Application of Powertrain and Fuel Technologies to Meet Emissions Standards

Conference Organizing Committee

C Ashley (Chair)
Cedric Ashley and Associates

C Bale
Tickford Limited

C Jones
RAC Motoring Services

A Osborn
Perkins Technology

M Russell
Lucas Diesel Systems

D Tidmarsh
University of Central England

IMechE
Conference Transactions

I MECH E
150th Anniversary
1847-1997

International Seminar on

Application of Powertrain and Fuel Technologies to Meet Emissions Standards

24–26 June 1996

Organized by the Automobile Division and the
Combustion Engines Group of the
Institution of Mechanical Engineers (IMechE)

IMechE Conference Transaction 1996 – 5

Published by Mechanical Engineering Publications Limited for
The Institution of Mechanical Engineers, London.

First Published 1996

This publication is copyright under the Berne Convention and the International Copyright Convention. All rights reserved. Apart from any fair dealing for the purpose of private study, research, criticism or review, as permitted under the Copyright, Designs and Patents Act, 1988, no part may be reproduced, stored in a retrieval system, or transmitted in any form or by any means, electronic, electrical, chemical, mechanical, photocopying, recording or otherwise, without the prior permission of the copyright owners. Reprographic reproduction is permitted only in accordance with the terms of licences issued by the Copyright Licensing Agency, 90 Tottenham Court Road, London W1P 9HE. *Unlicensed multiple copying of the contents of this publication is illegal.* Inquiries should be addressed to: The Publishing Editor, Mechanical Engineering Publications Limited, Northgate Avenue, Bury St. Edmunds, Suffolk, IP32 6BW, UK. Tel: 01284 763277. Fax: 01284 704006.

© The Institution of Mechanical Engineers 1996

ISSN 1356-1448
ISBN 0 85298 996 2.

A CIP catalogue record for this book is available from the British Library
Printed by The Ipswich Book Company, Suffolk, UK

The Publishers are not responsible for any statement made in this publication. Data, discussion and conclusions developed by authors are for information only and are not intended for use without independent substantiating investigation on the part of potential users. Opinions expressed are those of the Author and are not necessarily those of the Institution of Mechanical Engineers or its publishers.

Contents

Engine Design

C517/003/96	**A new Jaguar engine for the 21st century** M J Joyce, J Carling, M Clough, and W J Corkill	3
C517/038/96	**Potential of the new Ford Zetec-SE engine to meet future emission standards** R J Menne, G Heuser, and G D Morris	17
C517/041/96	**Piston assembly design for low emissions applications** M Willcock, S T Gazzard, and R B Ishaq	31

Aftertreatment

C517/009/96	**The prediction of fuel consumption and emissions during engine warm-up** P J Shayler, N J Darnton, and T Ma	53
C517/042/96	**Robustness design of experimental approach to the optimization of fast light-off of catalytic vehicles emission systems** B Rutter, R Hurley, D Eade, A Fraser, S Brett, R Shrieves, and J Kisenyi	65
C517/047/96	**Aftertreatment strategies to meet emission standards** R J Brisley, D E Webster, and A J J Wilkins	95

Diesel Control Technology

C517/044/96	**Impact of alternative controller strategies on emissions from a diesel CVT powertrain – preliminary results** M Deacon, R W Horrocks, C J Brace, N D Vaughan, and C R Burrows	115
C517/035/96	**Control technology for future low emissions diesel passenger cars** B Porter, T J Ross-Martin, and A J Truscott	125
C517/043/96	**Operating point optimizer for integrated diesel/CVT powertrain** M Deacon, R W Horrocks, C J Brace, N D Vaughan, and C R Burrows	147

EGR for Diesel

C517/028/96	**The effects of carbon dioxide in EGR on diesel engine emissions** N Ladommatos, S M Abdelhalim, and H Zhao	157

C517/034/96	**EGR technology for lowest emissions** R Baert, D E Beckman, and A W M J Veen	175
C517/027/96	**Experimental characterization of turbocharging and EGR systems in an automotive diesel engine** M Capobianco, A Gambarotta, and G Zamboni	191

Lubricity of Diesel Fuel Injection Equipment

C517/037/96	**The lubricity of hydrotreated diesel fuels** J N Davenport, M Luebbers, H Grieshaber, H Simon, and K Meyer	207
C517/021/96	**Vehicle tests for LSADO** A J Farrell and N G Elliott	219

Fuel Technology for Diesel Engines

C517/013/96	**Feasibility study towards the use of copper fuel additives for particulate reduction from diesel engines** J P A Neeft, M Makkee, J A Moulijn, J A Saile, and K L Walker	233
C517/032/96	**Fuel property effects on polyaromatic hydrocarbon emissions from modern heavy-duty engines** R C Doel	247
C517/023/96	**An experimental characterization of the formation of pollutants in DI diesel engines burning oxygenated synthetic fuels** C Beatrice, C Bertoli, N Del Giacomo, and M Lazzaro	261

New Generation Fuels

C517/022/96	**A fuel injection system concept for dimethyl ether** H Ofner, D W Gill, and T Kammerdiener	275
C517/046/96	**Dimethyl ether as an alternate fuel for diesel engines** S-E Mikkelsen, J B Hansen, and S C Sorenson	289
C517/008/96	**A new generation of LPG and CNG engines for heavy-duty applications** P C G De Kok	299

EGR for Spark Ignition

C517/030/96	**Improving fuel consumption of stoichiometric spark ignition engines by the use of high EGR rates and high compression ratio** C De Petris, S Diana, V Giglio, and G Police	311

C517/011/96	Enhanced exhaust gas recirculation using exhaust throttling for NOx reduction at WOT M Osses, P Clarke, G Andrews, A Ounzain, and G Robertson	321

Hydrocarbon Emissions from Spark Ignition Engines

C517/040/96	Evaluation of total HCs and individual species for SI engine cold start and comparison with predications J Edwards, D Tidmarsh, and M Willcock	337
C517/024/96	Emission of unregulated compounds from five different gasoline qualities A Laveskog	353
C517/033/96	Enhanced evaporative emissions requirements: a challenge to automotive industry, a chance for the environment N Raimann	363

Related Titles of Interest

Title	Author	ISBN
Gasoline Engine Analysis	J Fenton	0 85298 634 3
Gearbox Noise, Vibration, and Diagnostics	IMechE Conference 1995–5	0 85298 953 9
Computers in Reciprocating Engines and Gas Turbines	IMechE Conference 1996–1	0 85298 984 9
Automobile Emissions and Combustion	IMechE Seminar 1994–1	0 85298 892 3
Transient Performance of Engines	IMechE Seminar 1994–8	0 85298 935 0
Diesel Fuel Injection Systems	IMechE Seminar 1995–3	0 85298 980 6
Using Natural Gas in Engines	IMechE Seminar 1996–2	1 86058 033 5

For the full range of titles published by MEP contact:

Sales Department
Mechanical Engineering Publications Limited
Northgate Avenue
Bury St Edmunds
Suffolk
IP32 6BW
England

Tel: 01284 763277
Fax: 01284 704006

Engine Design

C517/003/96

A new Jaguar engine for the 21st century

M J JOYCE BSc, AIMechE, **J CARLING** BSc, **M CLOUGH**, and **W J CORKILL**
Jaguar Cars Limited, Coventry, UK

1. SYNOPSIS

In June 1996, Jaguar launched its latest all new engine; AJ-V8. This paper describes the design and function of the engine. The key aim of the engine was to meet current and future environmental demands, whilst simultaneously satisfying the customer requirements in the luxury passenger car market. The compact engine offers best-in-class specific performance, coupled with low friction, an efficient combustion system and particularly fast warm-up of both the engine and its catalysts. These features allow the engine to meet all world emission standards at launch including the Californian TLEV requirement.

2. INTRODUCTION

Jaguar competes exclusively in the luxury sports and saloon car market and since 1994 has used two engines, the six cylinder AJ16 and the V12. The car market is subject to both increasing competition and environmental legislation. The new V8 engine, AJ-V8, has been designed as a response to these twin challenges and represents a major step forward for the company both in terms of its design and manufacturing processes.

During the early 90's Jaguar actively researched the luxury car market in order to establish the key customer wants for a new engine, listed in Table 1. These were found to be broadly common across all world markets, with only minor variances. Also shown in Table 1 are the internal requirements which have to be satisfied if the engine is to be a viable business proposition. The task of balancing these requirements, some of which conflict, is the key task in the design of any new product.

Table 1. Customer Wants and Internal Needs

Customer Wants	Internal Needs
Good value	Meet emission legislation
Good acceleration	Meet noise legislation
Good top speed	Meet crash legislation
Good looking engine	Meet recycling standards
Freedom from faults	Meet fuel consumption legislation
Absence of vibration	Allow a profit to be generated
Low noise levels	Package acceptably in vehicles
Pleasing noise	
Low fuel consumption	

Using these requirements a large number of objective targets were defined for the engine and its 23 sub-systems. However, in order to provide emotional guidance to the design team, a subjective goal of 'Refined Power' was also defined.

3. DESIGN FEATURES

3.1. Engine configuration and capacity

The study into the most suitable configuration revolved around the need for low vibration levels coupled with good power outut. Layouts considered during the study included I6, V6, V8, V10 and V12. The need for balance shafts was reviewed but they were considered undesirable due to the friction and complexity penalties.

A 90° V8 configuration was selected as it offered a compact layout and provided the required balance between refinement and performance. It has an adequate number of cylinders both for low vibration levels and good power potential. The compact layout helps vehicle packaging and assists in weight control which in turn helps reduce vehicle fuel consumption.

The chosen configuration is extremely popular in Jaguar's class of engines (1, 2, 3). Great efforts were made to capitalise on the compact configuration in order to produce a physically small engine for its displacement. This was pursued primarily in order to assist in vehicle packaging but also helped promote the control of weight. The overall dimensions for a number of competitor engines are given in Table 2 and a cross-section of the engine can be seen in Figure 1.

Figure 1.   Sectioned AJ-V8 engine

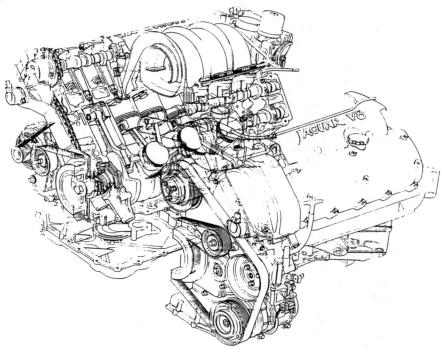

Table 2. Engine dimensions

Engine	Width [mm]	Length [mm]	Height [mm]	Mass [kg]
AJ-V8	680	570	680	200
AJ16	600	865	680	245
V12	750	900	685	340
Comp A	710	600	720	205
Comp B	725	730	710	220
Comp C	730	710	715	220

All competitor engines are 90° V8 engines

The choice of an V engine with a large number of cylinders is not optimal if one purely considers emissions. Small cylinder capacities result in a large combustion chamber surface to volume ratio which is disadvantageous for HC emissions. The V configuration also results in the need to use at least two catalytic converters with the resultant cost and complexity. It was however felt that the need to satisfy the overall customer requirement of 'Refined Power' justified the choice.

The selection of engine capacity, all other factors being equal, is a balance between the required output and fuel economy. For AJ-V8 the decision was made to maintain the capacity at that of our existing 4.0 litre AJ16 engine. The decision goes against the market trend of increasing engine capacity to offset vehicle weight increase. The targets set were therefore to both improve the specific output and the specific fuel consumption in comparison with the AJ16 which had in turn represented an advance on the previous AJ6 engine.

3.2. Combustion system

Once the cylinder capacity of 500cc had been determined, a study into the combustion system was commenced. It was quickly apparent from a market study that a minimum of 4 valves per cylinder would be required in order to achieve competitive output. From this base assumption the study split into two parallel branches; one to choose the bore/stroke (B/S) ratio of the engine and the other to compare 4 valve/cylinder and 5 valve/cylinder layouts.

A number of bore and stroke combinations were considered with the B/S ratio ranging from 0.83 to 1.0. The consideration of an over-square configuration was felt unnecessary for the target specific power output of 52 kW/litre and it was known to be directionally poor for both HC emissions and fuel consumption. HC emissions are increased with larger bores due to their greater crevice volume. Fuel consumption is degraded by the larger surface/volume of the chamber, coupled with the increased burn duration, resulting in greater heat transfer.

Test work on flow benches, a single cylinder engine and multi-cylinder engines were used in combination with calculations. It was found that the minimum B/S ratio which could satisfy the required output was 0.9 and this minimum should be used as it would give the best part throttle performance. Table 3 shows the calculated performance of engines with different B/S ratios with all other factors being held constant.

However, another design study on the crankshaft indicated that a long stroke would increase vibration levels. As a result the decision was taken to adopt a B/S ratio of 1.0 as it was felt to offer the best balance in our efforts to deliver 'Refined Power'.

Table 3. Calculated performance for different Bore/Stroke ratios

B/S ratio	Specific power [kW/l]	Relative fuel economy [%]	1500 RPM BMEP [Bar]	Piston ring diameter [mm]
0.83	48	96.5	10.5	81
0.91	52.5	98.5	10.25	83.3
1.00	57	100	10.0	86

The 5 valve investigation was carried out in a similar fashion and concluded that a 5 valve/cylinder configuration offered no functional advantage over a good 4 valve/cylinder system (Fig 2) whilst increasing engine complexity. This is in conflict with other work in the field (4,5) which was based on extremely high specific output engines but is consistent with the performance of 5 valve engines seen in the market when compared to equivalent 4 valve engines.

Figure 2. 4v and 5v single cylinder engines compared

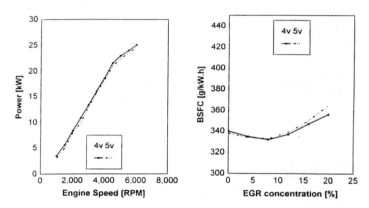

With the bore, stroke and valve number established, the design of the combustion chamber was now the subject of considerable detail development. A key aim of the development was a high dilution tolerance to either air or EGR in order to give flexibility in the calibration of the engine. The resultant combustion system is a relatively conventional pentroof chamber with a high velocity tumble inlet port. It does however have a number of unusual detail features. The included valve angle is narrow at 28°24', the chamber has no squish bands and the piston is flat topped. These features combine to provide an extremely low surface/volume ratio, which in combination with the high compression ratio of 10.75:1 gives excellent thermal efficiency.

Traditionally, a wide valve angle has been felt necessary for high output engines. (6) Through careful design of the inlet port and valve, adequate flow has been realised and any losses resulting from the valve angle are felt to be more than offset by the improved combustion chamber design. The compact design of the chamber has resulted in it being highly efficient at converting swirl into turbulence. As a result, the inlet ports have been designed to only generate a moderate level of swirl to ensure that the combustion does not become unduly harsh at high throttle openings.

The result is a simple combustion system with excellent part throttle dilution tolerance, good specific fuel consumption, low noise levels at wide open throttle and the flow capacity to support a specific power output of 54 kW/l in the installed condition.

3.3. Light weight valve gear

The valve gear of the AJ-V8 engine has the lowest reciprocating mass of any engine in its class as can be seen in Table 4. The general design strategy adopted for the valve gear was to minimise component mass without the need to use expensive 'exotic' materials such as titanium and ceramics, whilst maintaining excellent valve train functionality and reliability.

Table 4.   Valve Gear Masses

Engine	Year of introduction	Reciprocating valve train mass [kg/valve]	
		Inlet	Exhaust
AJ-V8	1996	0.096	0.096
Comp A	1994	0.109	0.100
Comp B	1992	0.149	0.147
Comp D	1987	0.146	0.141

All engines have bores in the range 86 to 89mm.

A direct acting design without hydraulic lash adjustment was selected for its low mass, high stiffness, low complexity and good cost effectiveness. The low valve train mass has been achieved despite the generous valve head sizes of 35mm on the inlet and 31mm on the exhaust. Both valves are a one-piece design, with a 5mm valve stem and neither requires the use of supplementary hard materials for the seat interface.

Cold-formed steel valve spring retainers are employed whose shape was 'sculptured' to minimise non-functional material. The 5mm collets are also a light-section cold formed design. Fluorocarbon valve stem seals are integrated with the steel spring seats to reduce cost and to aid engine assembly. The tappets are a 33mm diameter aluminium alloy design with an iron alloy spray-coated outer wall. Valve clearance is set during engine assembly with a top mounted hardened steel shim of 30mm diameter. The diameter of this shim was maximised to reduce the constraints on the cam profile design. Despite the absence of a hydraulic lash adjuster, no maintenance of the valve gear is required.

The low mass of the valve gear permits the use of relatively light valve spring loads with a high peak engine speed of over 7000 RPM. These light springs significantly reduce valve train loads yielding benefits in terms of low speed engine friction. The functional benefits of the low mass of the valve gear include reduced high speed engine noise, reduced valve train drive loads, improved fuel consumption and the direct benefit of reduced engine mass. All of these have been achieved without any loss of reliability or increase in cost and complexity.

3.4. Variable Cam Phasing system

The engine is equipped with a Variable Cam Phasing (VCP) system fitted to each of the inlet camshafts. This system is similar to that fitted to competitor engines (7,8) and provides two cam timing positions. The Jaguar system features an unusually large phasing range of 30°, chosen to give 35° of valve overlap in the advanced cam timing. This results in high levels of exhaust back flow at moderate throttle openings giving high internal Exhaust Gas Recirculation (EGR) levels. This results in low NOx emissions combined with improved HC emissions (Fig 3) and allows the engine to be sold in many markets without an external EGR system. In the more stringent markets, only a small amount of external EGR is required, giving very low NOx levels without the increase in HC emissions usually associated with external EGR systems.

Figure 3.   VCP effect on emissions at 2000 RPM, 2 Bar BMEP

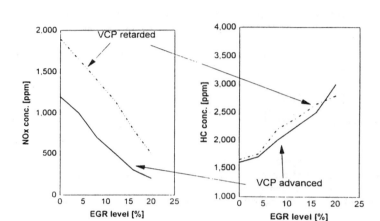

The system is calibrated to suit three main areas of operation. The timing is set in the retarded timing, with only 5° of overlap, to minimise residuals and give the best possible combustion quality at very low engine speeds and small throttle openings. At moderate throttle openings, at low to medium engine speeds, the timing is set advanced to minimise emissions by taking advantage of the internal EGR. At wide open throttle the timing is set to give the best volumetric efficiency; advanced up to 4500 RPM and retarded at higher speeds (Fig 4).

3.5. Air intake system

The use of composite intake systems is becoming the norm in the drive to reduce the weight of vehicles. The AJ26 intake system consists of a Polyamide air cleaner, a polypropylene induction tube and a Polyamide manifold. The only metal components are the throttle body and the throttle adapter which are manufactured in aluminium alloy. The throttle requires a thermally stable structure to allow idle speed control to be carried out by the main blade. The throttle adapter provides the dual role of a structural support for the throttle and a resilient mounting point for the EGR valve.

Figure 4.   WOT volumetric efficiency

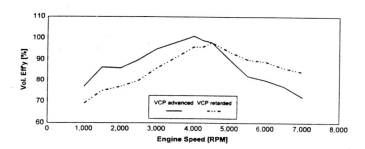

A further weight saving is achieved by the novel integration of the fuel rail into the manifold itself. Through the use of a suitable Polyamide grade and side feed injectors it has been possible to dispense with the need for a fuel rail by moulding it as an integral part of the manifold. As well as reducing weight and complexity, variability in the location of the injector tip is reduced thus increasing the consistency of the targeting of the twin spray injector.

Considerable effort was expended in the early stages of the program in order to develop a variable intake system (VIS) for the engine. Three main types of VIS systems are in use; port de-activation, runner length switching and plenum chamber switching (9). The first two were rejected primarily due to their complexity and the potential for unreliability. The third was explored using both computer simulations and prototype engines. However, achieving a performance benefit proved extremely difficult to deliver. The main problem was the tight package in combination with the odd firing on each cylinder head. The pipe runs required to link each cylinder to its appropriate plenum were very contorted. This resulted in high friction losses in the runners which more than offset the advantage gained from the linked plenums. It is perhaps for this reason that none of the engine's primary competitors feature a VIS system, despite such systems being found on 4 and 6 cylinder engines from the same companies.

The final, fixed geometry intake system has been tuned to compliment the VCP system. With two position VCP, the cam timing is optimised for both low and high engine speeds. The intake was therefore tuned for mid-range speeds to provide a broad spread of torque. The intake pipes are 430mm long with an effective diameter of 40mm which results in peak torque being achieved at 4250 RPM.

3.6. Cooling system

The main feature of the cooling system is a novel means of cooling (10) developed using CFD, flow visualisation, CAD and FEA techniques. The new method adopts precision cooling using high local coolant velocities and low coolant volume to offer both an improvement to control of metal temperatures and engine warm up (Fig 5) compared to traditional cooling methods. These improvements directly contribute to both the performance and fuel economy of the engine.

Figure 5.   Warm up during the FTP75 test

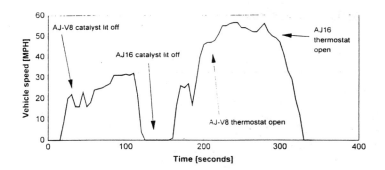

An end-to-end coolant flow strategy was adopted as this allows each cylinder to receive the same distribution of local coolant velocities. The cylinder block requires lower coolant velocity to achieve the same degree of control over metal temperature compared to the cylinder head. This is because temperature gradients generated through the cylinder wall are roughly a 1/3 compared to those of the cylinder head. Thus, to prevent overcooling of the cylinders, a bypass rail was added to both banks (Fig 6) to channel 50% of the coolant away from that directly cooling the cylinders. In addition to this feature, the block water jacket extends to only 2/3 of the piston stroke to prevent overcooling of the lower part of the bores.

In the cylinder head, the difficulty with end to end flow is to direct sufficient coolant between the exhaust ports as this requires the flow to be diverted locally from a longitudinal to a transverse direction. Coolant is manoeuvred by biasing the upstream and downstream passages so that the least restrictive path is between the exhaust ports (Fig 6). Another feature of the cylinder head water jacket is the reduced coolant volume around the exhaust ports to prevent overcooling of the exhaust gases and unnecessary heating of the coolant.

The coolant volume of the engine water jackets is 2.8 litres, this being only 57% that of a modern competitor engine. This degree of reduction in thermal mass has been shown to offer an 18% improvement in the time required to heat up the coolant to operating temperature (11). Table 5 compares the metal temperatures of the AJ-V8 with those of the conventionally cooled AJ16 Jaguar engine. Peak temperatures on the V8 are lower by over 80 °C and the cylinder-to-cylinder variation is within a few degrees due to the local coolant velocity distribution for each cylinder portion.

Figure 6. Cylinder block and head cooling

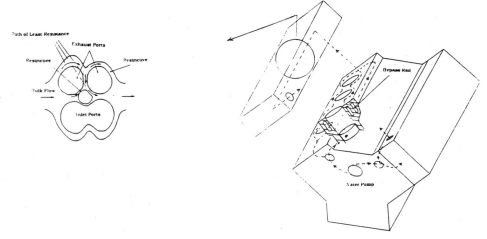

Table 5. Engine metal temperatures at rated power

	AJ-V8 [° C]	AJ16 [° C]
Cyl. head maximum	168	251
Cyl. head minimum	108	177
Cylinder block variation - max	8	25
- min	4	4

The twin outlet water pump was designed to fit in a confined space at the front of the engine. This required the water pump to be optimised in terms of package size. This resulted is a best in class overall efficiency of 62% at the system operating point. This was achieved using a shrouded plastic impeller which provides the means of attaining the most effective geometry. Careful attention was also applied to the design and development of achieving a volute profile which matched the performance of the impeller.

The thermostat is situated on the cold side of the radiator as this offers the best means of controlling top hose temperature that eliminates temperature cycling synonymous with top hose thermostats. Coolant is channelled from both cylinder heads via a water outlet pipe. This provides flow to the radiator, thermostat and heater as well as housing the sole water temperature sensor. The thermostat housing directs coolant to the thermostat from the engine and radiator, and also to the cylinder block via a 90 degree curved elbow to offer a good inlet condition to the water pump to offset pump cavitation. The major external cooling system components are all manufactured from Polyamide to further reduce the weight of the engine.

3.7. Engine Management System

The system is supplied by Nippondenso, an existing supplier to Jaguar and other manufacturers in the luxury vehicle market. It features an electronic throttle, cylinder selective ignition, fully sequential fuel injection with closed-loop AFR control based on air mass flow measurement. In addition, variable cam phasing, electronic EGR and fuel tank purge control are provided.

The electronic throttle carries out all air control functions including basic throttling, idle speed control, starting air, cruise control and power limitation. Throttle pedal progression is electronically varied in response to vehicle and driver requirements. Strategies are used to assist drivability, in particular tip-in and tip-out response. This enables torque converter lockup to be employed over a wider speed range, with fuel economy benefits. The throttle body also includes a mechanical backup for limp-home.

Other vehicle systems are interfaced via a vehicle communication network (CAN). The engine ECM continually transmits and receives information enabling it to play an active part in vehicle control systems. Examples include dynamic control of torque during gearshifts and power reduction in response to traction control demand.

The performance targets for the system were set to meet the relevant market emission and diagnostic legislation, in particular the Californian LEV and OBDII requirements. Advantage has been taken of the compact size of AJ-V8 to allow the entire catalyst volume to be mounted 'close coupled' in the exhaust down-pipe area of the engine. Two 2 litre Palladium/Rhodium catalysts are employed, one for each bank of the engine. Special fast warm-up oxygen sensors were developed to suit this application which feature high output heaters.

The combination of the close-coupled catalysts, fast warm-up sensors and sophisticated warm-up control have enabled hydrocarbon conversion efficiencies of up to 98% over the complete FTP75 cycle, including the cold start. During fully warm operation, efficiencies of up to 99.7% are attained. In order to prevent accelerated deterioration of the catalysts at high throttle openings, a combination of EMS strategies and new catalyst coatings are employed. These, in combination with the low exhaust temperatures resulting from the high thermal efficiency of the engine, provide good resistance to thermal degradation of the catalysts without a severe fuel consumption penalty.

4.0. RESULTANT PERFORMANCE

4.1. Friction

In order to achieve the combined objectives of good output with low fuel consumption, low friction levels were essential. The light weight valve gear contributes significantly to this achievement, particularly at low engine speeds. In addition the use of Nikasil bore plating, low load piston rings, high efficiency oil and cooling pumps and general attention to detail in the design of the bearing has resulted in an engine with highly competitive friction levels (Fig 7).

Figure 7.   Engine friction levels

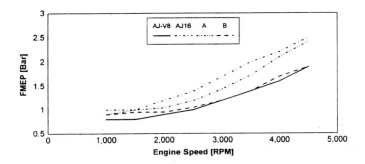

4.2. Engine output

The engine has good peak power and torque of 216 kW at 6100 RPM and 393 Nm at 4250 RPM. This is backed up by a broad spread of torque, thanks to the combination of VCP and the tuning of the intake and exhaust systems. 80% of peak torque is available between 1400 and 6400 RPM (Fig 8).

Figure 8.   AJ-V8 full throttle performance

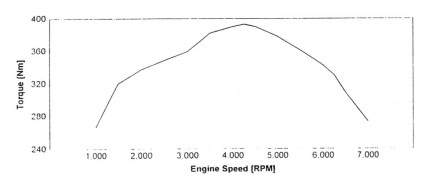

The resultant specific output is highly competitive as can be seen in Table 6. In combination with the low installed weight of the engine, the high power output results in a very high engine power/weight ratio of 1.08 kW/kg. This compares favourably with competitor engines in the class which are in the range 0.8 to 1.0, and with the AJ16 engine at 0.72 kW/kg.

Table 6. Specific full throttle performance

Engine	Specific power [kW/l]	Peak BMEP [Bar]	1500 RPM BMEP [Bar]
AJ-V8	54.0	12.4	10.1
AJ16	44.3	12.0	9.9
V12	37.5	10.1	8.8
Comp A	48.5	11.5	9.5
Comp B	47.7	12.0	9.5
Comp C	50.0	12.3	9.6

All competitor engines are 32v V8s.

4.3. Emissions and fuel consumption

The engine is able to meet all current emission requirements and there is great confidence that the basic architecture is in place for the engine to be developed to meet the more stringent legislation planned. Steady state, fully warm emission levels are competitive with other 'modern' luxury car engines as can be seen in Table 7, whilst the fast warm-up of both engine and catalyst provides excellent cycle and real-world emission control. This allows the European Stage 2 requirement to be met without external EGR and Californian TLEV requirement with external EGR.

Table 7. Specific emissions at 1500 RPM, 2.62 Bar BMEP

Engine	HC [g/kW.h]	NOx [g/kW.h]
AJ-V8; VCP advanced	5.3	5
AJ-V8; VCP retarded	5.7	15
AJ16	5.5	12
Comp A	6.3	13

The combination of low friction coupled with the high thermal efficiency combustion system, results in an engine with excellent specific fuel consumption. A minimum of 240 g/kW.h is seen at 2000 RPM, near WOT. Whilst fuel consumption did not rate highly in customer requirements, greater importance was attached to it during the development of the engine due to the likelihood of future CO_2 legislation.

5. FUTURE DEVELOPMENTS

With the impending Californian and European emission legislation, work has already begun on lower emission derivatives of the engine. The primary focus of these activities is on reduced engine-out emissions through the development of more sophisticated fuel delivery and variable valve timing systems combined with further developments in the control of the fuel, ignition and EGR systems.

.The simplicity of the engine heating the catalyst, linked with low engine-out emissions in the first seconds of the cycle, is felt to be highly attractive when compared to external catalyst heating systems. The primary attraction of such a system is the low complexity and hence high reliability. Prototype vehicles have already demonstrated compliance with Californian LEV standards and engine tests are showing very encouraging results from the fuel preparation system work in order that ULEV and Euro Stage 3 standards may be met.

Other planned developments are aimed at increasing specific output to provide higher performance without the impact of a capacity increase in terms of fuel consumption and emissions.

6. CONCLUSION

The AJ-V8 engine has been designed primarily to be highly competitive at launch, but also to provide a sound basis for Jaguar's future engines. It is a compact, high output engine with low specific fuel consumption and emissions. It both satisfies environmental legislation and customer requirements. As such it is well suited to meet the needs of the market as we move into the 21st century.

Acknowledgements

The design and development of this engine has been a true team effort and this paper is written on behalf of the entire team at Jaguar.

References

1 Nakamura et al, 'A new V8 engine for the Lexus LS400', SAE paper 892003, 1989.

2. Kinoshita et al, 'Development of a new generation high-performance 4.5 liter V8 Nissan engine', SAE 900651.

3. Foss et al, 'The Northstar DOHC V-8 Engine for Cadillac', SAE 920671.

4. Aoi et al, 'Optimization of Multi-Valve, four cycle engine design - The benefit of Five-Valve Technology', SAE 860032.

5. Downing & Bale, 'The design and development of a unique five valve cylinder head', SAE 905156.

6. Kimbara et al, 'Innovative standard engine equipped with 4 valve', SAE 870352.

7. Grohn & Wolf, 'Variable valve timing in the new Mercedes-Benz four-valve engines', SAE 891990.

8. Nishimura et al, 'Nissan V6 3.0 litre, 4-cam 24 valve high performance engine', SAE 870351.

9. Matsumoto & Ohata, 'Variable induction systems to improve volumetric efficiency at low and/or medium engine speeds', SAE 860100.

10. European Patent Specification No. 0461765B1. 7/12/94.

11. Clough M, 'Precision cooling of a four valve per cylinder engine', SAE 931123.

Potential of the new Ford Zetec-SE engine to meet future emission standards

R J MENNE, G HEUSER, and **G D MORRIS**
Ford Werke Aktiengesellschaft, Köln, Germany

Abstract

The key development targets of the All-New Ford Zetec-SE were improved fuel efficiency, high power and torque output, good **N**oise **V**ibration and **H**arshness characteristics, low emission levels and the capability to meet future emission standards.

As a result of reduced friction, optimised combustion chambers and improved tumble motion this engine provides very competitive specific fuel consumption with a minimum **B**rake **S**pecific **F**uel **C**onsumption of 245 g/kWh. However for optimum vehicle fuel consumption it is also critical to achieve good BSFC at part load. Therefore Ford put considerable effort into improving this operating regime without adversely affecting idle quality. The torque characteristic has been carefully balanced, in that the engine provides more than 90% of maximum torque in the speed range between 2000 and 5200 rpm.

Low tailpipe emissions are achieved due to the fast light-off characteristics of a **C**lose **C**oupled **C**atalyst. Consequently current emissions for HC+NO_x and CO are about 78% less than Stage II legislated levels.

One major factor reducing emissions is improved fuel preparation. The four-hole injector design, which is used in the Ford Zetec-SE engine, produces finely atomised fuel and therefore the hydrocarbon emissions are significantly reduced. However in order to achieve future emission standards capability advanced fuel preparation is necessary. Use of air-assist injectors generates even smaller fuel droplets and therefore improved combustion stability. As a result ignition timing can be retarded during cold start and warm up without affecting idle quality. As a consequence the higher exhaust gas temperatures, produced with retarded spark, allow the catalyst to light-off faster. This leads to a further 50% reduction in HC emissions during the **N**ew **E**uropean **D**riving **C**ycle established for Stage III emissions.

1. Introduction

Coincident with the launch of the new Fiesta, Ford has introduced an all-new engine family, the "Zetec-SE". This engine family, with a displacement range from 1.25 to 1.4 litres, reinforces Ford's commitment to four-valve technology in the B-car segment.

Development targets were established based on the results of an extensive analysis of customer requirements. As a result the new engine family's programme objectives were set to deliver excellent **N**oise **V**ibration and **H**arshness performance and fuel economy, allied to high power output and low engine weight. Extended service intervals were also a clear requirement, as was the need to meet Stage II emission levels. In fact the Zetec-SE engine does not just meet existing legislation but was specifically designed to have the capability to meet projected future standards.

2. Engine Concept and Main Dimensions

An aluminium alloy cylinder head with twin overhead camshafts, four valves per cylinder and near central spark plug mounts on to an aluminium alloy cylinder block with integral crankcase **(Fig. 1)**. The ribbed cylinder block is a deep skirt design with a separate bearing beam retaining the five main bearings. The combination of deep skirt block and bearing beam leads to significantly increased powertrain stiffness. A structural oilpan helps to reduce vibrations of the whole powertrain. To reduce hydrocarbon emissions pistons with a top land height of 5 mm are employed. Both cooling and oil systems are flow optimised to reduce parasitic work. The main dimensions of the 1.25 litre and 1.4 litre Zetec-SE are shown in **Table 1**.

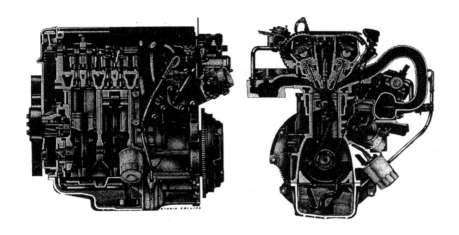

Figure 1: Engine Cross Section Views

		1.25-l-Engine	1.4-l-Engine
Displacement	cm^3	1242	1388
Bore	mm	71.9	76.0
Stroke	mm	76.5	76.5
Bore / Stroke Ratio	-	0.94	0.99
Bore Spacing	mm	87.0	87.0
Connecting Rod Length	mm	136.3	136.3
L / R	-	3.57	3.57
Compression Ratio	-	10.0	10.3
Max. Power	kW (hp) @ rpm	55 (75) / 5200	66 (90) / 5600
Max. Torque	Nm @ rpm	110 / 4000	125 / 4500
Torque at 1500 rpm	Nm	92	101

Table 1: Main Engine Dimensions

3. Functional Characteristics of the Zetec-SE Engine

NVH

As mentioned above, Best in Class (BIC) NVH was one of the major drivers during the development of this engine. In order to optimise the sound and vibration levels of the powertrain a number of sophisticated analytical techniques such as Eigenmode-analysis, **F**inite **E**lement calculations and sound pressure measurements were applied.

In order to achieve good NVH performance a stiff engine design is essential. During the advanced design phase about 20 different cylinder block and crankcase concepts were evaluated. The main differences between the various concepts considered were short skirt versus deep skirt block, the material type, the main bearing configuration and the oilpan design.

The selected concept has a deep skirt closed deck cylinder block and an aluminium bearing beam. The ribs on the surface of the structural oilpan are optimised to increase powertrain stiffness and to reduce sound pressure levels. The valvetrain was optimised using **C**omputer **A**ided **E**ngineering analysis to reduce valve spring load and valvetrain noise.

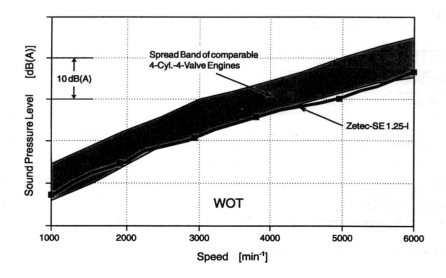

Figure 2: Sound Quality at Full Load

The increased powertrain stiffness results in the first powertrain bending frequency being close to 300 Hz, which is higher than the frequency of the second order shaking forces. In addition, vibrations were reduced by the use a relatively long con-rod giving an L/R ratio of 3.57. NVH characteristics were also taken into account when designing the **Front End Accessory Drive**. As a result most parts of the FEAD are direct mounted to the cylinder block.

The outstanding NVH behaviour of the Zetec-SE is shown in **Figure 2** where it is compared to other recent four-cylinder four-valve engines. The figure shows the radiated sound pressure levels versus engine speed at full load conditions. The 1.25 litre Zetec-SE is very quiet over the entire speed range and for the majority of speeds the sound pressure level is lower than that of any of its competitors.

Friction

Every engine sub-system was carefully optimised to minimise friction. **Figure 3** shows the **Friction Mean Effective Pressure** of the Zetec-SE compared to engines of similar configuration. The upper part of the figure shows the overall engine

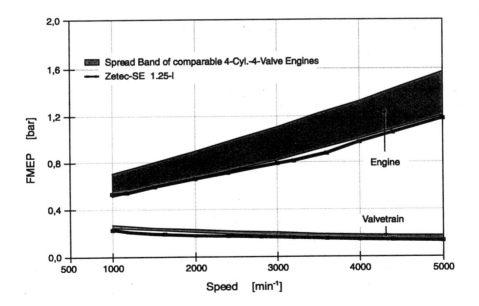

Figure 3: Engine and Valvetrain Friction

friction, the lower part the valve train sub-system friction. The comparison clearly shows that both friction characteristics are commendably low and hence aid the achievement of good fuel economy. In order to minimise friction a considerable amount of detailed work was undertaken. The piston rings were carefully optimised and matched to the cylinder bore in order to minimise tangential forces. Through detailed modelling and optimisation of the valve train dynamics the valve spring loads were considerably reduced. The wear between cam and mechanical bucket tappet was also analysed in great depth, so that an optimum combination of materials and surface quality could be selected. The combination chosen allows up to 150000 km operation without readjustment of the valve clearance. In comparison to a valvetrain concept with hydraulic elements the dimension of the oil pump was considerably reduced, this also contributes to reduced engine friction.

Power and Torque

The torque characteristic and the power output is strongly affected by the intake manifold design. During the development phase of this new engine family some new development tools, such as **C**omputational **F**luid **D**ynamics were used to complement the test bench investigations. As a result the optimised intake plenum has a conical shape, with the cross sectional area reducing towards the fourth cylinder. The inlet ports induce a medium tumble motion, which leads to a fast and highly efficient combustion without unnecessary pumping losses. In addition the selected tumble level limits the maximum in-cylinder pressure rise. This assists in the achievement of excellent NVH behaviour under full load conditions.

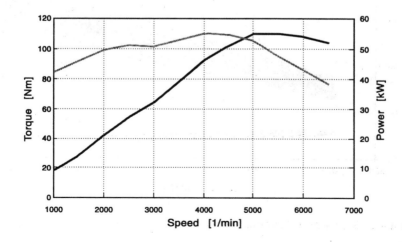

Figure 4: Power and Torque of the 1.25-l Zetec-SE Engine

The power and torque curves of the 1.25 litre Zetec-SE are shown in **Figure 4**. The engine torque curve is balanced to provide both optimum low and high speed performance. Maximum power is 55 kW at 5200 rpm and maximum torque is 110 Nm. This is a **B**rake **M**ean **E**ffective **P**ressure of more than 11 bar, which is impressive for a small displacement engine in this high volume vehicle segment. In the speed range from 2000 to 5200 rpm the engine delivers 100 Nm or more, which is over 90% of maximum torque.

Fuel Economy

The optimised combustion system is to a significant degree responsible for the good fuel economy, low emission levels, high power output and excellent NVH of the Zetec-SE engine. As previously mentioned, the intake ports induce a medium intensity of tumbling air motion, which leads to fast combustion with high thermal efficiency.

The compact combustion chamber includes a near central spark plug location. The squish area is about 10% and the valve angle is 41.45°. The bore/stroke ratio for the 1.25 litre Zetec-SE is designed to be "under square" for optimum thermal efficiency and low speed torque. It was not necessary to employ **E**xhaust **G**as **R**ecirculation to meet the 1996 Stage II emission standards. However use of a closed loop EGR system provides benefits in terms of fuel consumption and the flexibilty to meet potential future emission levels with the minimum of further internal modifications. The 1.25 litre production engines maximum EGR rate is about 12%.

Figure 5 shows a **B**rake **S**pecific **F**uel **C**onsumption map of the 1.25 litre Zetec-SE. The minimum BSFC is 245 g/kWh, which is very competitive for an engine of this displacement. Of particular significance is the excellent fuel consumption of the engine in the low speed low load operating range. This ensures that the new Zetec-SE delivers good fuel economy in the customer critical area of urban driving. **Figure 6** shows the BSFC of both Zetec-SE engines for such an operating point (speed= 1500 rpm, BMEP= 2.62 bar, AFR= 14.3, EGR= 0%) in comparison to competitor engines. The excellent BSFC measurement for each Zetec-SE engine can be clearly seen.

Considerable effort was put into the optimisation of the idle quality of this new engine family. This was another of the key development targets, identified during customer surveys. The standard deviation of **I**ndicated **M**ean **E**ffective **P**ressure, which is used to quantify the combustion stability at idle, is 4 kPa. Compared to similar engines this is an outstanding value, particularly when recognising the good torque characteristic and low fuel consumption of the Zetec-SE.

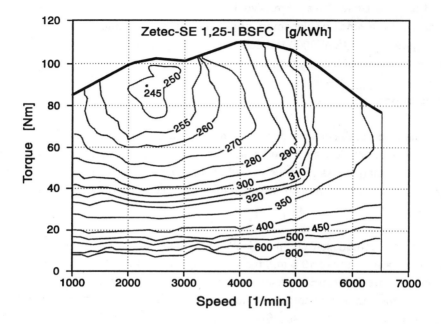

Figure 5: BSFC-Map of the 1.25-l Zetec-SE Engine

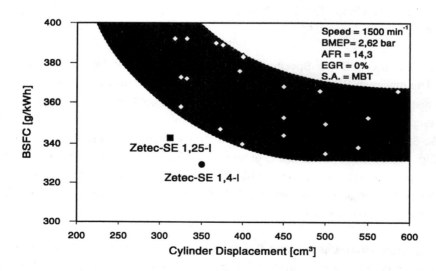

Figure 6: Part Load BSFC of the 1.25-l Zetec-SE Engine

Exhaust Emissions

As previously mentioned, this engine family was designed to meet both current and future emission levels. Due to the optimised combustion system and reduced crevice volumes, for example, the piston top land, the Zetec-SE has intrinsically low feedgas emissions. **Figure 7** shows the Feedgas HC emissions at part load (speed= 1500 rpm, BMEP= 2.62 bar, AFR= 14.3, EGR= 0%) versus the cylinder displacement. Highlighted in the diagram are the 1.25 litre and the 1.4 litre Zetec-SE, which are compared to similar engines. In both cases the very low engine-out HC emissions are easy to recognise. This clearly demonstrates that this new generation engine has a very high potential to meet future emission levels.

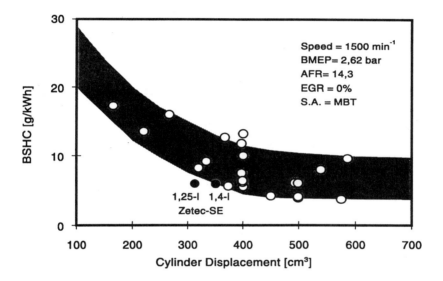

Figure 7: Comparison of Part Load BSHC

However in order to achieve consistently good tailpipe emissions additional parameters, other than the base engine design, also have to be taken into account. One of the major factors, which influences the exhaust emissions is the exhaust aftertreatment system in combination with the engine management system, or more specifically the engine calibration. In order to optimise the aftertreatment system CAE tools such as CFD modelling have been extensively used. This helped to locate the **H**eated **E**xhaust **G**as **O**xygen sensor at a location where the exhaust gas stream from all four cylinders is well mixed and where the temperatures are not excessive. These modelling techniques were also used to achieve an almost uniform exhaust gas distribution across the entire catalyst cross section.

It is well known that about 80% of the total HC emissions are emitted during the engine warm-up period, when the catalyst efficiency is low. To improve the light-off duration of the aftertreatment system Ford has applied a **C**lose **C**oupled **C**atalyst (CCC) for the first time in this vehicle segment. **Figure 8** shows the HC conversion efficiency of a CCC compared to an under-body catalyst during a cold-start within the **N**ew **E**uropean **D**rive **C**ycle (NEDC). After about 40 seconds the CCC shows visible HC conversion and after 65 seconds a 50% conversion rate is reached. In the case of the under-body catalyst there is almost no measurable HC conversion at this point in time. 90% conversion efficiency are reached after about 130 seconds in the case of the close coupled catalyst; the under-body catalyst needs nearly 200 seconds to reach the same conversion efficiency. As expected, the faster light-off performance of the CCC leads to reduced tailpipe emissions.

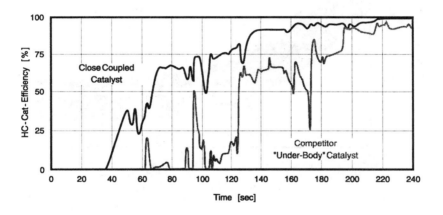

Figure 8: Comparison of Catalyst Efficiencies

The combination of the optimised combustion system, the excellent fuel preparation, the dynamically optimised EEC V engine management system and the fast light-off of the exhaust aftertreatment system provides the Zetec-SE with very low exhaust emissions. **Figure 9** shows emission levels of HC+NO$_x$ and CO of the 1.25 litre Zetec-SE compared to Stage I and Stage II legal emission levels. With 0.11 g/km for HC+NO$_x$ and 0.48 g/km for CO the 1.25 litre Zetec-SE emits about 78% less than the Stage II emission standards would allow.

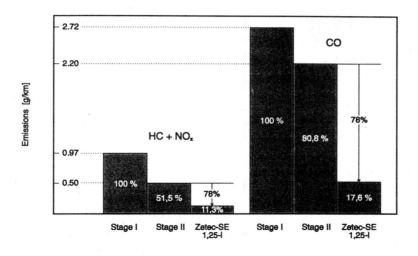

Figure 9: NEDC-Emissions of the 1.25-l Zetec-SE Engine

4. Improved Fuel Preparation

Considerable effort has been put into the optimisation of fuel preparation and consequently of mixture formation. An extensive evaluation of injector location and installation angle was undertaken with a number of alternative injector designs.

The best installation was determined to be in the cylinder head, close to the inlet valves. The injector spray of the 2-spray injector was optimised at an angle of 8 degrees, which ensured that the fuel is accurately targetted on the back of the valve. This minimises the wall film thickness in the inlet port and hence provides very good dynamic engine behaviour combined with low engine-out HC emissions.

A number of recent technical publications have indicated that improved fuel preparation, which means smaller fuel droplets (the optimum is fully vaporised fuel), helps to reduce HC emissions especially under cold-start conditions /1,2/. Extensive testing on the Zetec-SE engine confirmed these findings and as a direct result Ford made the decision to use 4-hole injectors for the first time. These injectors deliver substantially reduced droplet sizes when compared to conventional 2-hole designs. **Figure 10** shows the injection sprays of both a 2-hole and a 4-hole injector. The injection duration is 5 ms and the pictures are taken 5 ms after start of injection.

Two-hole Injector

Four-hole Injector

Figure 10: Injection-Spray of Two-hole and Four-hole Injector

In the case of the 2-hole injector single droplets are visible. In comparison the 4-hole injector delivers a combination of smaller droplets within a fuel "mist". This leads to a decrease in HC "engine-out" emissions. A comparison of the HC emissions for both injector types for different injection timings shows that, in the case of the 4-hole injector, the HC emissions are lower for all injection timings. In particular, with open valve injection, the 4-hole design delivers very significant improvements in HC emissions.

Further improvements in cold-start, HC feedgas emissions and catalyst light-off duration can be achieved by the use of even more advanced fuel preparation such as air-assist injectors. Compared to conventional designs, these injectors offer the potential to achieve very fine fuel atomisation. This can result in further combustion improvements and therefore reduced HC emissions. They also offer the potential for further reductions in wall film thickness within the intake manifold and ports, which can lead to further enhancements in dynamic engine behaviour /3,4/. The use of air-assisted injectors allows the engine calibration to be modified, by the use of retarded ignition timing, to enhance light-off performance without degrading idle quality or driveability. This leads to higher catalyst conversion efficiencies during cold-start and therefore to lower tailpipe emissions.

Initial investigations, applying air-assist injectors to a 1.4 litre Zetec-SE running an NEDC (Stage III) emission test, demonstrated a dramatic decrease in HC emissions. When using the standard calibration, the HC emissions during the NEDC-cycle (Stage III) were reduced by about 30% **(Figure 11)**. If, in addition, the calibration was modified to generate fast catalyst light-off, then the HC emissions could be reduced by about 52% compared to a more conventional 4-hole injector.

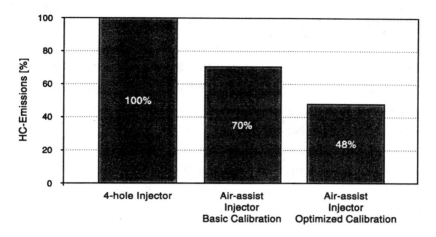

Figure 11: NEDC HC emissions (Stage III, 1.4-l Zetec-SE)

5. Conclusions

The all-new Zetec-SE was developed to meet future emission levels. The benefits of an optimised combustion system and enhanced fuel preparation, combined with the fast light-off performance of a close coupled catalyst, enable this engine family to deliver exhaust emissions which are far below Stage II legislated levels. Additional improvements, especially in HC emissions, can be achieved by further enhanced fuel preparation. The use of air-assist injectors combined with optimised engine management strategies can significantly shorten catalyst light-off duration. Clearly, improved fuel preparation will be a major factor in reducing engine emissions to meet Stage III emission legislation.

6. References

/1/ Boyle, R.J., Boam, D.J., Finlay, I.C.: Cold Start Performance of an Automotive Engine Using Prevaporised Gasoline; SAE 930710

/2/ Moser, W., Lange, J., Schürz, W.: Einfluß der Gemischaufbereitungsqualität von Einspritzventilen auf den stationären und instationären Motorbetrieb, 13. Internationales Wiener Motorensymposium 1992, Band 1

/3/ Wittig, S., Himmelsbach, J., Hallmann, M., Samenfink, W., Elsäßer, A.: Gemischaufbereitung und Wandfilmverhalten in Saugrohren von Ottomotoren; Teil 1: Experimentelle und numerische Grundlagenuntersuchungen, MTZ 55 (1994), Nr. 3

/4/ Müller, H., Bellmann, H., Himmelsbach, J., Hallmann, M., Samenfink, W., Elsäßer, A.: Gemischaufbereitung und Wandfilmverhalten in Saugrohren von Ottomotoren; Teil 2: Untersuchungen am Vollmotor, MTZ 55 (1994), Nr. 4

C517/041/96

Piston assembly design for low emissions applications

M WILLCOCK BEng, PhD, **S T GAZZARD** BSc, CEng, ARSM, MIM, and **R B ISHAQ** BEng, PhD
ATC, AE Goetze Automotive, Bradford, UK

ABSTRACT
The impact of reducing hydrocarbon, particulate, NOx and CO_2 emissions on the design of the piston assembly is discussed, in the light of the need to meet emissions legislation for both gasoline and automotive diesel applications. Particular attention is given to the durability issues associated with reducing the crevice volumes between the piston and cylinder in gasoline applications, and the increased engine powers of modern diesel applications. The development of appropriate enabling technologies is reviewed, including the introduction of a new range of piston alloys for improved performance and durability.

Ring and groove interaction arising, due to the development of low oil consumption in gasoline engines, is described in detail and the effect on performance discussed.

The consequences of low oil consumption required to meet particulate targets in diesel engines, is also discussed and attention is given to piston skirt coatings, developed to achieve a reduction in assembly friction and a consequential reduction in specific fuel consumption.

1.0　INTRODUCTION

The major driving force in engine and component development continues to be the demand for a reduction in exhaust emissions. Piston manufacturers are having to adopt strategies of developing many design solutions to enable piston assemblies to operate in the harsher operating conditions.

The emissions which are currently monitored are

　　i.　　Hydrocarbons (HC),
　　ii.　　Carbon Monoxide (CO),
　　iii.　　Oxides of Nitrogen (NOx), and
　　iv.　　Particulates (diesel only).

Fuel economy is a selling factor for vehicle manufacturers and a benefit of this is reduced Carbon Dioxide (CO_2) emissions. As an inevitable by product of combustion, there are limiting factors, but improved fuel utilisation and better efficiency can lead to improvements. For the piston assembly friction is the main factor that needs to be addressed in order to reduce fuel consumption. It is claimed that up to 44% or more of the total frictional energy can be assigned to the piston assembly.

Emissions of hydrocarbons (HC) indicate the combustion efficiency in internal combustion engines and they arise when vapourised unburnt fuel or partially burned fuel products, leave the combustion region and are emitted with the exhaust. Indeed, several piston parameters can be adjusted to benefit the HC emissions, including reduction of the crevice volume and oil contribution to the emissions.

CO emissions arise due to poor air/fuel mixing and fuel rich regions, better controlled by improvements in injection and combustion. Piston design features are not currently accountable for the control and completion of combustion, hence alterations to effect these are not the primary concern to piston manufacturers.

Diesel engine manufacturers are calling for higher rated engines, which in turn increase operating pressures and temperatures, giving rise to the formation of NOx. Exhaust Gas Recirculation (EGR) is one method employed to control these NOx emissions, which affects the piston manufacturer.

Oil consumption control can reduce particulate emissions. In current technology engines, oil contribution to the particulate mass is up to approximately 30% under certain operating conditions. This is an area which is receiving much attention for the reduction of overall particulate mass.

This paper discusses some of the current piston assembly design features for low emissions and describes in some detail the problems which arise in their manufacture and operation. It also outlines some of the cost-effective solutions to these problems.

2.0 GASOLINE PISTON ASSEMBLY STRATEGIES

2.1 HYDROCARBONS

HC emissions arise from several sources. Two of which relate directly to piston assembly features are:

  i.   Crevice volume: The air/fuel mixture is compressed into crevices between cylinder bore and piston where it is unable to undergo combustion due to flame quenching at the narrow entrances to these regions, hence contributing to the HC emissions.

  ii. Oil contribution: Lubricating oil film on cylinder liner/bore surfaces has been shown to absorb fuel, which is later emitted into the exhaust stream during the expansion stroke, giving rise to further HC emissions.

A reduced top land height is the favoured design strategy adopted to reduce crevice volumes in order to reduce the flame quench areas. This also tends to lead to lower lubricating oil consumption resulting in further HC emissions reduction. However, this can lead to wear and damage at the ring and groove interface.

2.2 RING/GROOVE INTERACTION

A phenomenon often referred to as ring welding, describes the damage on the top ring groove due to interaction with ring side faces. This can be attributed to:

(i) reduced levels of lubrication,
(ii) increased temperature of top ring and groove due to the reduction in top land height to limit the HC emissions, and
(iii) abnormal engine operation adopted for emissions control (mainly over fuelling and detonation).

The problem is application specific and only occurs in a few engines. However, results to date show that ring welding may be an early life problem and does not always lead to failure. There is damage during early life, which ceases at a point so that it may go unnoticed for the remainder of the life of the engine. Figure 1a shows a schematic of the presence of ring weld on a groove side face, while Figure 1b demonstrates excessive ring pounding against the grooves in a tested piston. Figure 1c is a scanning electron microscope image showing localised damage on the groove surface.

Although not fully understood, it has been observed that ring weld occasionally includes some material transfer between the top ring groove and the ring side face, but mainly excessive wear and localised regions of severe damage on groove surfaces. This could be caused by localised fatigue during the engines early life.

When lubricant is present in sufficient quantities, the piston and ring do not come into direct contact due to the presence of the lubricant film. However, with low oil consumption engines, the decreased level of lubrication leads to point contact of the protrusions, on the piston material with the ring, due to the surface finish of the piston. This leads to fatigue damage due to high loads, over a reduced number of cycles [1]. Cracks propagating from these regions of damage, allow small particles from the ring groove to break away. Engine failure does not always occur, because oil consumption may increase and eventually allow satisfactory lubrication between ring and groove. At the end of a long test damaged regions will be packed by hard carbon deposits forming a satisfactory surface. In such cases wear of the ring groove would be insignificant. However, in extreme cases the damage develops into severe deformation of the groove (ring pounding) resulting in engine failure. The early indication of this is increased oil consumption and blowby.

2.2.1 Materials Development For Resistance to Ring Weld

In order to develop materials for resistance to ring weld, a series of tests were conducted to compare different ring surfaces. The apparatus used for achieving wear, was a reciprocating wear machine. A pin made up of the piston material (a standard Lo-Ex surface (AE413)), including any coatings, was made to rub against a sample of the ring material. The load and temperature for the teat could be varied and lubricant supplied. The results of these tests are shown in Figure 2a. The superiority of phosphated nodular iron and carbon steel surfaces for rings, relative to a nitrided high-chromium steel is self-evident.

There is further evidence, confirmed also by engine tests, that an improvement in the ring side-face finish can enhance the resistance to ring-welding. As a result of these tests a surface finish of $2.5\mu m$ Rz max. has now been introduced for nitrided steel rings in high top ring positions.

Of the alternative side-face coatings available, a resin-based coating has proved to be the most beneficial, provided that the groove temperature is below about 250°C. Physical Vapour Deposition (PVD) coatings are currently being developed for use at higher groove temperatures.

A range of piston materials have also been subjected to some wear rig tests. The high copper alloy, AE135 and ceramic fibre reinforced materials showed better wear characteristics compared to a standard nitrided high chromium steel surface (Figure 2b). The benefit of a phosphate coated surface is also clearly evident. An anodised surface has also performed exceptionally well in rig tests and has been successfully employed in the USA for several years now, to counteract potentially serious ring welding problems.

2.2.2 High Top Ring Design

When the top land width is reduced below about 3.5mm and particularly when detonation is regularly experienced, the strength of the top land becomes the controlling factor. In these extreme cases, copper or ceramic fibre reinforced alloys which offer high temperature fatigue strength, become necessary to ensure land survival.

Alternative piston assembly designs, which lower the crevice volume with minimised risk of land failure, include a half-keystone ring. These designs are in later stages of development (Figure 3) and can enable the top land width measured at the root of the groove to be increased by around 20%.

If the top land width is reduced below about 2.5mm, then additional reinforcement will almost certainly be required. Localised alloying (with nickel for instance) is feasible, but ceramic reinforcement is technically and commercially the more attractive solution for high volume production.

2.2.3 Piston Materials (Gasoline engines)

A new range of aluminium-silicon alloys (designated 'Æxelsius') have been developed, specifically to meet the durability needs of low emission assemblies. These new alloys are characterised by an increased copper content and high purity (Table 1).

The Æxelsius alloys also combine the enhanced high temperature strength associated with the traditional 'Y' alloy, with the favourable physical and wear characteristics of the conventional Lo-Ex alloys (Table 2). Special metallurgical controls have been introduced in order to achieve a consistently refined structure. Together with the use of sophisticated solidification modelling techniques, these have permitted the attainment of excellent fatigue properties at high temperature, without detriment to the general integrity of the piston or deterioration in the low temperature fatigue strength.

The coefficient of thermal expansion and other physical characteristics of the Æxelsius alloys are close to those of the conventional Lo-Ex alloys (AE413, AE109 and A-SI2UN). However, the hardness is increased somewhat and the wear resistance is significantly enhanced, particularly in the case of Æxelsius 135, where the highest copper level (5%) is combined with a small increase in the silicon level (to 13%). The wear characteristics of this gasoline piston alloy are, in fact, similar to those of the hyper-eutectic alloy (AE425) already well-established in the USA for high-duty gasoline applications.

3.0 DIESEL PISTON ASSEMBLY STRATEGIES

With each new generation of engine, the power ratings of diesel engines are being increased. The drive for this is the need to reduce emissions. The effect of increased engine power on the piston can be assessed by the thermal loading it is exposed to. The trend for increased thermal load is illustrated in Table 3. In addition to the increased load the light duty engines are adopting direct fuel injection (DI) technology and turbocharging (TC) which requires deep re-entrant combustion bowls in the piston crown. When the thermal load is greater than the 3.8 MW/m^2 the temperatures obtained at the bowl edge can exceed 320°C. Around this temperature there is a significant reduction in the fatigue strength of conventional piston alloys. This has a significant impact on the design of the piston.

3.1 RING GROOVE FEATURES

The general trend for European diesel engines is to raise the top ring to allow for the improved utilization of combustion air. An additional benefit is the possible reduction in hydrocarbon levels. The presence of a high top ring does, of course, impose additional durability demands on the ring groove reinforcement. With the gravity die cast piston, these should be satisfactorily met by on-going refinement of the groove reinforcement process.

With the squeeze cast piston two alternative forms of groove reinforcement are possible:

i. For mid-range applications, the Ni-resist insert can be plasma-sprayed with stainless steel, prior to incorporating, in order to enhance substantially the bond strength and permit the employment of very high top rings [2].

ii. For the light duty sector, a ceramic fibre reinforcement (at 15% density) can be used in conjunction with a high-copper piston alloy and a nitrided high-chromium steel ring. Although not in series production this solution is cost effective. Durability tests in a turbo-charged, direct injection engine have indicated that acceptable wear rates are probably achievable with a top land of around 6% of the bore diameter (Table 4).

3.2 CROWN DURABILITY AND BOWL DESIGN FEATURES

The need to improve combustion efficiency has made it necessary to introduce combustion bowl features, such as reduced edge radii and re-entrance. These tend to be prejudicial to the crown durability. Also with the increase in the thermal loading of the piston, as discussed above, the crown experiences higher temperatures. To achieve the durability requirements in these conditions, materials with improved creep and fatigue strength at the maximum crown operating temperature (typically 350°C), have therefore become imperative.

The control of particulate emissions requires a sharp bowl edge upper radius to ensure effective air / fuel mixing. However, the radius at the bowl edge is critical for the crown stresses in both planes and must, therefore, be a compromise between the needs of combustion control and piston durability. Radii greater than 2mm would be preferred for the latter. The current value generally adopted varies from 1 to 2mm.

The top land height also has an influence on the bowl edge. A raised top ring increases the positive hoop stress at the bowl edge due to the pressure balancing effect from the top land. This is demonstrated in Figure 4.

Another bowl design feature which has an influence on durability, is the degree of re-entrancy (the overhang of the bowl lip). Higher temperatures and stresses at the bowl edge are experienced with a large bowl edge overhang. This effect is greater than the bowl edge radius.

3.3 PISTON COOLING

A method for improving the crown durability is to cool the crown and ring groove region of the piston. To achieve this, oil cooling galleries are being developed for the higher rated LVD piston. By using this method, the temperature can be reduced by 20 or 30°C in this region of the piston. The improvement in the resistance to bowl edge cracking has been confirmed by both suitable engine testing and by analysis based on FE simulation of cyclic loading during the tests. Figure 5 plots the predicted effects of an oil cooling gallery in an automotive diesel piston and shows that the shortest life is at the bowl edge. This can be improved by a factor of 5 with gallery cooling.

3.4 MATERIAL / ALLOY DEVELOPMENT

Pistons with ceramic fibre reinforced crowns are in series production for a number of highly rated diesel applications, with undoubted durability benefits [3]. However, Æxelsius 113 may provide a more cost-effective solution. In one turbo-charged application, the substitution of Æxelsius 113 for conventional Lo-Ex has provided an improvement in bowl-edge life by a factor of 3 to 4. As the trend for peak cylinder pressures continues to increase, the mechanical and thermal loadings associated with these pressures will push conventional aluminium materials even closer to their limits, so that piston design will have to be modified in conjunction with alternative high strength materials. Single piece ferrous pistons are currently being developed for medium duty applications. However, during preliminary tests cast iron pistons in an IDI TC light duty engine, struggled to meet the weight and durability targets. The high development and production cost of ferrous pistons for this class of engine, will seriously limit the potential applications which are likely to be confined to a niche market.

3.5 E.G.R.

The NOx from diesel engines is controlled by giving attention to the combustion pressure and temperature profiles. To this end , the use of EGR is becoming widely accepted. However, there are abrasive and corrosive gases in the exhaust such as sulphur dioxide and the nitrogen oxides, which when brought into the cylinder can lead to corrosion problems of the bore, ring and ring groove. Although the sulphur content of fuel has been reduced to 0.05%. This only partially helps to diminish the corrosion problems. In order to improve the resistance of the rings to peripheral scuffing wear, resistant base materials and scuff resistant coatings are being adopted.

3.6 PARTICULATE EMISSIONS

Oil consumption is still a major contributor to particulate emissions and its control is a major issue in minimising diesel exhaust emissions. The piston - cylinder system has been shown to be the main route through which oil enters the combustion chamber. Oil consumption is also the primary source of carbon deposits in the engine. These deposits can result in cylinder liner polishing wear and consequently, an increase in oil consumption rates.

Piston assembly system designs, facilitate adequate lubrication, minimise oil consumption and control the blowby gases. Blowby gases can enhance the deterioration of the lubricating oil, both in the ring groove and in the oil sump. When the piston assembly becomes starved of lubricant, scuffing of the ring peripheries and the piston can occur.

3.7 Ring Design for Low Oil Consumption in Engines

Oil consumption is effected by many parameters of the piston assembly. Most are specific to individual applications, but of major importance is the specification of the ring pack. Current developments for low oil consumption include multi-piece oil control rings, which have improved conformability to the bore [5]. The optimisation of the ring pack to meet the stringent oil consumption requirements, is being aided by predictive and analytical techniques, involving an understanding of the whole system.

4.0 SPECIFIC FUEL CONSUMPTION

A major factor in the engine friction is the piston assembly, which accounts for approximately 44% of the total and the greatest contributor to this is the ring pack [6] , although the influence of the piston increases with engine speed. Figure 6 illustrates the proportions of these mechanical losses. One method used to evaluate friction is to measure the specific fuel consumption. From this several aspects of the piston assembly design and engine architecture have been found to influence the reduction in friction. The most critical are:

- reciprocating mass,
- skirt contact area,
- operating clearance,
- piston skirt coating,
- ring pack specification.

4.1 RECIPROCATING MASS

Energy lost due to piston side loads, can be reduced by the development of low mass pistons. An additional benefit is the reduced load on the crankshaft, due to inertia of the piston and this can allow the down-sizing of the con rod and crankshaft. A shorter compression height (the dimension between the crown and pin centre), can also be achieved to reduce the total height of the engine. However, all piston design factors are inter-related in achieving variations in engine refinement with respect to noise, vibration and harshness. Due consideration must be given to skirt guidance and inter-ring volumes in order to achieve satisfactory noise levels and ring pack performance.

4.2 SKIRT FRICTION

4.2.1 Contact Area

Although piston skirt operation is mainly hydrodynamic, boundary lubrication can occur at engine start up, piston reversal around top and bottom dead centre and sudden engine transients [7]. At these conditions, reductions in friction can be achieved by a reduction in the area of piston skirt in contact with the cylinder bore.

Two methods of achieving this were developed:

i. the Æconoguide skirt form [7], which consists of three raised regions on the skirt and
ii. the X-piston (Figure 7).

Although reductions in friction are achieved with these designs (Figure 8), the Æconoguide cannot achieve the noise performance of a conventional skirt design and the X-piston is more expensive to manufacture.

4.2.2 Skirt Surface Finish and Coatings

The specific fuel consumption can be reduced by adopting a truncated surface finish on the skirt, (Figures 9a and 9b). An additional benefit is the stabilisation of piston to bore operating clearance.

A low friction coating can have a significant effect on reducing the specific fuel consumption (sfc). The coating initially specified for bedding-in during the engine's early life, was graphite. However, with the perceived benefits, further development in material technology have resulted in the production of low friction skirt coatings, with more wear resistant characteristics. Coatings such as AE072 have demonstrated superior improvements in both sfc and durability, surviving extreme gasoline engine tests. Figures 10 shows the fuel economy benefits of this coating during an ECE-based test cycle.

4.2.3 Piston Rings for Low Friction

There have been many developments of ring materials and design, which can reduce piston assembly friction further. Reductions in the axial width of both compression and oil control rings, resulting from the use of steel materials [2], facilitate improvements in piston assembly mass and reduced ring pack friction. Table 5 lists typical ring pack axial widths specified in volume production.

Lower ring widths enable a reduction in piston compression height and hence piston mass. The associated benefit of reduced frictional losses, results from a reduction in both the natural peripheral ring/bore contact load and the gas force on the inner surface of the narrower rings. Further improvements can be achieved with the three piece oil ring with a low radial width which, with a low wall pressure will still achieve oil consumption and blowby performance.

5.0 CONCLUSIONS

The continued pressure for low emissions engines has posed some serious challenges to the piston manufacturer. There is some commonality between the gasoline and diesel pistons in the requirement for improved emissions, low oil consumption and reduced friction. These objectives are being achieved with cost effective and durable components.

For gasoline pistons low mass, high top ring designs have been developed and tested achieving low emissions. Also, ring pack developments including a highly conformable oil control ring, have achieved low friction and low oil consumption.

For diesel engines the increased power ratings have required the development of various design and reinforcement features to achieve crown and ring groove durability.

The various design solutions adopted for each type of engine, has required the development of improved piston and ring materials and these have achieved significant improvements in durability.

6.0 REFERENCES

1 Sagawa, J., Ikeda, Y. and Onuki, T. MoS_2 Coated Piston Ring to Prevent Pistons Aluminium Sticking. SAE 911281, (1991).

2 Neuhauser, H.J., Jijina,,H.D., Plant, R. and Nicolini, G. Steel Piston Rings - State of Development and Application Potential. T&N Technical Symposium, 1995. Paper 16.

3 Jakobs, R., Gazzard, S.T. and Ongetta, R. Mid Range Diesel Piston Systems. T&N Technical Symposium, 1995. Paper 22.

4 Andrews, G.E. Diesel Engine Processes That Influence Particulate Formation. A Short Course on Diesel Particulates and NOx Emissions. University of Leeds. April 1995.

5 Avezou, J.C., Ongetta, R., Philby, J.D. and Risch, H.J. Development of Light Vehicle Piston System. T&N Technical Symposium 1995. Paper 21.

6 Furuhama, S., Tahiguchi, M. and Tomizawa, K. Effect of Piston and Piston Ring Designs on the Piston Friction Forces in a Diesel Engine. SAE 810977, (1981).

7 Rhodes, M.L.P. and Parker, D.A. Æconoguide - The Low Friction Piston. SAE 840181, (1984).

TABLES

ALLOY	GRAVITY DIE CAST	SQUEEZE CAST	NOMINAL COMPOSITION (%)			
			Si	Cu	Ni	Mg
Lo-Ex	*	*	12	1	1	1
Y Alloy	*		<0.5	4	2	1.5
Æxelsius 113	*		12	3	1	1
Æxelsius 121		*	13.5	3	1	1
Æxelsius 135	*		13	5	1	1
Hypereutectic (AE 425)	*		16	2.5	0.5	0.8

TABLE 1. The Composition of Piston Alloys.

CHARACTERISTICS	LO-EX (AE413)	ÆXELSIUS 135
Thermal expansion : 20-200°C ($\times 10^{-6}$ K^{-1})	21	20.5
Hardness (HB)	115	125
0.2% Proof stress* : 200°C	145	180
(MPa) 250°C	85	105
300°C	50	65
10^7 Fatigue strength* : 350°C	32	37

* Following 100 h prior exposure at temperature

TABLE 2. Comparison of Typical Properties for Alternative Gasoline Piston Alloys.

ENGINE TYPE (TURBOCHARGED)	THERMAL LOAD (MW/m^2)	MAX. CYLINDER PRESSURE (MPa)
CURRENT IDI	3.0	13
MID 90'S DI	3.5	13.5
DEVELOPMENT DI	4.0	15

TABLE 3. Typical Operating Conditions for Light Vehicle Diesel Engines.

Top land width (% bore dia.)	9.2	5.3
Groove wear (μm per 100h)	< 0.5	< 10
Ring Wear (μm per 100h)	< 1	< 1

TABLE 4. Durability Test Results for Fibre Reinforced Ring Groove Pistons in a 74 kW Turbocharged DI Engine.

RING	AXIAL WIDTH (mm)		
	LATE 1980's PRODUCTION	EARLY 1990's PRODUCTION	CURRENT PRODUCTION
TOP	1.75 / 1.5	1.2	1.2 / 1.0
2nd	2.0 / 1.75	1.5	1.2
OIL	3.0	2.5 / 2.0	2.0

TABLE 5. Trend for Ring Axial Widths.

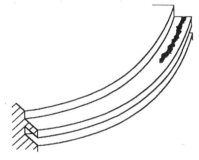

Figure 1a Schematic of ring groove damage

Figure 1b Ring pound damage on top ring groove

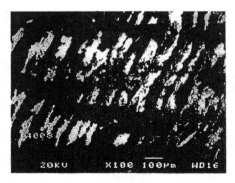

Figure 1c Scanning electron microscope image of damage to ring groove bottom face

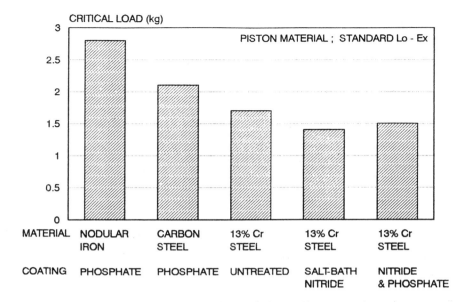

FIGURE 2a RING MATERIAL AND COATINGS TESTED AGAINST Lo - Ex PISTON MATERIAL

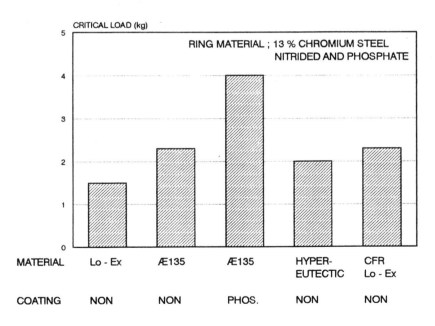

FIGURE 2b RING WELDING TESTS FOR PISTON MATERIALS AND COATINGS AGAINST 13 % CHROMIUM STEEL RING WITH NITRO. AND PHOSPATE COATING

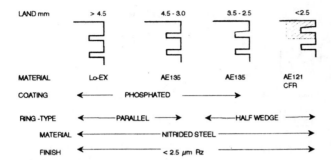

Figure 3 Preferred piston assembly designs for various top land widths

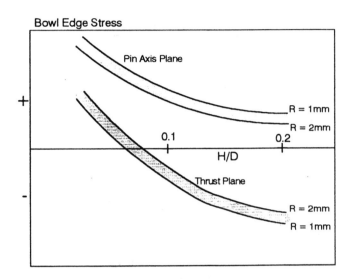
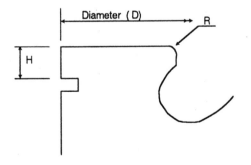

Figure 4 Influence of Top land height and bowl radius on bowl edge stress

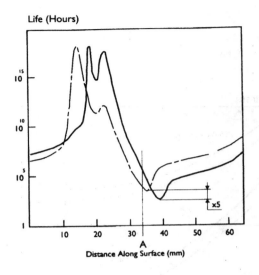
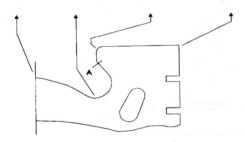

Figure 5 Effects of an oil cooling gallery

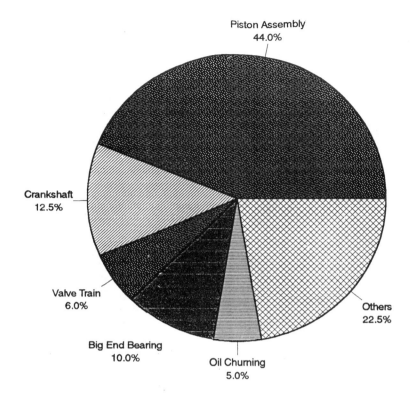

Figure 6 Factors Contributing to Mechanical Losses

Figure 7 The X-Piston

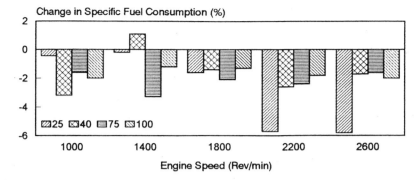

Figure 8a Improvements in Specific Fuel Consumption due to AEconoguide

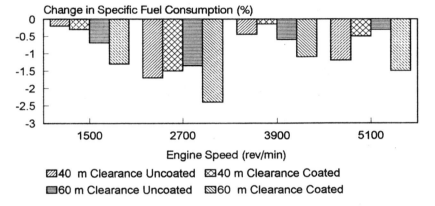

Figure 8b Improvements in Specific Fuel Consumption due to X - Piston

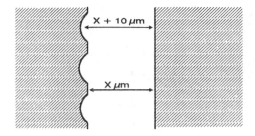

Figure 9 sketch of truncated skirt finish

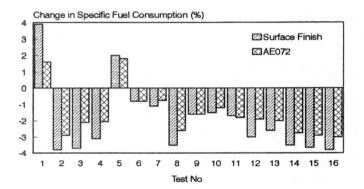

Figure 10 Fuel economy benefits of truncated surface finish and a low firction coating

Aftertreatment

The prediction of fuel consumption and emissions during engine warm-up

P J SHAYLER MSc, PhD, MIMechE, MSAE and **N J DARNTON** BEng
Department of Mechanical Engineering, University of Nottingham, UK
M A
Ford Motor Company Limited, Basildon, UK

SYNOPSIS

Given the availability of test-bed data for fully-warm engine conditions, the prediction of fuel consumption and emissions of HC, CO and NO_x over the ECE+EUDC drive-cycle with a 20°C start-up has been investigated. Predictions of fuel consumption are based upon the assumption that fuel consumption is dependent on indicated operating conditions, but independent of engine temperature. The indicated and brake operating conditions are related through an analysis of engine warm-up and friction characteristics. A similar approach is described for the prediction of emissions. In this case, indicated specific emissions are shown to depend on indicated operating conditions and engine coolant temperature. The effect of engine temperature is shown to be independent of operating condition for a given engine.

NOMENCLATURE

α	Friction correction factor
η_{trans}	Transmission efficiency
ECT	Engine Coolant Temperature (K)
EGR	Exhaust gas recirculation
N_{engine}	Engine speed (rpm)
$\dot{m}_f$	Fuel mass flow rate (kg/s)
P_a	Atmospheric pressure (bar)
P_i	Intake manifold pressure (bar)
$\dot{P}_f$	Total friction power loss (W)
$\dot{P}_{prf}$	Predicted total friction power loss (W)
$\dot{P}_{rl}$	Road load power (W)
T_{engine}	Engine brake torque (Nm)
S_p	Mean piston speed (m/s)

1. INTRODUCTION

Recent proposals to revise specifications for emissions test drive cycles reflect the importance of the early minutes of cold-engine operation. These have been reviewed by Hadded et al [1]. The California Air Resources Board (CARB) have adopted cold start CO legislation at a test temperature of -7°C for current- technology vehicles. In Europe, the EC Directive 91/441/EEC brought European emissions legislation in line with US 1983 standards and the more recent Stage 2 legislation and Stage 3 proposals to be introduced at the turn of the century are planned to bring European emissions legislation in line with that in California. The current European light duty test cycle is also under review in connection with the current 40 second conditioning period before emissions sampling starts, and the specified soak/test temperature, currently 20°C to 30°C. Stage 3 legislation is likely to remove the 40 second idle period and reduce the soak/test temperature to between 0°C and 5°C to reflect patterns of vehicle use and operating conditions in northern Europe more accurately. Although levels of pollutant emissions are the main focus of legislation, fuel economy penalties associated with cold running conditions are related and also of concern, and drive cycle fuel consumption figures provide a useful bench mark for vehicle performance appraisals.

The aim of the work reported in this paper has been to investigate the possibility of predicting drive-cycle emissions and fuel consumption from cold-startup, given the availability of test bed data for fully-warm engine conditions and details of engine brake load and speed variations during the cycle. In general, engine maps are produced for fully-warm operating conditions, which are most easily controlled and provide the bulk of the information required for engine development. Extending mapping conditions to cover the range of possible states encountered as the engine warms-up poses practical difficulties and although coolant and oil conditions can be maintained at lower temperatures by forced-cooling arrangements, these are not routine procedures.

The methodology examined for fuel consumption predictions makes use of indicated specific fuel consumption data for fully-warm engine conditions, mapped as a function of indicated mean effective pressure and engine speed. This map is assumed to be independent of coolant temperature and thus, provided engine friction loads can be predicted, fuel consumption can be determined from indicated loads during warm-up. For application to the drive cycle with the engine cold started at 20°C, the interactions between friction losses, fuel consumption, and coolant and oil warm-up behaviour are taken into account using an engine warm-up model. The application of the procedure to predict fuel consumption was described in detail in [2]. In the following, the formulation and details of the procedure are described and the possibility of extending this to predict emissions is examined. Experimental data is presented to support the formulation.

2. FUEL CONSUMPTION PREDICTIONS

For a given vehicle and drive-cycle specification, engine speed variations with time over the cycle can be defined from knowledge of rolling tyre radius and transmission ratios, and the main component contributions to the engine brake load can be described by the following equation:

$$T_{engine} = \frac{60\ (\dot{P}_{rl}+\dot{P}_{KE})}{2\pi\ N_{engine}\eta_{trans}} \quad (1)$$

where:

$$\dot{P}_{rl} = F_0 + F_1 v + F_2 v^2 \quad (2)$$

v is vehicle speed, F_0, F_1 and F_2 are coefficients determined by experiment. $\dot{P}_{KE}$ is the power required at the road wheels to increase the kinetic energy of the vehicle during acceleration. This term is set to zero when decelerating, as under these conditions energy is dissipated through the vehicle brakes and the engine is motored. The over-run periods when the engine is motored are treated in the following as an idle running condition with fuel consumption proportional to engine speed.

When the engine is operating at conditions other than fully-warm, indicated conditions will differ for a given brake operating condition, due primarily to higher engine friction loads during warm-up. Mixture preparation conditions in the intake port will also differ and this will affect emissions particularly, but indicated specific fuel consumption is much less sensitive. Here, indicated specific fuel consumption is assumed to be simply a function of imep, speed, calibration settings for spark timing and air fuel ratio and, if appropriate, exhaust gas recirculation rate. An example of data showing that the dependence of isfc upon imep is weakly dependent upon coolant temperature for a given calibration is given in Fig 1. In this case, the data are for a 2.0l DOHC 8v engine operating at stoichiometric mixture conditions. The engine was cold started at -25°C, and initial transient effects associated with start-up were found to have decayed by the time the coolant temperature reached -10°C. These effects give rise to a further increase in fuel consumption over and above that associated with poor mixture preparation directly, but over the period of an engine drive cycle the additional increase is a small proportion of the total.

If the fully-warm engine data are available only as a function of brake operating conditions then some post-processing is required to relate brake loads to indicated loads. This requires calculations of mechanical friction losses which the authors compute using the friction model for spark ignition engines developed by Patton et al [3]. The friction model should account for mechanical friction contributions only (excluding throttling losses for example). The total friction power loss $\dot{P}_f$ is split into four components such that:

$$\dot{P}_f = \alpha(a+b+c+d)\dot{P}_{prf} \quad (3)$$

where the coefficients a to d represent the proportions of the steady state friction losses for the valve train, the bearings, the piston, and the auxiliary components respectively, and these coefficients sum to a value of unity. The factor α, taken here to be 1.34, is used to compensate for discrepancies between friction losses measured on several four-cylinder in-line engines with capacities in the range 1.1 to 2.0l and the friction losses predicted by the model. Generally piston friction accounts for the largest proportion of the total friction losses, and is dependent on gas pressure. The contribution to the piston mean effective pressure friction loss (in kPa) due to gas pressure loading is defined by [3]:

$$\text{fmep}_{gas} = 6.89 \frac{P_i}{P_a} \left[0.088 r_c + 0.182 r_c^{(1.33-kS_p)} \right] \quad (4)$$

where P_i is the intake manifold pressure, r_c the compression ratio S_p the piston speed and $k = 2.38 \times 10^{-2}$ s/m. The ratio P_i/P_a is described as a function of mass fuel flow rate, air/fuel ratio, and swept volume.

Predicting fuel consumption as a function of time during the drive cycle entails determining friction changes as the engine warms up in parallel with a time marching calculation of fuel consumption from the corresponding indicated operating conditions. The rate at which the engine structure and lubricating oil warm-up influences friction losses over the cycle, because these depend on oil viscosity which in turn depends upon temperature. The effect of viscosity changes is taken into account by scaling the instantaneous friction loss in proportion to values at the same running condition for fully-warmed up conditions. The correction is given by [4]:

$$\dot{P}_{warm-up} = \left(\frac{v}{v_{ref}} \right)^n \dot{P}_f \quad (5)$$

where the reference viscosity value is for a temperature of 90°C and v is the viscosity of the oil at any time during warm-up. $\dot{P}_{warm-up}$ is the friction loss at the lower oil temperature and $\dot{P}_f$ is the friction loss at the fully-warm operating state. The index n has been determined experimentally to be typically 0.19 - 0.24. The warm-up behaviour of the engine over the drive cycle is determined using a computer program for modelling engine thermal systems (PROMETS), which is described in detail in [4]. For each time step through the cycle, an iterative procedure is used to obtain a self-consistent variation of friction, indicated operating conditions, and oil warm-up characteristics. Generally, not more than three iterations are required to establish the final fuel consumption variation. Once these calculations have been carried out, the fuel consumption over the period of interest is obtained by summing the fuel used during each time step.

4. APPLICATION

The following illustrates the application of the procedure to a Ford 2.0 1 DOHC 16v engine installed in an Escort sized vehicle driven over the ECE+EUDC drive cycle in the first place with the engine operating fully-warm throughout and in the second with a cold-startup from soak conditions of 20°C. The initial 40 seconds of idle running has been eliminated. In both cases the engine has been run at stoichiometric mixture settings throughout, and the EGR rate and spark timing conditions were retained as provided by the corresponding steady state fully-warm dynamometer test data. A comparison of cumulative fuel-used results is shown in Fig 2. In the fully-warm case, the predicted fuel used was within 0.6% of the measured value. For the 20°C cold-start cycle, the fuel consumption penalty associated with cold-start up was 4.98%. This is primarily associated with the higher fuel flow rate in the early part of the drive cycle, when the friction levels are most significantly different from the fully-warm values, although the effect of the warm-up period extends over the entire duration of the drive cycle. The size of the penalty will depend upon vehicle details, and is likely to be higher for smaller vehicles because friction work will represent a higher proportion of the total indicated work done over the drive cycle.

Although the main test of interest in this study entails cold starts at 20°C, the effect of start temperatures down to -20°C has been examined. Fig 3 shows the fuel penalty associated with cold operation over the drive-cycle for the range of start/soak temperatures. The fuel penalty of 4.98% for a 20°C start increases to 10.53% for a start/soak temperature of -20°C. The predictions are based on stoichiometric running conditions throughout the cycle, and hence are likely to represent the minimum penalty at the lower end of the temperature range, when fuel enrichment is normally required. The ratio of total friction work to total indicated work for the drive cycle (ie friction ratio) increases as the engine start temperature is lowered, from 0.39 at 80°C to 0.49 at -20°C. The generally high value of this ratio is indicative of the importance of frictional losses under the relatively light-load operating conditions imposed by the drive cycle. For the fully-warm engine case, 26% of the total friction work occurs during idle and over-run periods. The friction ratio increases from .39 to .44 when the engine is started cold at 20°C rather than fully-warm. The increase in fuel consumption, of 5%, is a direct consequence of the increase in friction, but not in direct proportion to this because the higher indicated loads affect a change in indicated specific fuel consumption (isfc). For the engine map examined, the isfc was improved, and partly offsets the effect of the increased friction work, as indicated by Fig 4.

5. EMISSIONS PREDICTIONS

The procedure for predicting fuel consumption can, in principle, be extended to predict exhaust emissions. This task is more difficult because emission concentrations are more sensitive to mixture preparation changes during warm-up and because over-run conditions, air/fuel ratio excursions during transients and the early seconds of engine running after start-up may, in each case, give rise to significant contributions to the cumulative total emissions over the cycle. The importance of the latter is considered in Section 6.

To examine if the effects of mixture preparation on emissions might be characterised by a relatively simple function of coolant temperature, experimental data were recorded on a Ford 2.0l DOHC 8v engine during a series of cold start warm-up tests. The tests were carried out for a range of engine speeds, indicated loads and tail-pipe air/fuel ratios. In all cases mass flow rates of emissions were calculated from air mass flow rate and tail-pipe air/fuel ratio, together with concentration measurements taken with a Signal 3000 series FID HC analyser and 4000 series chemiluminescence NO_x analyser. CO measurements were recorded using a Horiba MEXA-324 GE analyser. In the first instance, the early seconds of engine running after cold start were excluded from the measurement period to eliminate initial transient effects which influence HC emissions particularly.

Figure 5 shows the experimental data for ISCO, ISHC, and $ISNO_x$ plotted as a function of engine coolant temperature(ECT). At each ECT value, the absolute emission value is divided by the corresponding fully-warm value to give an ECT Correction Factor. The data were recorded from tests covering engine speeds from 1250 - 2500rpm, IMEP values of between 2.5 and 7.5 bar, and air/fuel ratios from 12-1 to 17.5-1. As can be seen from the figure, when normalised to values obtained at fully warm operating conditions, the indicated specific emissions are strongly correlated to engine coolant temperature. For each of the pollutants of interest, a simple function of ECT can be defined which characterises the correction factor required to define indicated specific emissions to within plus or minus 5% of the observed values. This function is independent of speed, load and air/fuel ratio to within this level of accuracy.

In an earlier paper, Shayler et al [5] described studies of the relationship between emissions and operating conditions and fuel injection system details. They showed how many of the effects on emissions could be explained on the basis of mixture inhomogeneity within the cylinder at the time of combustion. In the current work, ECT provides an indication of mixture preparation conditions in the intake port, which in turn give rise to varying degrees of mixture inhomogeneity within the cylinder and consequently to variations in emissions from values obtained at the fully warm engine state. For the engine used in the current study, emission concentrations against air/fuel ratio at one operating condition are shown in Fig 6. With reference to this, the lower indicated specific CO emissions observed during warm-up are attributable to the bulk of the cylinder mixture producing CO being leaner than the overall mixture air/fuel ratio. Fig 5 suggests that CO emissions at 20°C are between 5%-10% lower than those for the corresponding fully warm operating condition, irrespective of engine speed, load and air/fuel ratio when normalised.

Indicated specific HC emissions are relatively high during the warm up period. Fig 5 shows that the indicated specific HC emissions are approximately 30% higher when the coolant temperature is 20°C, compared to the fully warm case. Attributing this increase to the inhomogeneity resulting from poorer mixture preparation would require a stratification of air/fuel ratio in the mixture of about 10% between the rich and lean portions.

Normalised data for indicated specific NO_x emissions are lower at low ECT values, by around 30% at an ECT of 20°C, again independent of speed, load and air/fuel ratio when normalised to the fully-warm emission levels at a given operating condition. The NO_x emissions are influenced by the air/fuel ratio of the lean part of the stratified mixture, although the changes in NO_x emissions cannot be entirely due to this, since the same dependence on ECT is observed regardless of whether the engine was run lean or rich of the air/fuel ratio producing peak NO_x in Fig 6. It is likely that the reduction in NO_x emissions is significantly influenced by lower combustion temperatures during warm up.

The mechanisms producing these trends are likely to arise in any engine but whether or not the simple ECT correction functions used in the current work are adequately representative for general application is an open issue and requires further work.

6. ILLUSTRATION

The procedure has been applied to the drive cycle with a 20°C soak/test temperature, and without the initial 40 seconds of idle before emission sampling begins. As before, the engine has been run at stoichiometric mixture settings throughout, with EGR and spark timing settings provided by the steady state fully-warm dynamometer test data. The data are for a Ford 1.8l Zetec engine installed in a Mondeo sized vehicle. In the first instance, to assess the effect of mixture preparation changes during warm-up, the cumulative variation of engine-out emissions were compared to variations assuming the engine is fully warm at the start of the drive cycle. These variations were computed using the assumption that start-up effects on emissions are negligible, that perfect mixture control maintains stoichiometric conditions throughout the drive cycle, and that during the over-run periods of the cycle, emissions mass flow rates for no-load apply at a given speed. The computed changes in cumulative totals associated with the deterioration in mixture preparation during warm-up represented a 2.1% increase in HC and decreases of 0.7% and 1.6% in CO and NO_x respectively. Poor mixture preparation increased the HC penalty but actually reduced the CO and NO_x penalty associated with a cold start. For the engine/vehicle combination examined, emissions during the over-run periods of the cycle accounted for approximately 11% of the HC and CO totals and 8% of the total NO_x emissions.

To take into account the effect of air/fuel ratio excursions during transient changes in operating conditions requires a prediction of excursions. This requires knowledge of fuel transfer behaviour in the intake port and a representation of the fuel supply control strategy which is beyond the scope of the current work. However, the potential influence of such variations has been examined by imposing excursions of ± 2 air/fuel ratios for two seconds at points in the cycle where rapid changes in charge flow rate or gear changes occur. This enabled an approximate percentage change in each of the emission concentrations during the transient points to be calculated and hence the percentage change of the total drive cycle results to be determined. Fig. 7 shows that the imposed excursions give rise to increases of 1.7% and 33.23% in HC and CO emissions, respectively, and a 7.25% decrease in NO_x emissions. The large increase in CO emissions is due to the sensitivity to air/fuel ratio variations on the rich side of stoichiometric.

A further area of concern is the period immediately after start-up when HC emissions are higher than would be associated solely with the mixture preparation changes characterised by the ECT correction function. An additional increase in HC occurs which is due in part to fuel deposited on the combustion chamber surfaces prior to these reaching high temperatures sufficient to vaporise the fuel. During this initial running period, there is a difference between the fuel injected and the fuel burned, as accounted for by exhaust gas analysis. Experimental results indicate that up to 30% of the difference may be transported into the exhaust gas stream as unburnt hydrocarbons. To model this, a function derived from experimental results for a Ford 1.8l Zetec engine has been used to relate 'lost' fuel to fuel supplied, start temperature and ECT. The 'lost' fuel has an almost negligible influence on the cumulative fuel used over the cycle ($\sim 0.1\%$), but if 30% of this appears as unburnt HC emissions, it increases total feedgas HC emissions by 2.2% and, unless very rapid catalyst light-off is achieved, makes a potentially large contribution to tail-pipe HC emissions because it occurs early in the drive cycle as shown in Fig 8.

To examine the effects on tail-pipe emissions, computations have been carried out for a catalyst with an assumed conversion efficiency of 98% and light-off times typical of various catalyst types and locations. These times were assumed to be 23 seconds for an electrically-heated catalyst (EHC), 70 seconds for a close-coupled catalyst (CCC), and 130 seconds for an underbody catalyst location (UBC). Fig. 9 shows that in the best case, and excluding the effect of the 'lost' fuel contribution to HC, tail-pipe emissions are reduced to around 2.7% of the feedgas total. Including the 'lost' fuel contribution to HC increases the best-case from 2.7% to 3.5%. This illustrates the importance of transient phenomena immediately following engine start-up.

Finally, the effect of air/fuel ratio excursions on tail-pipe emissions has been examined. The same pattern of excursions during the cycle was imposed as before when considering the effect on feedgas emissions, and the catalyst efficiency was assumed to be zero for HC and CO during rich excursions and for NO_x during lean excursions. The result is shown in Fig. 10. As for the feedgas emissions (Fig. 7), the CO increase reflects the strong dependence upon air/fuel ratio. The HC and NO_x tail-pipe emissions account for about 15% of the feedgas totals when the simulated mixture excursions are imposed, compared to 2.5% over the same intervals of the cycle when mixture control is assumed to be perfect.

7. CONCLUSIONS

A method of predicting fuel consumption and emissions over drive cycles starting with the engine at cold-soak temperatures has been investigated. The procedure requires fully-warm test bed data to be available for the engine over a range of indicated loads and speeds imposed by the cycle.

Experimental data supports the assumptions made. Indicated specific fuel consumption is a function of indicated operating conditions independent of coolant temperature. Indicated specific emissions of HC, CO and NO_x during warm-up can be normalised to the levels of the corresponding fully-warm indicated operating conditions and related to simple functions of engine coolant temperature to account for the deterioration in mixture preparation.

Illustrative applications of the procedure have been presented. Typically, fuel consumption over the ECE+EUDC drive cycle is increases by approximately 4-5% when the engine is cold-started at 20°C and compared to results for fully-warm start values. The effects of cumulative emissions over the cycle of mixture preparation changes, start-up contributions, imperfect mixture control during transients, and over-run conditions have been explored and their relative importance examined.

Poor mixture preparation and cold-start phenomena prior to catalyst light-off have substantial affects on cumulative tail-pipe emissions of HC. Mixture ratio excursions throughout the cycle have a potentially large influence on HC and, particularly, on CO emissions.

ACKNOWLEDGEMENT

The authors wish to express their thanks to the Ford Motor Company for support for this work and permission to publish the results given in this paper.

REFERENCES

[1] O Hadded, J Stokes, D W Grigg, 'Low Emission Vehicle Technology for ULEV and European Stage 3 Emission Standards', IMechE Paper C462/18/126, Autotech '93, 1993.

[2] P J Shayler, N J Darnton, T Ma, 'Predicting the Fuel Consumption of Vehicles for Drive Cycles Starting from Cold Ambient Conditions', SIA Paper SIA9506A27, Proc. EAEC 5th Int Congress Conference A, Powertrain and the Environment, Strasbourg, 1995.

[3] K J Patton, R G Nitschhe, J B Heywood, 'Development and Evaluation of a Friction Model for Spark Ignition Engines', SAE Paper 890836, 1989.

[4] P J Shayler, S J Christian, T Ma, 'A Model for the Investigation of Temperature, Heat Flow and Friction Characteristics During Engine Warm-up', SAE Paper 931153, Reprinted in SAE Transactions, J of Engines, 3, pp 1588-1597, 1993.

[5] P J Shayler, J Turner, K Ford, 'Exhaust Emissions: The Influence of Fuel Injection System Details', IMechE Paper C394/015, IMechE Int Conf on Automotive Power Systems, Chester, 10-12 September, 1990.

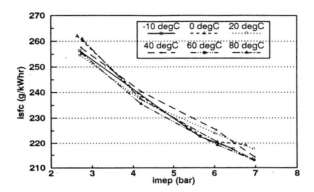

Fig 1   isfc plotted against imep for four fixed throttle warm-up tests at 1750 rpm for 2 litre DOHC 8v engine. Tail-pipe air/fuel ratio stoichiometric throughout.

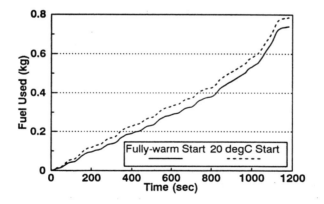

Fig 2   Cumulative fuel used over the ECE+EUDC drive cycle for cold start (20°C) and fully-warm (90°C) engine initial conditions.

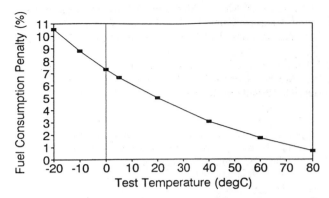

Fig 3   Fuel consumption penalty (% increase compared to value for fully-warm initial engine state) as a function of test-temperature.

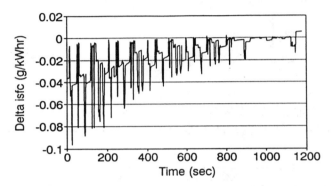

Fig 4   Change in isfc (Delta isfc) when start/soak temperature is reduced from fully-warm to 20°C, as a function of time from drive cycle start.

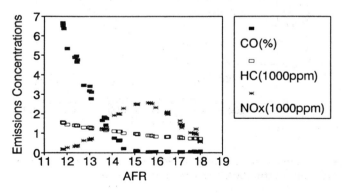

Fig 6   2 litre DOHC 8v engine-out emissions concentrations plotted against tail-pipe air/fuel ratio. Steady-state fully-warm test conditions: 1250 rpm, 80 Nm brake load.

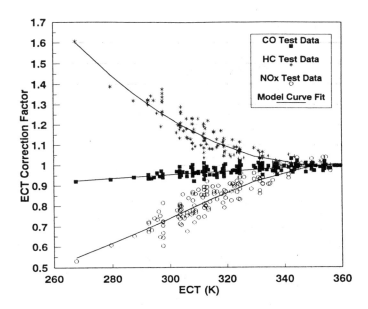

Fig 5 Indicated specific emissions divided by fully-warm indicated specific emissions (ECT correction factor) against ECT. 2 litre DOHC 8v warm-up test data with range of speeds, imeps and AFRs (1250-2500 rpm, 2.5-7.5 bar, 12:1 - 17.5:1)

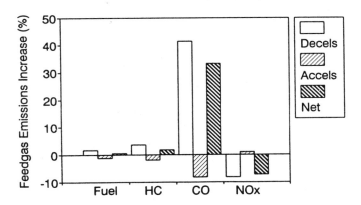

Fig 7 Increase in total fuel consumption and emissions for ECE+EUDC drive cycle with 20°C start due to air/fuel ratio excursions at transient operating points.

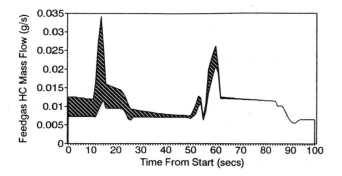

Fig 8   Predicted HC mass flow rate increase due to lost fuel (shaded) for ECE+EUDC drive cycle with 20°C start temperature and without 40 second idle period.

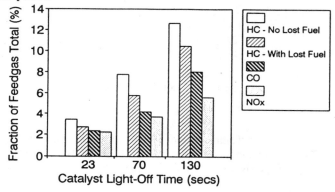

Fig 9   Tail-pipe emissions as a fraction of feedgas emissions for ECE+EUDC drive cycle with 20°C start for various catalyst light-off times. HC emissions shown with and without affect of accounting for 'lost' fuel.

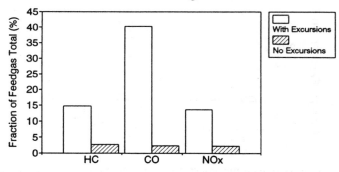

Fig 10   Effect of transient air/fuel ratio excursions on total tail-pipe emissions as a fraction of total feedgas emissions for ECE+EUDC drive cycle with 20°C start.

# Robustness design of experimental approach to the optimization of fast light-off of catalytic vehicles emission systems

RUTTER, R HURLEY MSAE, **D EADE, A FRASER, S BRETT, R SHRIEVES,** and **J KISENYI**
Ford Motor Company Limited, Essex, Basildon, UK

Abstract

The first 200 seconds of the MVEURO automotive drive cycle produces about 80% of the overall hydrocarbon (HC), carbon monoxide (CO) total emissions during that period. The current MVEURO drive cycle permits an initial idle of 40 seconds during which time the exhaust after treatment system has sufficient time to warm to near an exothermic temperature. The first 40 second idle has been deleted from the proposed new MVEURO drive cycle. The implication of this deletion are that more unconverted HC and CO will be measured during the total cycle, making regulatory compliance very difficult.

In order to meet this legislative compliance challenge and become 'best in class' (BIC) a technological innovation is required. There are in fact two current technologies in Ford which reduce these gas emissions during the first 200 seconds after engine start. These technologies are referred to as:

a. EGI - Exhaust Gas Ignition
b. CSSRe - Cold Start Spark Retard (enleanment)

This paper specifically looks at the application of Design of Experiments on the CSSRe technology and how these statistically designed processes enable the prime characteristics in the design to be successfully optimised to ensure that all forms of variability are taken into account before the vehicle is sold to the consumer.

The Design of Experiment chosen is often referred to as a central composite design since it requires each factor (design characteristic) under investigation to be set at three levels in a planned controlled way. The output from this experimental design enables the effects of the factors at these three levels and takes into account interactions between them. Resulting prediction equations provide an opportunity to develop three dimensional plots referred to as response surfaces which visually indicated the effects of changing these factors against the responses being measured. Further these prediction equations enhance computer aided engineering models bring them close to 'real life' working conditions.

Introduction

There are six major aspects to robustness considerations. All need to be taken into account to fulfil FAO's mission to provide 'Best in Class' (BIC), cost competitive products. These developments are:

- obtain and confirm customer wants
- fulfil customer wants at a system level
- analyse the system
- assess and confirm effects of 'noise' on the system
- conduct investigative and confirmatory tests
- conduct key life tests to validate the design

Obtain and Confirm Customer Wants

The latest Corporate quality definition states:

"Quality is defined by the customer. The customer wants products and services that throughout their lives meet and exceeds his or her expectations at a cost that represents value".

If we focus our design activities on satisfying the customer we must first find out what the customer wants. Market research provides the necessary processes. Having conducted our market research activities and acquired customer wants, a translation process is required to enable the engineering activities to find design solutions to meet these wants. In the first instance engineering activities should concentrate efforts at a conceptual system design level. This is in keeping with the customer approach to defining quality since customers see systems and functions of a product.

Fulfil Customer Wants at a System Level

The Ford System Design Specification Handbook states:

Systems Engineering is a process to transform customer's needs into effective designs. The process enables product engineers to optimise designs within and across systems.

Systems Engineering is very closely linked to Customer Focused Engineering (CFE) which is the process of taking Market Researched customer wants and translating them into engineering terminology. Knowing customer wants enables engineers to 'design in' quality. Customers will also convey what they do not want. These take the form of failures in function generally. This data should be added to a System Failure Modes and Effects Analysis (FMEA). Both the 'customer wants' and 'don't wants' are the major inputs into engineering System Design Specifications (SDS). The distinct domain of customer and engineering are illustrated on the next page in Fig 1.

The engineered after treatment system - customer and engineering domains

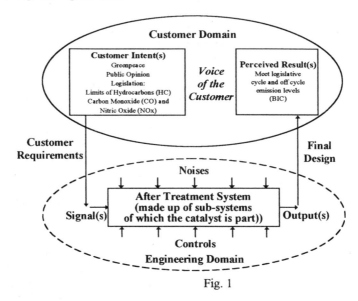

Fig. 1

Customer Domain

The customer or user domain centres on the expectations of those who come into contact with the product. Some customers will be internal to the organisation, those who are recipients of previous department output: for example, the manufacturing organisation is a recipient of the design output, the service organisation is recipient of both design and manufacturing outputs.

External customers include the dealers, the consumer, pressure groups (eg. Greenpeace) and the legislators. The latter are making increasing demands on the product for improved safety and the ever decreasing permissible emissions into the atmosphere.

Each group of customers will have their particular requirements and it is important that all these are captured before design solutions are considered. Indeed, this activity of 'gathering the voice of the customer' is the basis of defining the 'design problem' to which solutions are to be found.

Customers will indicate their requirements in terms of expectations. With careful questioning and observation customer perceptions of performance and satisfaction can be deduced and measured technically by instrumenting vehicles during customer competitive appraisals. This data is particularly important during the process of translating wants into an engineering specification.

Concerning reliability, the expectations of customers are changing. Many vehicle manufacturing organisations are offering 60,000 mile or 3 year warranties. Legislation on emissions is rapidly moving towards 150,000 mile or 10 year compliance. Indeed Tom Cackett, deputy executive director of California's Air Resources Board (CARB) stated the intention to:

> *"assure that every car meets the very low emissions vehicle (LEV) standards being adopted throughout their useful life".*

'Corning Incorporated Auto Emissions June/July 1993 No. 1'

Within the CFE methodology, the Quality Function Deployment (QFD) process provides a structure on which to operate on these wants. Other quality tools and techniques provide a means of identifying those wants which relate to ideal function and those usages to which the final system design must be robust. The legislators demand lower emissions which on reaching and anticipating new requirements will be 0.1 gms Hydrocarbons (HC) during the Motor Vehicle European Drive Cycle (MVEURO Emissions Cycle). The engineers will be required to meet these requirements as a minimum.

In the future, the principle customer, the legislators will require lower emissions for a range of temperatures and humidities and over an ever increasing period of service life. The engineers will need to test a range of ambient temperature and humidities ensuring lower emissions under all conditions. These conditions of ambient temperature humidity, etc. are referred to as noise factors and are factors over which the engineering has no control, but for which the final solution must be robust. Robust in this context is defined as little or no variation in function in the presence of these 'noises'.

Engineer Domain

The engineer translates customer requirements into engineering terms known as Technical System Expectation (TSE) with target values. These target values are established using instrumentation on vehicles perceived by customers to best provide for their needs during vehicle appraisal. The engineering terms become the characteristics of the system design specification (SDS) for which the engineer is in control. At this stage the engineer can begin to include 'noise factors' which should be considered and included in any planned experimental work to ensure ideal function under all conditions including time. The FMEA and Cause and Effect Diagrams are a fertile source of noise information.

The Engineered System

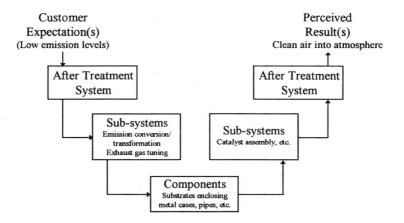

Focus on system and sub-system solutions to meet customer expectations rather than permitting components to define the system.

Fig. 2

Typically the system design solution is made up of several sub-systems, which in turn comprise of one or more components. A design selection process using the customer focused engineering system design specification as the determining process will assist in deciding the most appropriate solution system, one which has the lowest variability of function and meets and betters the target emission requirements in the catalyst's case. This is the analytical process where each requirement is traced from concept through to components of the system. The components are then recombined into the final system solution

At each stage system, sub-system, or component, an optimising process can take place using knowledge based methodologies such as design of experiments, FMEA and Process Capability studies.

During the optimisation from component to total system a complete design is generated which takes into account reliability. The focus is not just on the initial optimisation considering the function when parts are first manufactured and assembled, but in-depth consideration is given to the function over time. It is the deterioration or degradation, % loss of function or time to failure, of each component and sub-system that will have an effect on the performance of the total system and customers perception of the product over time. Further, a deterioration of one system is likely to have a detrimental affect on other systems.

The principles of robustness extend to ideal function over time. Ideal function over time and the maintenance thereof is the subject of reliability. Traditionally, reliability has focused on tests to bogie or failure. Reliability in the robustness arena focusses on maintaining low variability of function and in the case of the catalyst high percentage efficiency over time.

Analysing the System - Catalyst Design

To understand how to analyse a system it is useful to remind ourselves that the focus should always centre on ideal function, the system intent/purpose. If we consider an exhaust catalyst, the signal (input) are the untreated harmful emissions from the engine. The ideal function of the catalyst is to transform these harmful gases by oxidation of hydrocarbons (HC), carbon monoxide (CO) and a reduction of Nitric Oxides (NO_x) into carbon dioxide (CO_2), nitrogen (N) and water (H_2O).

The catalyst comprises of a metal container with an inlet and outlet into which are placed ceramic or metal substrates. These substrates (metal or ceramic - see Fig. 3 below) are coated with a preparatory wash coat followed by a coating of precious metals (palladium, platinum and rhodium). It is the contact interaction of the exhaust gas as it passes through the substrate with these precious metals that stimulates the conversion/reduction processes.

Over a period of time however, the exhaust temperature, together with various contaminants including sulphur and lead from the fuel, damage the wash coat and precious metals reducing their efficiency (see Fig. 4). Eventually a point is reached when unconverted gases 'break through' (pass through) these substrates unconverted and escape into the atmosphere possibly resulting in failure to meet legislation.

Substrates

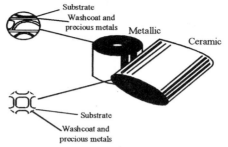

Fig. 3

Effects of Deterioration Over Time

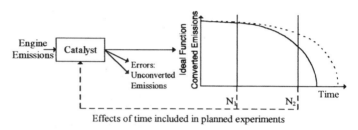

Effects of time included in planned experiments

Fig. 4

Assessment of Noises

There are five main areas of 'noise' that an engineer must consider when analysing a system. This will be referred to as a noise tree in this paper. These are combined in Fig 5. It is these noises that are likely to affect the ideal function of the system. Any component, sub-system or system test should account for these noises and find the optimum design solution that is robust to them.

Institutionalisation of Robustness

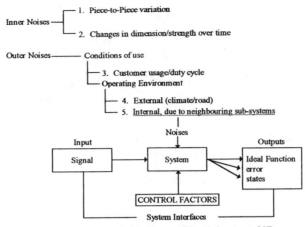

Fig. 5

All noises should be included in the System Design Specification. All 5 noise types will be considered with particular attention to the 5th, that of outer noises due to internal interactions with neighbouring sub-systems.

Analytical Perspective

If we were to consider the catalyst as a component then the tests to prove robustness would take the following into account using the noise tree from the previous page. This list is not exhaustive but serves to illustrate the thought processes.

Piece to piece: manufacturing variability
- loading of precious metals
- thickness of the washcoat
- length of the substrate

Changes in dimension/strength over time - reliability given that degradation has occurred - % conversion efficiency
- available surface area of precious metals
- free % of substrate available for gas flow

Conditions of use/duty cycle
- different types of fuel
- modes of driving

External: operating environment - differing climatic conditions
- temperature and humidity
- altitude
- road conditions

Internal - sub-system to sub-system
- engine exhaust gas temperature variability
- engine exhaust gas emissions composition

Testing might take two forms:

1. multiple regression of a set of measured parts against a response to establish any correlation
2. one factor at a time experiments taking each extreme value separately
3. making several changes at once in a planned Design of Experiment which can vary from a screen experiment to a detailed consideration of a few factors and the establishment of interaction.

The principle of system engineering is to consider all causes of variability and prove robustness of function to them. It would be easy to evaluate this component in isolation from the rest of the vehicle and believe that tests built around the above sources of noise would prove robustness, particularly if DoE were used to explore and discover relationships between the control and noise factors. Control factors are characteristics of the design over which the engineer has control/influence.

Illustrated over the page in Fig. 6, it can be shown that the catalyst is required because the engine process of converting the chemical energy in the air/fuel mixture to power at the crankshaft is less than ideal. The error states of unburned hydrocarbons, heat and lead/sulphur poisoning are noises to the catalytic process. Design settings can be chosen to provide ideal function with minimum variability given these error states occur in practice.

System Engineering with Robustness as the Focus

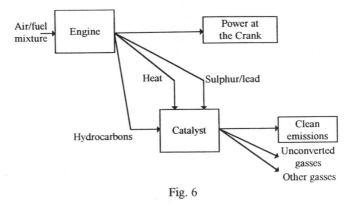

Fig. 6

Customers measure our vehicles against engineering's ability to provide function under all conditions. In the case of the catalyst the main customer is the legislator.

There are some key messages in the above diagram which are worth considering, namely:

1. Error states from the engine form both signal input into the catalyst and noises. This is not always the case since some sub-systems have signal inputs from completely separate functions.

2. Error states take the form of Things Gone Wrong (TGW) and are usually inter-linked in some way, for example, if we were to reduce the exhaust gas temperature either by over fuelling or positioning the catalyst a greater distance away from the engine to allow the gases to cool, we would require a more expensive larger catalyst to deal with the greater emissions or the likely greater deposits of sulphur and lead respectively. It is important to focus on the ideal function of each part of the system. In general terms this leads to cost efficient solutions.

3. Lastly, whatever the design chosen and however well optimised, laws of physics will always apply.

Associative Perspective: System to system interaction

The catalyst is fitted to the exhaust system and forms part of the 'after treatment system'. Exhaust gases pass from the engine after combustion through the catalyst and out through the expansion and resonator systems to the atmosphere. There are other systems which have an impact on the ideal function of the catalyst.

Team work, inter-departmental communication and co-operation, is essential if all sub-system interactions are to be identified. The first step is to detail associative relationships to any system. As illustrated in Fig. 7 the catalyst has many interactions. Each one of these interactions will have greater or smaller impact on the catalyst performance (ideal function). Each of the inputs is subject to all or some of the five noises types discussed earlier, but now we have a bigger picture. For example we can note that degradation on the ignition system is likely to create a greater amount of unburned hydrocarbons which in turn causes increased stress on the catalytic process.

Systems Affecting the after Treatment System Ideal Function

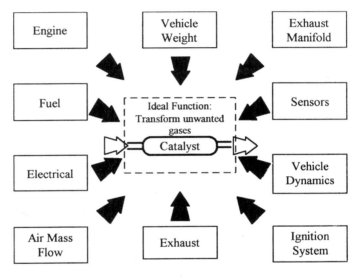

Fig. 7

The Overall Affect of System Interactions

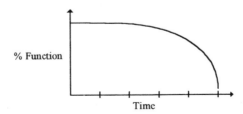

The system will comprise of many sub systems (which may be individual components) whose functions will interact.

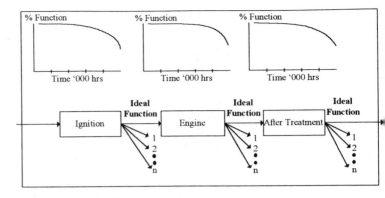

The output (either ideal or error) from one sub-system either:

a. provides the input signal for another sub-system, or
b. is a source of noise for another sub-system

Systems Engineering is about understanding how each of the engineering models for each sub-system fit together.

Fig. 8

In general, contributing sub-system will have an impact on the overall function of the whole system as illustrated in Fig. 8. The effect of each sub-system's contribution will either impact on the variability or mean strength of the overall system's ideal function and links to the stress/strength distributions illustrated in Fig. 9. The target of any tests conducted will include the verification of the design's robustness over time. Emission components in the future will have to function for much greater periods of time. It is possible to fit the largest catalyst possible within the package constraints of a given vehicle to take account of all the noises encountered in the life of the system. This is likely to be an expensive solution.

Optimisation, minimising variability in spite of the noises at each stage of the system is a more cost effective approach. Fig. 9 shows that at zero time the stress, amount of emission requiring treatment and the capability of the catalyst to convert the gases is separate. The consequences are that the catalyst converts the legislated gases however, the catalyst becomes less efficient due to degradation and eventually the distributions of engine and catalyst variability begin to overlap. At this point failure occurs, that is, the engine output is no longer converted by the catalyst.

Effects on Ideal Function and Dispersion Over Time

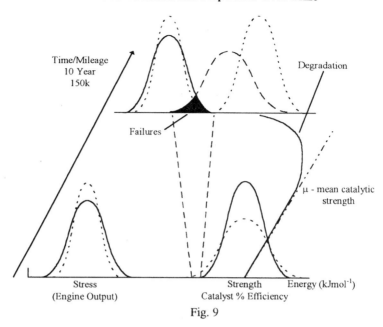

Fig. 9

Noises associated with the various sub-systems in Fig. 7 are listed below. Once the noises have been identified, effort should be concentrated on the effect of the noise, not on the noise itself. This gives us the opportunity to either simulate the effect or to choose an appropriate accelerated ageing test to degrade functions in a shorter time frame. These effects are listed at the end of each sub-system together with an indication of the simulation processes.

Engine

1. **Piece to Piece:**
   - eg. Engine capacity, clearances around pistons, compression ratio, valve seating, timing, etc.

2. **Changes in dimension/strength over time:**
   - eg. Wear of piston in bore, reductions in compression ratio, spark plugs degradation, etc.

3. **Customer usage/duty cycle:**
   - eg. Incorrect oil, type of driving, lack of servicing, etc.

4. **External (climate/road):**
   - eg. Very hot/cold, high/low altitude, etc.

5. **Internal neighbouring sub-systems:**
   - eg. Fuel type, engine cooling system, air induction system, calibration/strategy, etc.

Effects/Simulation:

Quantity of hydrocarbons stressing the catalyst
   - simulated by increasing fuelling to the engine - air/fuel ratio

Exhaust temperature
   - simulated by degrading catalyst surfaces. This ageing process must correlate with 'real world' degradation. The accelerated ageing process is required since it is impossible to mimic the effect of ageing.

Poisoning of the substrate (less precious metal area available)
   - simulated by making parts with less surface area to mimic degradation

Vehicle Weight

1. **Piece to Piece:**
   - eg. Overall weight differences between vehicles of the same class spec and variants (base to top spec), etc.

2. **Changes in dimension/strength over time:**
   - N/A

3. **Customer usage/duty cycle:**
   - eg. Additional accessories, baggage, towing, ice and mud

4. **External (climate/road):**
   - eg. Very hot/cold, high/low altitude

5. **Internal neighbouring sub-systems:**
   - N/A

Effects:

Quantity of hydrocarbons stressing the catalyst
   - simulated by increasing fuelling to the engine

Exhaust temperature
   - simulated by degrading catalyst surface as poisoning

Electrical

1. **Piece to Piece:**
   - eg. Wiring thickness, connector contact area, contact forces, etc.

2. **Changes in dimension/strength over time:**
   - eg. Corrosion, fretting, and contact force, etc.

3. **Customer usage/duty cycle:**
   - eg. Disconnections, vibration, acceleration, etc.

4. **External (climate/road):**
   - eg. Very hot/cold, humidity, salt spray, thermal expansion, etc.

5. **Internal neighbouring sub-systems:**
   - eg. Other electrical systems' magnetic fields

Effects:

Weak signals simulated by
   - lower electric current and voltage
   - small contact surfaces
   - introduction of EM interference (simulate starter and alternator
   - weaker contact force

Fuel

1. **Piece to Piece:**
   - eg. Winter/summer fuel, fuel variation across customer territory, volatility, mixing (blending), etc.

2. **Changes in dimensions/strength over time:**
   - eg. Changes in composition if left for prolonged periods in the vehicle

3. **Customer usage/duty cycle:**
   - eg. Leaded fuels used, methanol ethanol fuels, changes in formulation by the oil companies, etc.

4. **External (climate/road):**
   - eg. Very hot/cold, high/low altitude

5. **Internal neighbouring sub-systems:**
   - eg. fuel injection system, calibration and strategy

Effects:

Quantity of hydrocarbons stressing the catalyst
   - simulated by increasing fuelling to the engine

Exhaust temperature
   - simulated by degrading catalyst surface (as poisoning)

Poisoning of the substrate (less precious metal area available)
- simulated by making parts with less surface area to mimic degradation

Exhaust

1. **Piece to Piece:**
   - eg. Back pressure, etc.

2. **Changes in dimension/strength over time:**
   - eg. Back pressure increase/decrease due to deposits/leaks, degradation of baffles and muffler material

3. **Customer usage/duty cycle:**
   - eg. Rate of carbon and other particle deposition, etc.

4. **External (climate/road):**
   - eg. Very hot/cold, high/low altitude, vibration from road type, etc.

5. **Internal neighbouring sub-systems:**
   - eg. Catalyst, body

Effects:

Increased engine airflow for given power causes changes in gas velocity through the substrate affecting the emission conversion process.
- simulated by control of back pressure during tests

Air Induction

1. **Piece to Piece:**
   - eg. Filter porosity, surface area, etc.

2. **Changes in dimension/strength over time:**
   - eg. Less air due to deposition on the filter surface, air leakage after MAF sensor

3. **Customer usage/duty cycle:**
   - eg. Dusty track driving, infrequent air filter changes

4. **External (climate/road):**
   - Very hot/cold, high/low altitude, dusty environment

5. **Internal neighbouring sub-systems:**
   - Electrical sensors (MAF) induction system, engine, EGI secondary air

Effects:

Poor airflow leading to loss of power, air mass flow is compensated for.

Quantities of hydrocarbons to the catalyst simulated by
- decreased fuel into the engine - changes in air/fuel ratio
- more temperature for a given engine speed and load

Exhaust Manifold

1. **Piece to Piece:**
   - eg. Mass of material, etc.

2. **Changes in dimension/strength over time:**
   - eg. Leaks at gasket faces, etc.

3. **Customer usage/duty cycle:**
   - N/A

4. **External (climate/road):**
   - eg. Very cold, snow packed, etc.

5. **Internal neighbouring sub-systems:**
   - eg. Engine, catalyst

Effects:

Cooling of the exhaust gases - preventing catalytic processes
Design issue simulated by
- forced cooling of the manifold during the first 200 seconds from engine start
- testing in a cold facility/climate

Sensors

1. **Piece to Piece:**
   - Sensor variability and sensor set-up

2. **Changes in dimension/strength over time:**
   - Sensor surface contamination and degradation widening gaps

3. **Customer usage/duty cycle:**
   - Disturbance, vibration

4. **External (climate/road):**
   - eg. Very hot/cold, high/low altitude, vibration from road type, etc.

5. **Internal neighbouring sub-systems:**
   - eg. Electrical (EMF interference), engine (heat), induction, exhaust, body

Effects:

Poor signal leading to incorrect fuelling

Quantity of hydrocarbons stressing the catalyst
   - simulated by increased fuelling to the engine
   - poisoning of the substrate - less precious metals available
   - simulated by making parts with less surface area to mimic degradation

Exhaust temperature
   - simulated by degrading catalyst surface

Ignition System

1. **Piece to Piece**
   - eg. Spark plug gap, high tension lead resistance, etc.

2. **Changes in dimension/strength over time:**
   - eg. Gap growth on spark plug electrodes, plug contamination, HT lead corrosion, etc.

3. **Customer usage/duty cycle:**
   - eg. Type of vehicle use, length of journey, etc.

4. **External (climate/road):**
   - eg. Very hot/cold, high/low altitude, vibration from road type, etc.

5. **Internal neighbouring sub-systems:**
   - eg. Electrical, air charging, etc.

Effects:

Poor signal leading to incorrect fuelling

Quantity of hydrocarbons stressing the catalyst
   - simulated by increasing fuelling to the engine

Exhaust temperature
   - simulated by degrading catalyst surface (as poisoning)

Poisoning of the substrate - less precious metal area available
   - simulated by making parts with less surface area to mimic degradation

If we sum up the effects of the incoming signals and noises from other sub-systems it can easily be reduced to variability in unburned hydrocarbons from the engine, including liquid fuel hitting the substrate, and reduction in the effective surface area in the substrate as the surfaces become contaminated. The former is easy to simulate, but the reduction in surface area is more complex since the degradation may be non-uniform across the face of the substrate. Degraded parts are best derived from an accelerated ageing test. Any tests should be conducted using 95 percentile available fuels and at the extremes of operating temperature.

It will be necessary to validate the accelerated ageing test pieces against the same degraded 'real life' parts from customers. Failure modes specific to the accelerated test technique can make the test work void. Comparisons can be made using post-mortem examinations in the laboratory of both 'real life' and accelerated tested parts.

Application: Design of Experiments

To meet the more stringent Stage III emission requirements, a greater focus of attention has been made on the first 200 secs. stage of the MVEURO emission test cycle. It is during this part of the cycle that some 80% of the hydrocarbon emission are generated. This is largely due to the cold surfaces within the engine absorbing heat reducing the effectiveness of combustion and the fact that the engine is over-fuelled initially during the starting process.

Two technologies are under investigation currently to reduce the emission during this period.

1. Exhaust Gas Ignition (EGI)
2. Cold Start Spark Retard (CSSR)

1. Exhaust Gas Ignition

This technique of extracting energy from the engine exhaust gases by adding secondary air into the exhaust prior to the catalyst and then igniting the resulting gases was 'discovered' about three years ago. The gases are ignited within a combustion chamber between a front and rear substrate internally to the catalyst (see Fig. 10). The resulting heat raises the rear substrate temperature to a level where gas conversion can take place. The optimising experiment for the chamber length, front brick length and cells per inch (tubes through the substrate) is shown on the following page.

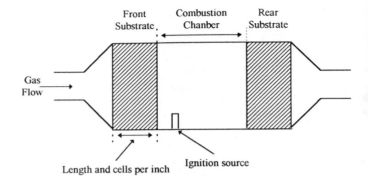

Fig. 10

The experimental array used is known as a central composite design with three factors at three levels. The exhaust gas excess hydrocarbons are controlled by adjusting the EEC IV module output to the injectors. The fuelling level is set as a ratio against stoichiometric. The experiment was repeated twice for each run. The levels of fuelling, 0.57 and 0.62, were set in line with the calculated variability of the air/fuel ratio at start up and during the initial period of engine running.

Results:

The three dimensional predictive plots on pages 21, 22 and 23 indicate that to maximise signal-to-noise ratio the system design required a long front brick, short combustion chamber and 400 cells per inch. This is in opposition to the requirements to set the hydrocarbon target to minimum where a long combustion chamber and short front brick are required. The engineering requirements now can be directed towards improving the combustion process and minimising variability.

Central Composite Design for Three Factors at Three Levels

	Abbreviate for 1st Factor			Abbreviation for 2nd Factor						Abbreviation for 3rd Factor					
	A							**B**				**C**			
	linA	linB	linC	QuadA	QuadB	QuadC	AxB	AxC	BxC	CC1	CC2	CC3	CC4	CC5	Response
1	-1	-1	-1	1	1	1	1	1	1	-1	-1	-1	-1	-1	
2	1	-1	-1	1	1	1	-1	-1	1	1	-1	-1	-1	-1	
3	-1	1	-1	1	1	1	-1	1	-1	-1	1	-1	-1	-1	
4	1	1	-1	1	1	1	1	-1	-1	1	1	-1	-1	-1	
5	-1	-1	1	1	1	1	1	-1	-1	-1	-1	1	-1	-1	
6	1	-1	1	1	1	1	-1	1	-1	1	-1	1	-1	-1	
7	-1	1	1	1	1	1	-1	-1	1	-1	1	1	-1	-1	
8	1	1	1	1	1	1	1	1	1	1	1	1	-1	-1	
9	0	0	0	-2	-2	-2	0	0	0	0	0	0	0	-16	
10	-1	0	0	5	-4	-4	0	0	0	4	4	-4	0	4	
11	1	0	0	5	-4	-4	0	0	0	-4	-4	4	0	4	
12	0	-1	0	-4	5	-4	0	0	0	0	4	4	-4	4	
13	0	1	0	-4	5	-4	0	0	0	0	-4	-4	4	4	
14	0	0	-1	-4	-4	5	0	0	0	0	0	0	-4	8	
15	0	0	1	-4	-4	5	0	0	0	0	0	0	8	0	
Divisor	10	10	10	18	18	18	8	8	8	8	8	8	8	8	
Contrast value	0	0	0	0	0	0	0	0	0	0	0	0	0	0	

Factors for the DoE on EGI

Factor	Levels			
	-1	0	1	
A - Front Brick Length	80	110	150	mm
B - Cells/inch	200	400	600	cells/inch
C - Chamber Length	165	215	265	mm

Response Surface for Front Brick Length Versus Catalyst Chamber Length ((B) Cells/Inch=0) - First 200 Secs

Predicted Signal to Noise Ratio

- 55-60
- 50-55
- 45-50
- 40-45
- 35-40
- 30-35
- 25-30

(A) Front Brick Length

(C) Chamber Length

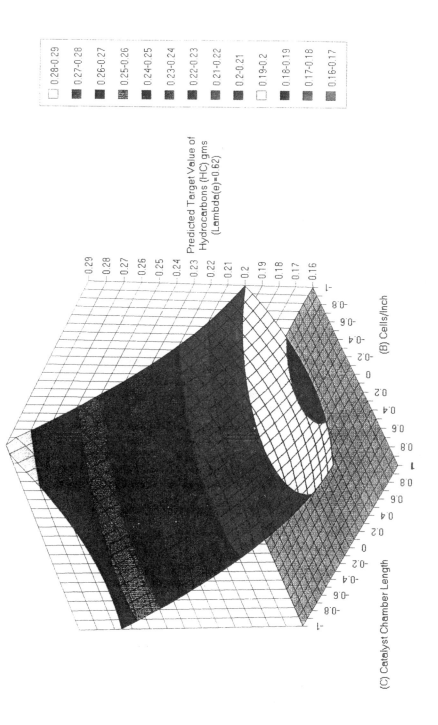

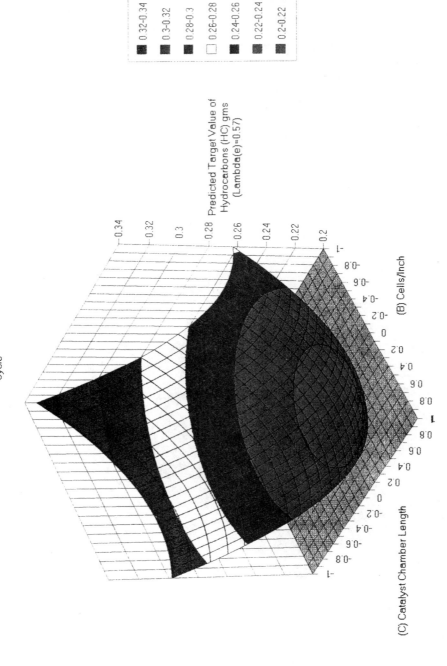

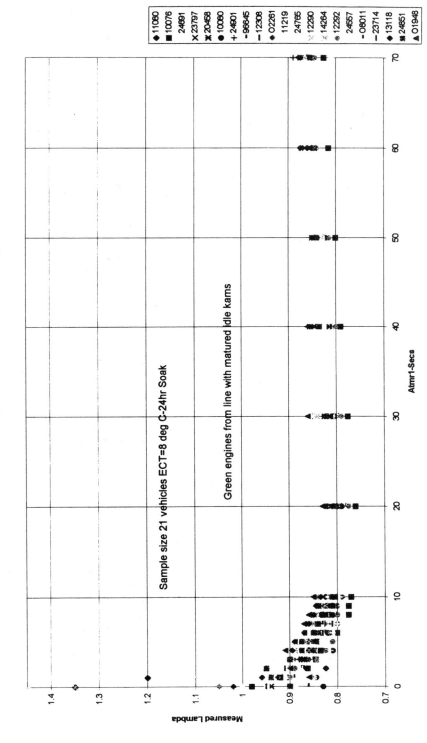

Investigative work is continuing and it is planned to test the EGI system using different fuels (signal inputs) at varying ambient temperatures. Shell Oil Company will be supplying fuels of different mixtures. This work is scheduled for year end.

Second Experiment:

In parallel to the robustness study on EGI, a robustness experiment is underway to optimise the HC, CO and NO_x using substrates with different precious loading ratios (Pt, RL, Pd) and overall total loadings. The experiment is being conducted at ambient 20^0C temperature using aged substrates. Indeed tests will be conducted at 4k and 50k aged substrates. Using these aged substrates it will be possible to test using different fuels before further ageing to 100k and then to 150k (Corporate target). These tests will enable optimisation of expensive metals over time.

The results of the above experiments will enable the function and costs to be optimised in the presence of 'noise'.

2. CSSR

An alternative process to EGI is one which retards the ignition timing in such a way that the engine combustion process continues as combustion gases pass from the engine into the manifold. If the catalyst is an integral part of the exhaust manifold, close coupled, then the heat from these engine gases raises the substrate temperature quickly to the point at which the conversion process will take place.

Background investigation of end of line vehicle emissions in Genk Plant Belgium, shown on page 24, indicates that the air/fuel ratio is likely on average to drift approximately $\pm$ 10% during the cold start phase after engine start.

Other work conducted in the calibration activity at Dunton England, on drift of sensors, substantiated the plant investigation and further indicated that the spark could drift $\pm$ 10%.

Armed with this data, it is possible to deduce those variations that need to be considered in any experimental work and fulfil the need to simulate the effects deduced from the noise tree mentioned earlier with the exception of degradation. A thirty vehicle fleet is currently undergoing tests to gain sufficient degradation knowledge to ascertain the drifts in these sub-systems and enable and experimental investigation into function given that these variations occur over time.

It is believed currently that all the variations will boil down, in the case of CSSR, to changes in spark and air fuel ratio. The outcome of the precious metal ratio/loading experiment mentioned above will enable knowledge based reductions in the system costs whilst maintaining minimum variability with the lowest emissions.

Conduct Key Life Tests to Validate the Design

At the commencement of the design process, system solutions to customer requirements are first evolved. These soon become sub-systems and components with appropriate characteristics. During this phase, robustness experiments are conducted to discover interactions and understand the best design and manufacturing settings for the total system (given the noises are present as already discussed). The process is interactive as indicated in Fig. 11, each time refining knowledge of the system and its' parts, and providing test data evidence on which to base further optimising improvements.

Once the design becomes more settled, with interactions and all noises identified, it is not only possible to analyse the effects of the noise and mimic these states as part of the tests, but also a more simplified test procedure related back to the customer usage profiles is possible. The implication is faster testing and sign-off of systems since each component/sub-system can be tested and optimised to ensure its' ideal output, input to the next sub-system, is tuned to suit the next system's needs and eventually that of the customers.

System Engineering Perspective on Life Testing

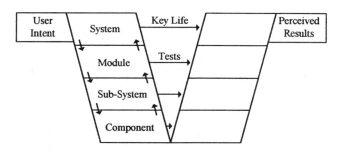

Fig. 11

The left hand side of the life tests concentrates on verification, conducted using planned experimentation whilst the right hand side concentrates on validation using customer focused key life tests.

Tests to bogie (set number of hours/miles) or tests to failure simply provide data to consolidate life expectation. Test to failure data are more meaningful as far as life expectation prediction is concerned. These form in a limited way the tests required during the sign off period. The system should of course be signed off at all levels before mass production commences.

Since in-depth knowledge of the system is available, only a limited number of tests are required. Also, planned design of experiments permit prediction of control factor settings to provide longer life if required since the important factors and their affects are known.

Linkages to other initiatives

Systems engineering is associated with all the EQUIP modules and other quality tools. Already mentioned is the strong linkage to Quality Engineering (ideal function and robustness) and Experimentation (planned designed experiments).

At the beginning of this section Customer Focused Engineering was mentioned, the process of providing a high quality product using all the quality tools in a structured way. The customer provides the necessary information to initiate this process. The Quality Function Deployment provides the framework for designing in quality on a 'right first time' basis, including the reliability requirements. System Failure Modes and Effects Analysis (FMEA) looks at the things gone wrong and assists in defining the required tests and conditions necessary at the system level to ensure that failures are detected within the company and not with the customer.

During the design and manufacturing process selection, the Design and Process FMEA will aid the test planning processes. Team Oriented Problem Solving (TOPS) output is also a contributor to ensuring failures of the past don't re-occur on the next model. Lastly, Process Management contributes greatly to the identification and control of sources of variation within the manufacturing processes and during any experimental work.

Fig. 12 indicates some of these linkages, but is by no means exhaustive. All the above relate strongly to Value Management providing high quality products at the lowest costs.

Linkages to Other Initiatives

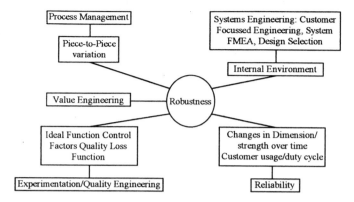

Fig. 12.

Aftertreatment strategies to meet emission standards

J BRISLEY, D E WEBSTER, and A J J WILKINS
Johnson Matthey Catalytic Systems Divisions, Royston, UK

1. INTRODUCTION

The pending introduction of Californian low emission vehicle, Federal Tier 2 and European Stage III standards by the end of this decade has led to substantial levels of research into a number of possible approaches towards meeting these targets. While the main body of this work has primarily been in the context of conventional 4-stroke gasoline engines, there is an accelerating interest in the development of engines which run part or all of their time at air-fuel ratios substantially lean of stoichiometry. Vehicles equipped with these lean burning engines - for example, diesel, lean burn 2- and 4-stroke gasoline, and CNG fuelled engines - have the potential for improved fuel economy (and hence lower CO_2 emissions), but must also meet the new regulations for hydrocarbon (HC), carbon monoxide (CO), nitrogen oxides (NOx), and particulate emissions. Emission control systems and catalysts developed for stoichiometric control may not be appropriate for the leaner burning engines; in particular, control of NOx under these conditions presents a different set of problems, and requires different types of catalyst technologies and systems control to that used for TWC/stoichiometric operation. In contrast, for engines running at, or close to, stoichiometric air-fuel ratio, the challenge - particularly in meeting the Californian regulations - is hydrocarbon control, especially under the conditions of cold start.

Recently, the requirements for on-board diagnostics (OBD) applied to the emissions system has highlighted the sometimes contrasting requirements of the OBD system with the performance requirements in the formulation of the catalyst. This represents yet another challenge to meeting all the requirements of the emission package.

The objective of this paper is to review how the various legislative standards are being met with a range of strategies which include not only the development of new catalyst technologies - important as that is - but also the development of systems which, in concert with the catalyst technology, will achieve the necessary performance.

2. HYDROCARBON CONTROL

In California, the 1993 Clean Air Act (CAA) required vehicles to meet a standard of less than 0.25 grams of hydrocarbon (excluding methane) per mile travelled, using the FTP drive cycle as the criterion. However, a series of even tighter levels are being progressively introduced over the period to the beginning of the next century, so that the Ultra Low Emission Vehicles (ULEV), for example, will have to meet 0.04 grams of non-methane hydrocarbons per mile.

In contrast, the CO and NOx emissions will only halve between the CAA and ULEV standards (3.4-1.7 g/mile CO, 0.4-0.2 g/mile NOx). While this does not mean that the CO and NOx standards are "easily" achievable, it highlights the greater importance attributed to the control of HC by the regulatory bodies in the USA. Similarly, although not finalised, it is expected that the ECE European standards for hydrocarbon will progressively reduce at each stage of the legislation. In addition, for the first time the HC and NOx will be separated rather than be combined as at present. It has long been recognised that the key to HC control at this level is the cold start, that is the period before the catalyst reaches operating temperature. It is thus essential to find means by which this period can be reduced to be as short as possible. A number of options have been investigated which, either alone or in combination, seek to meet this aim.

The first of these is to provide a secondary source of heat, in addition to the hot gases coming from the engine, to enable the catalyst to reach operating temperature more quickly. These include providing direct heat to the catalyst via electrical energy - generally called electrically heated catalysts or EHC's (1, 2, 3) - or by burning a small quantity of fuel injected after the engine, but before the catalyst (4, 5) thus raising the temperature of the exhaust gas entering the catalyst. Although both systems would be designed to work only during the cold start period, they add significantly to the complexity of the systems, and have raised a number of questions concerning, for example, durability.

An alternative approach which avoids or minimises increased complexity is to use catalyst bricks only. In recent times a number of advanced three-way catalysts have been developed to give improved performance with stoichiometrically controlled spark ignition gasoline engines. These, which also have improved high temperature durability, enable catalysts to be located much closer to the engine, thus reducing catalyst warm-up time (6). Perhaps the most significant advance has been the realisation that high loaded palladium-containing catalyst systems can be made which have superior performance to the traditional platinum-rhodium catalyst at the same precious metal cost. These improved catalysts can be based on palladium as the only precious metal, or in combination with lesser amounts of rhodium, or in some cases with a combination of rhodium and platinum to form a "trimetal" catalyst. The total precious metal loading and ratio of the elements can readily be varied to give the required performance at optimum cost (7, 8). Similarly, there are some options for the location of the catalyst piece : those most commonly considered are:

- Underfloor only
- Starter + Underfloor
- Close-coupled only

Figure 1(a) shows the benefits of high loaded palladium catalysts, in comparison to Pt/Rh, in the underfloor position on a 1.2 litre European car run on the proposed modified Euro III drive cycle. In each case the catalysts (1.66 litre volume) had been pre-aged equivalent to 80,000 km. From this figure it can be seen that with the palladium-containing systems, the hydrocarbon emission levels are significantly lower than for the corresponding Pt/Rh catalyst. Nevertheless, all systems would fail the likely Euro year 2000 standard for hydrocarbon. Figure 1(b) shows that this is because the catalyst does not warm up to light-off quickly enough, and the putative standard is exceeded before full light-off is achieved.

The inclusion of a starter catalyst in front of the underfloor unit should help the situation by lighting off more quickly itself, and through the exotherm produced assist the underfloor catalyst to light-off earlier. For the results in Figure 2 and Table 1, a trimetal starter catalyst (0.6 litre) was inserted 30 cm from the manifold, with the volume and location of the underfloor catalyst being maintained as before. Table 1 shows that the starter alone has better performance than the underfloor catalyst alone, and would meet the requirements for HC and CO presently under discussion for Euro year 2000. NOx, however, although also lower is still likely to exceed the standard. Combining the starter catalyst with an underfloor catalyst gives substantial further improvement in all three pollutants, especially HC and NOx, so that now the data are well within the likely Euro year 2000 standards, and are approaching the proposed levels for discussion for year 2005. It might be argued that this excellent result is obtained at the expense of a substantial increase in catalyst volume : accordingly, a further test on the same vehicle was conducted in which the volume of the underfloor catalyst was halved, thus giving a total catalyst volume slightly less than that of the original underfloor catalyst. The results (shown in Table 1) show almost identical results for HC and CO, but the NOx performance - although still better than either the starter or full-sized underfloor alone - has deteriorated. Finally, given the encouraging results with the starter catalyst alone, the effect of a larger volume in the close-coupled position was evaluated. Because of space constraints only 1.2 litres of catalyst volume could be accommodated. No underfloor catalyst was used in this case. The data was obtained for the same series of formulations as had been used for the underfloor catalysts, and these are shown in Figure 3(a). The results clearly show the same trends seen for the corresponding underfloor catalysts, namely that the palladium-containing catalysts do confer an advantage with respect to hydrocarbon performance compared to Pt/Rh, and in spite of the lower volume these systems give the best HC figures in the whole series of tests. Indeed, the hydrocarbon emissions are comfortably inside the Euro year 2005 targets as presently suggested. The CO and NOx figures generally meet Euro year 2000 targets, and in the case of Pd/Rh and trimetal come close to meeting Euro year 2005 on this car. Figure 3(b) shows that the catalyst has lit-off much earlier than in the underfloor case, and has thus made substantial inroads into the "cold start" hydrocarbon emissions.

One further means of reducing cold start hydrocarbons is the use of a material acting as a trap for the hydrocarbon species. This works by adsorbing hydrocarbon in the cold start phase, before the catalyst is lit-off, then desorbing the stored hydrocarbons once the catalyst has lit-off, and thereafter being passive when the catalyst is working. Early forms (9) of this system were in a dual brick format in which the first brick is the hydrocarbon trap, and the second brick a three-way catalyst. In order to protect the trap material, a by-pass valve was incorporated so that the hydrocarbon trap was not exposed to very hot exhaust gas. A potential problem with this concept is that the thermal inertia of the trapping brick delays light-off of the rear catalyst : thus the trap desorbs before the TWC is working. In addition, in many smaller cars packaging constraints limit the use of extra bricks. A better option is to combine the adsorbing material and the catalyst formulation into a single brick, and to maximise the overlap of the desorption of stored hydrocarbon and catalyst light-off temperatures. A key factor is the stability of the trapping material after ageing; also important is the potential effect of the trapping material on the performance of the catalyst function. Such a system is under development (10) and at the present time suitable trap materials have been found, with good stability after ageing and good overlap of the desorption temperatures and TWC light-off.

3. NOx CONTROL UNDER LEAN CONDITIONS

In principle, HC and CO control under lean operating conditions should be straightforward, but reduction of NOx under these strongly oxidising conditions is not. Nevertheless, especially in Europe, NOx levels in the atmosphere are a major concern, and legislative proposals now in discussion will require some level of removal of NOx from lean burn engines, whether run on diesel, gasoline or other fuels.

So far, three approaches have been tried; two of these have met with some level of success.

i) Direct decomposition of NOx. Since the early reports by Iwamoto (11), who showed that copper exchanged ZSM-5 zeolite had stable steady state activity for the decomposition of nitric oxide selectively to dinitrogen, much work has been put into seeing if this system can be made to work under the high space velocities, low NOx levels encountered in automotive applications. The influence of degree of copper exchange in the zeolite, excess oxygen levels, and the presence of sulphur oxides and water may all contribute to the poor levels of conversion achieved in real applications.

ii) NOx reduction under oxidising conditions. For many years NOx emissions from stationary sources (power stations, chemical operations containing an excess of oxygen) have been controlled by the so-called SCR process, using added ammonia as the reducing agent over a range of catalysts:

$$4NH_3 + 4NO + O_2 \rightarrow 4N_2 + 6H_2O$$

$$4NH_3 + 2NO_2 + O_2 \rightarrow 3N_2 + 6H_2O$$

Held et al (12) showed that very promising levels of NOx reduction under automotive conditions could be achieved by the use of reductants (for example, urea, hydrocarbons) over a Cu-ZSM-5 catalyst. Since that time a large number of catalyst concepts have been screened, but many initially encouraging systems have proved to be lacking in durability, and ways of adequately stabilising their performance have yet to be found.

Two systems have been more thoroughly researched and refined, namely platinum on modified alumina or zeolite for low temperature (150-250°C) and Cu-ZSM-5 for higher temperatures (300-450°C). These temperature ranges are quite narrow compared with the operating window of a three-way catalyst, and at the upper end are limited by the competing direct oxidation of the reductant over the catalyst. Nevertheless, they fall into a suitable range for use in diesel applications which form a major part of the current lean burning vehicle fleet. Ideally, since engine-out emissions contain species (hydrocarbons, CO) capable of reducing NOx, these would be used to achieve the removal of NOx. However, in many cases the level of hydrocarbon in the exhaust gas is not sufficient to achieve a necessary level of NOx control. Alternatively, small amounts of added hydrocarbon - most conveniently using the fuel which is used in the engine - can be injected into the exhaust gas prior to the catalyst to effect the necessary reaction with the NOx.

Figure 4 shows the NOx performance with three generations of platinum- based lean NOx catalysts after a simple conditioning cycle (3 hours at 300°, 400° and 500°C, 33% of time at each level), and after high temperature ageing (four cycles of a 13-hour programme in which the maximum temperature is 5 hours at 750°C in each cycle), tested on a 1.9 litre TDI bench engine at a space velocity of 45,000 hr^{-1}. In Figure 4(a), the reductant is that issuing from the engine, and it can be seen that NOx conversion is of the order 20-30%. If now the same catalyst and engine is used, but with low levels of injected fuel, a further 20% on average NOx conversion results (Figure 4(b)).

Over the MVEG test cycle using a car equipped with a similar 1.9l TDI engine, the catalyst inlet temperature was in the range 130-200°C for the major part of the ECE cycle, but rose to the range 250-350°C for most of the EUDC section. Thus, parts of the operating cycle fall outside the optimum operating temperature range for both the platinum and copper catalysts individually. Therefore, a future challenge is to develop catalysts with a wider effective operating temperature range. A limiting factor with individual catalysts arises when the removal of the reductant by oxygen is more effective than the reductant NOx reaction, that is the selectivity towards reductant NOx reactions needs to be high over as broad a range of temperatures as possible. Some bi-metallic systems, comprising at least two catalytic elements, are showing some promise - at least in laboratory tests - as is shown in Figure 4(c). However, the search continues for a lean NOx catalyst showing better performance over a wider temperature range.

iii) NOx storage and release. A further possible means of reducing NOx in high oxygen-containing exhaust gas is to incorporate adsorbent materials into the catalyst formulation, together with a catalytic element capable of oxidising nitric oxide to NO_2. The NO_2 is then taken up by the adsorbent (oxides of barium, calcium, cerium, lanthanum, strontium, potassium and zirconium are all possible adsorbents). If the exhaust gas is then switched to reducing for a very short period, the surface complex nitrate becomes unstable at lower temperatures, and releases NO and oxygen (13, 14, 15). If the catalyst system then also contains elements capable of reducing the released NO - for example, a rhodium component - then the tailpipe NOx will be reduced compared to engine-out NOx. The relative efficiency of this system can be high, but is very dependent on a number of factors. For example, the capacity of the NO_2 adsorbing component, the rate at which the NOx can be adsorbed, and the rate at which NO is desorbed when the system is periodically switched to reducing conditions are key parameters. However, the first step seems to be the oxidation of NO to NO_2. Platinum has been shown to be a very efficient catalytic element for this reaction (13, 14), and temperature programmed desorption measurements (15) have shown that on the adsorbent the capacity for NO_2 adsorption is substantially higher than for NO. By changing the atmosphere over the catalyst to a richer composition, the adsorbed NOx (probably in the form of a nitrate species) decomposes releasing NO and O_2:

$$[MNOx]_{ads} == NO + \tfrac{1}{2} O_2 + MOx$$

Under the richer condition the released NO can react with, for example, CO over a suitable catalytic element:

$$NO + CO \rightarrow CO_2 + \tfrac{1}{2} N_2$$

A major advantage of the lean burn engine over the conventional stoichiometric engine is the potential for increased fuel economy/lower CO_2 emissions. Clearly, making the exhaust gas richer to regenerate and react the adsorbed NOx will offset some of the fuel economy gain, and so it is important to minimise the time of regeneration and maximise the period spent lean. Taking the stoichiometry during the regeneration phase back to lambda=1, as shown in Figure 5(a) results in a NOx breakthrough, implying that the rate of release of NO from the adsorbent is faster than that of subsequent reduction. However, if the mixture is richened still further the breakthrough of NOx is progressively lessened. Furthermore, as shown in Figure 5(b) the regeneration time required decreases markedly as the regeneration lambda becomes richer, and so these parameters can be optimised. Two factors still need resolution with this system. The first is that many of the best NOx adsorption components from the viewpoint of kinetic and adsorption capacity considerations are also excellent for the adsorption of SOx - also present in the exhaust gas - yielding sulphates which are often substantially more stable than the corresponding nitrates. Consequently, the proportion of adsorbent which is available for NOx adsorption progressively reduces as the material is sulphated. Secondly, there is the possibility of thermal deactivation of the adsorption component in long-term use. Both of these aspects are being reduced in criticality by further developments presently in hand.

4. DIESEL OXIDATION CATALYSTS

Recent developments in advanced combustion systems for diesel engines, including better control (precision, accuracy and repeatability) of fuel injection, robust design of combustion chamber geometry and optimised swirl, together with improved injection timing and turbocharging have all contributed to better performing, cleaner engines. Nevertheless, as the emission standards progressively tighten there is an increasing need to reduce engine out emissions further. The problem of reducing the levels of NOx emissions under the lean conditions in which diesel engines operate have already been discussed under 3. above. But, future emission levels, especially in Europe, call also for tight hydrocarbon and particulate control. Since exhaust gas temperatures are generally quite low, highly active catalysts with low light-off temperatures are required to effect the desired conversion of these pollutants. However, diesel fuel contains sulphur compounds which, during combustion, are converted to SO_2 and may in turn be converted by the active catalyst to SO_3. This sulphur trioxide can react with certain components in the washcoat to form a sulphate species (sulphate storage), or contribute to the weight of particulates emitted. While it is possible to devise catalyst formulations which inhibit sulphate formation, these often also exhibit inhibition of the desired HC/CO/VOF function with the effect of - for example - increasing light-off temperature. In addition, while diesel exhaust gas temperatures under start-up and low load conditions can be quite low they can be substantially higher under frequent high speed/full load driving which might be experienced on motorways. Accordingly, the catalyst also needs to retain its low temperature performance after having previously been exposed to high temperatures, ie, thermal durability is important. This summarises some of the key parameters required for these catalysts as:

a) good stable low temperature performance for HC/CO/VOF

b) low sulphate production and storage/release

c) high thermal stability/durability

By optimising the loading of the principle catalytic elements and washcoat components, progress towards these targets can be made. Figure 6 shows some results on aged catalysts tested on a 6 litre engine on the ECE 49 test. The fuel used contained 0.05% sulphur, and the maximum temperature during the cycle was 540°C. In this case the principle catalytic element was chosen to be platinum, at two different loadings, 10g ft^{-3} and 2.5g ft^{-3}. From Figure 6(a) and (b) it can be seen that the lower loading is not so efficient for CO removal, but gives almost equivalent performance for gaseous hydrocarbon removal. This is mirrored in Figure 6(c) which shows a very similar efficiency for the removal of the soluble oil fraction of the particulate over the two catalysts. However, the higher platinum loaded catalyst still shows some ability to generate SO_3 and hence sulphate, whereas the low loaded system does not, as indicated in Figure 6(d). The net effect on particulate emissions (Figure 6(e)) shows that due to the contribution of sulphate the 10g Pt ft^{-3} has increased particulate emissions compared to the baseline, whereas due to the reduction in the soluble oil fraction the 2.5g Pt ft^{-3} catalyst shows an overall reduction in particulate emissions.

5. NATURAL GAS VEHICLES

There is a growing interest in compressed natural gas (CNG) as a fuel for both passenger and heavy duty vehicles. In terms of emissions, this fuel would seem to have a natural advantage in the context of US legislation where the principal component of CNG - methane - is specifically excluded. However, relative cost, ease of distribution and availability are additional factors to be considered in order for its use to be widespread. In European legislation methane is not excluded from the hydrocarbon measurement, and since methane is one of the most unreactive hydrocarbons, it would initially appear that it would be more difficult for CNG fuelled vehicles to meet the required levels, especially when HC and NOx are not combined. However, CNG vehicles generally give lower emissions, especially CO, than their gasoline fuelled counterparts (16). Palladium-containing catalysts are generally good for methane removal, but platinum and rhodium also have good activity, and so in common with three-way catalysts a combination of some or all of these elements is normally incorporated into the catalyst. Figure 7 shows some results from a CNG fuelled Escort van, with an aged Pd/Rh catalyst in the underfloor position. This vehicle runs with retarded ignition over the ECE portion of the European cycle. The results show that the vehicle meets the presently proposed Euro year 2000 standards when run over both the Euro II and Euro III test cycles, and for CO and NOx gives results better than are currently outlined for Euro 2005 standards.

6. ON-BOARD DIAGNOSTICS (OBD)

The objective of emission systems and emission standards is to reduce the incidence of pollutants from motor vehicles in the atmosphere. It is therefore important to know whether, in use, the system is operating to an acceptable level, and whether all or part of the emission system needs replacement. In order for the monitoring systems to be effective, it is critically important that the measuring system itself is extremely stable, and is only affected by the parameter it is set to measure. Thus, for US legislation, the ideal system would be a hydrocarbon sensor; but presently available HC sensors are not sufficiently stable, and may be affected by other components in the exhaust gas varying during normal operation. The most frequently used system is therefore the dual oxygen sensor, which measures the voltage generated from sensors in front of and behind the catalyst (17). The difference between the two sensor outputs is intended to be a measure of oxygen used over the catalyst for oxidation of pollutants, eg, HC. However, in order to improve the performance of three-way catalysts it has been normal practice since the earliest days to include in the composition materials which behave as oxygen storage components. A typical example is ceria : this works by storing excess oxygen from the gas phase during periods of lean operation and releasing it during more oxygen deficient modes, thus encouraging oxidation reactions of HC and CO during these periods (18). This ability will clearly affect the gas oxygen content measured by the rear sensor. If the oxygen storage component deteriorates faster than the real rate of deterioration of hydrocarbon removal (18), then the oxygen content of the gas measured by the rear sensor will imply a faster rate of catalyst deterioration than actual. Furthermore, the sensor voltage signal is not linear with oxygen content. These factors now require the catalyst components to be adjusted to meet the requirements, not of catalyst performance but of the system for measuring it.

7. CONCLUSIONS

Recent improvements in catalyst science together with appropriate engine management systems has enabled emission systems to be developed to meet increasingly stringent legislation. The requirements for improved fuel economy has encouraged development of a number of lean burn concepts in Europe, for which a primary challenge is the development of improved catalysts for NOx control.

There now exists an array of technologies at various stages of development for meeting even the toughest proposed emission standards.

Table 1   The effect of catalyst volume and location on residual emissions from a 1.2l European vehicle over the modified ECE + EUDC drive cycle.

	HC	CO	NOx
Underfloor (1.66l)	0.246	1.996	0.470
Starter (0.60l)	0.117	1.600	0.237
Starter (0.60l) + Underfloor (1.66l)	0.076	1.179	0.089
Starter (0.6l) + Reduced Underfloor (0.80l)	0.082	1.242	0.171

References:

(1) Kubsch, J.E., SAE Paper 941996, (1994).

(2) Mizuno, H., Abe, F., Hashimoto, S., and Kondo, T., SAE Paper 940466, (1994).

(3) Reddy, K.P., Gulati, S.T., Lambert, D.W., Schmidt, P.S., and Weiss, D.S., SAE Paper 940782, (1994).

(4) Ma, T., Collings, N., and Hands, T., SAE Paper 920400, (1992).

(5) Hepburn, J.S., Adamczyk, A.A., and Pawlowicz, R.A., SAE Paper 942072, (1994).

(6) Bartley, G.J.J., Shady, P.J., D'Aniello Jr, M.J., Chandler, G.R., Brisley, R.J., and Webster, D.E., SAE Paper 930076, (1993).

(7) Brisley, R.J., Chandler, G.R., Jones, H.R., Anderson, P.J., and Shady, P.J., SAE Paper 950259, (1995).

(8) Punke, A., Dahle, U., Tauster, S.J., Rabinowitz, H.N., and Yamada, T., SAE Paper 950255, (1995).

(9) Hochmuth, J.K., Burk, P.L., Tolentino, C., and Mignano, M.J., SAE Paper 930739, (1993).

(10) Brisley, R.J., Collins, N.R., and Law, D., European Patent Application 95308884.

(11) Iwamoto, M., Yokoo, M., Sasaki, K., and Kagawa, S., J. Chemical. Soc. Farad. Trans., 77, (1981), 1629.

(12) Held, W., Koenig, A., Richter, T., and Puppe, L., SAE Paper 900496, (1990).

(13) Mijoshi, N., Matsumoto, S., Katoh, K., Tanaka, T., Harada, J., and Takahara, N., SAE Paper 950809, (1995).

(14) Bögner, W., Krämer, M., Krutsch, B., Pischinger, S., Voigtländer, D., Wenninger, G., Wirbeit, F., Brogan, M.S., Brisley, R.J., and Webster, D.E., J. Appl. Catal. B., (1995), 153.

(15) Brogan, M.S., Brisley, R.J., Walker, A.P., Webster, D.E., Bögner W., Fekete, N.P., Krämer, M., Krutzsch, B., and Voigtländer, D., SAE Paper 952490, (1995).

(16) Fricker, N., Janikowski, H.E., and Stover, G.P., The Institution of Gas Engineers 57th Autumn Meeting, Harrogate (1991) Communication 1474.

(17) Clemmens, W., Sabourin, M., and Rao, T., SAE Paper 900062, (1990).

(18) Hepburn, J.S., and Gandhi, H.S., SAE Paper 920831, (1992).

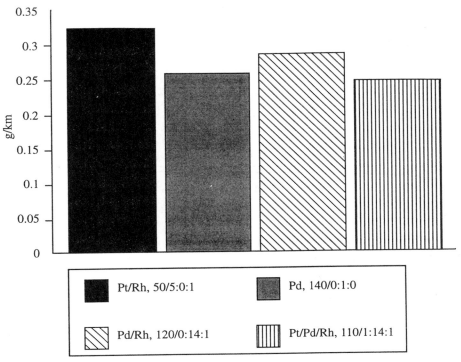

Figure 1 (a)
HC emissions, 1.2L car, stage III ECE/EUDC cycle

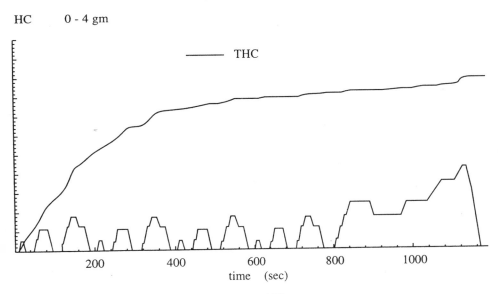

Figure 1(b)
Cumulative THC emissions with an underfloor catalyst in the ECE3 + EUDC cycle.

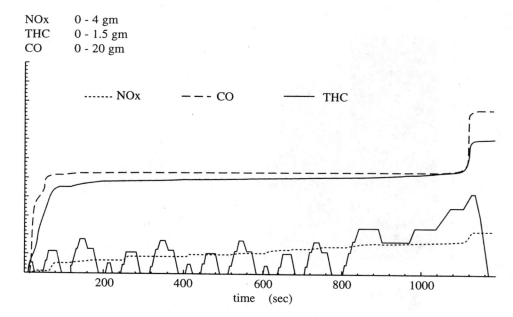

Figure 2
Cumulative NOx, CO & THC emissions Starter & underfloor catalyst.

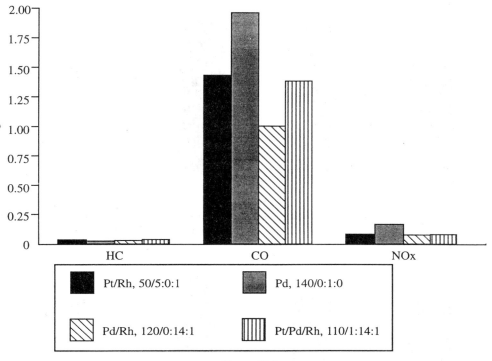

Figure 3(a)
HC/CO/NOx emissions, close coupled catalyst, 1.2 L vehicle over the ECE3 & EUDC cycle

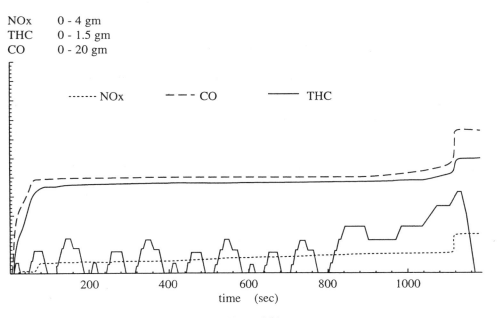

Figure 3(b)
Cumulative NOx, CO & THC emissions, Close Coupled catalyst.

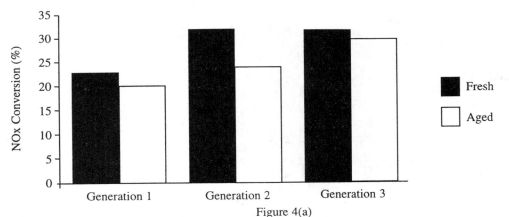

Figure 4(a)

NOx Performance, without fuel injection, after conditioning (fresh) and high Temperature Ageing (ag
Engine: 1.9 L TDI. Space Velocity = 45,000 hr^{-1}

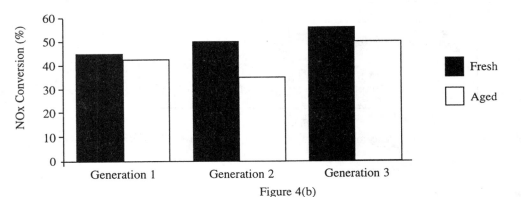

Figure 4(b)

NOx Performance, with fuel injection, after conditioning (fresh) and high Temperature Ageing (age
Engine: 1.9 L TDI. Space Velocity = 45,000 hr^{-1}

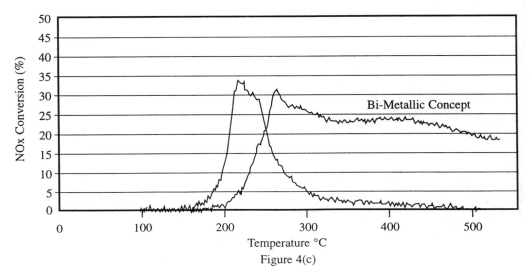

Figure 4(c)

Preliminary results on a wide window lean NOx catalyst at SV = 30,000 hr^{-1}

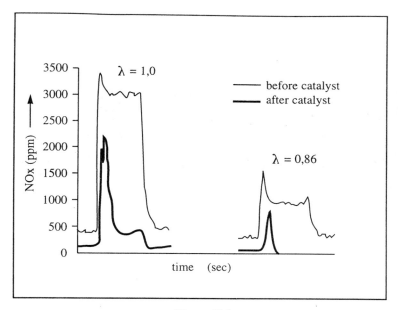

Figure 5(a)
Regeneration behaviour with different air-fuel ratios. T= 390°C

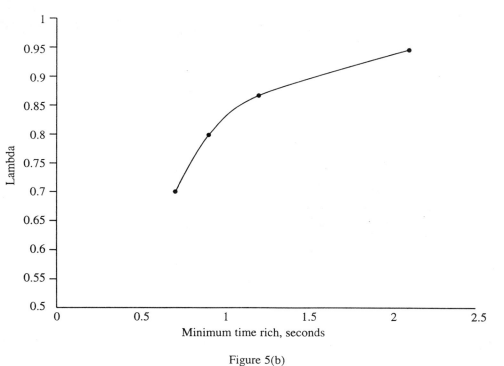

Figure 5(b)

Minimum regeneration time at rich air-fuel ratios.

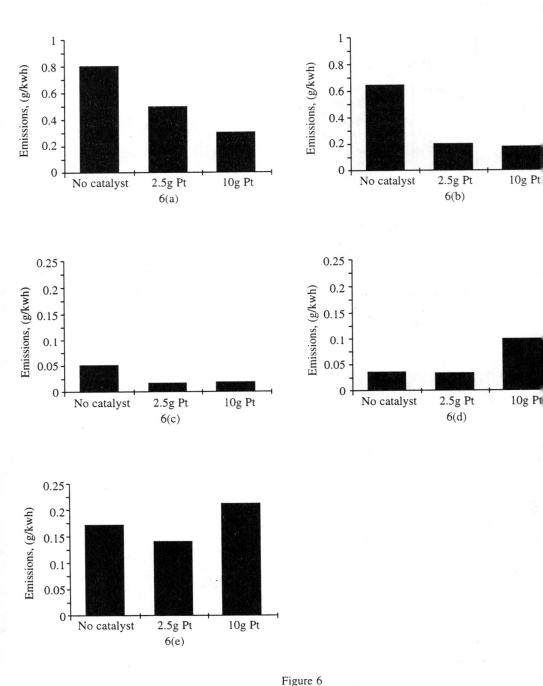

Figure 6
Low PGM loaded diesel oxidation catalyst (loading in g ft^{-3}) Effect on 6(a) CO, 6(b) HC, 6(c) SO, 6(d) SO$_3$, 6(e) particulates. ECE R49 test procedure. Tmax 540°C, 0.05% sulfur.

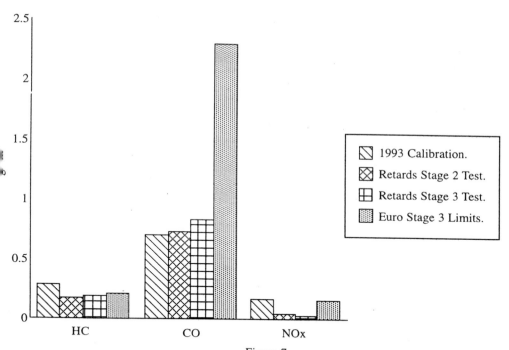

Figure 7
Natural Gas fuelled vehicle: effect of engine retard on emissions with a Pd/Rh Catalyst.

Diesel Control Technology

Impact of alternative controller strategies on emissions from a diesel CVT powertrain – preliminary results

S P DEACON BEng, MIMechE and **R W HORROCKS**
Ford Motor Company Limited, Essex, UK
C J BRACE BEng, CEng, MIMechE, **N D VAUGHAN** BSc, PhD, MIMechE, and **C R BURROWS**
School of Mechanical Engineering, University of Bath, UK

SYNOPSIS

The need for improved economy and exhaust emission levels of passenger cars is emphasised by the current contribution they make to atmospheric pollution. This paper gives an overview of work aimed at a substantial improvement in emissions and economy in the passenger car application, made possible through integrated control of a powertrain incorporating a continuously variable transmission. Several alternative approaches to the task of CVT driveline control were considered. These included artificial intelligence and more traditional and intuitive methods. Computer models of the powertrain components were validated against experimental data from a transient test rig and used to aid controller design. Preliminary results show the newly designed controller strategies to have significant impact on vehicle emissions.

1. INTRODUCTION

1.1 Alternative approaches to the reduction of emissions

Future improvements in emissions will be made by means of better engine design but immediate improvements can be made by alternative approaches. In addition to improving engine design, legislation has caused vehicle manufacturers to look more widely at the whole vehicle powertrain rather than at the engine alone. Improvements have been made by better control and matching of the engine and transmission (1) (2). There is further scope for work in this area.

Considerable improvements in both emissions and economy can be obtained by the adoption of an engine control strategy such that the required output power is always produced in the most advantageous region of the torque - speed map. This implies the use of an engine ideal operating line and requires the use of a continuously variable transmission to decouple the fixed relationship which would otherwise exist between vehicle and engine speed with a stepped ratio transmission. Ensuring strict adherence of the engine to an ideal operating line, chosen for emissions and economy, would not, however, be satisfactory in terms of the vehicle's transient performance. It would thus be necessary to allow the engine to deviate from the ideal operating line during vehicle transients, the length and duration of the deviation being dependent mostly on the driver's demand.

Early controllers for both stepped ratio and continuously variable transmissions were implemented hydraulically. Input signals were used from the accelerator pedal, selector lever position, engine speed and vehicle speed sensors. With the general trend towards the use of electronics and microprocessors in the control of vehicle systems, controllers which have been dependent upon hydraulics alone have been superseded. Hydraulics have remained for the actuation of clutches, whilst electronics are being used in determining ratio and for the modulation of hydraulic clutch pressures or electromagnetic clutch actuation. Electronics used in this way have enabled the production of more flexible and refined strategies for all types of transmission.

2. DESCRIPTION OF THE DIESEL CVT POWERTRAIN

2.1 Engine

The engine used for this work was a pre-production Ford 1.8L Direct Injection Diesel engine. The choice of Diesel as opposed to petrol was made for reasons of efficiency. Over recent years there has been growing interest in the High Speed Direct Injection (HSDI) version of the Diesel engine for passenger car applications owing to its inherent ability to produce higher efficiencies than the Indirect Injection (IDI) Diesel engine. Turbocharging and intercooling were used to increase the efficiency and power to weight ratio of the unit.

The Lucas EPIC (Electronic Pumping Injection Control) system was used on the engine for both the fuel injection and EGR (Exhaust Gas Recirculation) control.

To enable compatibility with the transmission it was necessary to derate the engine from a maximum torque production of 180 Nm to one of 130 Nm. In order to produce a similarly shaped torque curve on the derated engine, this meant a reduction in rated engine power from 65 kW to 50 kW.

2.2 Transmission

A modified Ford CTX transmission was used for the work presented here. The transmission utilises the Van Doorne push belt design variable speed unit described by Hendriks et al (3). The original hydromechanical control unit was upgraded to enable electro-hydraulic control by the addition of two proportional solenoid valves which were used to control the transmission primary and secondary actuator pressures. The parts of the hydraulic circuit used to control the clutch pressure and hence engagement remained as in the hydromechanical version of the CTX. The modified electronically controlled CTX transmission became known as the CTXE.

2.3 The test rig

The rig dynamometer comprised a fixed inertia in the form of a steel flywheel and a separate hydraulic loading system. It was developed at the University prior to the work described here and its operation is explained by Dorey and Guebeli (4). The rig enabled the testing and developing of the control software and the measurement and investigation into the dynamics of the engine and transmission. In addition, emulations of vehicle legislative emissions tests were possible together with in depth analysis of engine emissions production. The operating strategy enabled both static and dynamic frictional drag terms of the vehicle to be emulated.

3. NEWLY DEVELOPED ALTERNATIVE CONTROL STRATEGIES

3.1 Modelling of the powertrain

The powertrain comprising engine, transmission and vehicle were modelled using the Bath*fp* simulation software developed at the University and described by Richards et al (5). The modelling is described by Deacon et al (6). This enabled the early development and comparison of controller strategies. The simulation work enabled investigations into both dynamic system performance and into vehicle emission formation during transient manoeuvres represented by the legislative test. It also made possible the observation of those variables not measurable on the vehicle or test rig.

3.2 Application of the new control strategies

The task of controlling an automotive transmission can be considered to fall into two areas. Firstly there is the strategic function of determining the most appropriate response of the powertrain to the driver's demands. Secondly there are the tasks associated with achieving this response. The first task is concerned with interpreting the drivers requirement from the accelerator pedal and its relationship to the current operating conditions. The strategy may adapt to the driving style and also be influenced by other vehicle and powertrain sensors. Finally a choice must be made about the amount of power demanded from the engine and the ratio at which to operate the transmission. The demanded rates of change of these variables during transients may also be determined.

The decision was taken to design the controller architecture in a centralised hierarchical manner. This was because the hardware supplied by both Lucas and Van Doorne Transmissie b.v. made the use of a supervisory controller more appropriate. The controller architecture, shown in Figure 1, is therefore divided into two distinct areas, firstly the supervisory controller and secondly the powertrain controller. A logical boundary existed between the two areas and each had its own tasks. The supervisory controller contained the software responsible for the choice of current engine operating point in terms of torque and speed. In determining the final steady state operating point following a transient, the supervisory controller referenced data supplied by a module known as the operating line optimiser. The optimiser is described in detail by Brace et al (7). It used a model of the engine to determine a best operating line depending on a chosen weighted compromise between the different engine emissions. Other inputs used by the supervisory controller (depending upon the particular strategy in place) include the pedal position, a measure of time, and some powertrain feedbacks.

The outputs of the supervisory controller are passed to the inputs of the powertrain controller where the closed loop control of the engine speed (via control of the transmission ratio) and control of the transmission secondary pressure are performed. The powertrain controller sets the values of the EPIC demand signal and the transmission primary solenoid current in order to achieve the demanded engine torque and speed. The engine speed control loop is closed by feeding back the engine speed. The secondary pressure demand is also set by the powertrain controller to a level depending upon the engine torque, speed and the ratio. This loop is closed by using a signal fed back from the secondary pressure transducer.

The VDT controller is very similar to the powertrain controller in that it can be used to control engine speed externally via transmission ratio. Since the VDT controller gains were already set up and its performance was evaluated and seen to be satisfactory it was used in conjunction with the powertrain controller for the tasks of engine speed control and transmission secondary pressure control.

In the following section some sample results are presented from three of the five alternative designs of supervisory controller which were evaluated. All of them required the accelerator pedal position as an input in addition to the reference values of 'ideal' engine torque and 'ideal' engine speed produced by the operating point optimiser. Each controller was designed to generate a demanded engine operating point in terms of torque and speed. The five different approaches were pursued initially because they were all viable options and each was believed to have different strengths in the areas of emission control, drivability and ease of controller set up and tuning.

3.3 Results of predictive and experimental, transient manoeuvres and emissions tests

Controller 1 simulated performance

Figure 2 shows the predicted path followed by the engine in the torque speed domain as the result of a step increase followed by a step decrease in the accelerator pedal position. The results were created using the Bathfp simulation package mentioned in Section 3.1. Two ideal lines were used in this work, one optimised for low NOx emissions the other for high economy. These two particular lines were chosen because they represented the extremes of ideal line operation in terms of high torque, low speed and low torque, high speed operating points. The NOx ideal operating line was used during the simulation shown in Figure 2. The powertrain starts from a position of near steady state on the NOx ideal operating line and moves smoothly towards the higher power point on the ideal NOx line. The torque demand is increased at a rate such that the final torque level is demanded long before the final speed is demanded. The actual engine speed lags the demanded by a small amount. Following the decrease in pedal position, the demanded torque and speed are decreased rapidly towards the engine rest position. The fuel delivery is reduced to a minimum and the transmission is pushed into overdrive.

Controller 1 rig test example

Figures 3 to 5 show a test of controller 1 completed on the rig. In this test the power demand was generated linearly from the accelerator pedal signal level. In this paper preliminary experimental results using only the NOx ideal line are presented.

The results show the response of the powertrain to a step increase in the accelerator pedal signal. During the first seconds of the test, conditions are stabilised with the pedal at about one quarter of its full scale value. The engine and vehicle speeds are steady. The measured torque signal exhibited cyclic variations although these were not matched by any such variations in the EPIC fuel signal. There are small steady state errors between the demanded and actual engine torques and speeds.

Following the step increase in accelerator pedal signal, the demanded values of engine speed and torque start to change immediately. The demanded and actual values of engine torque move quickly towards the new ideal value with no overshoot. The settling time to the new torque value is less than two seconds. As before there is a small steady state error. The engine speed takes five or six seconds to reach a new steady state value, again with a small error. The actual value lags the demand slightly and there is no overshoot. The vehicle speed increases gradually following the pedal signal change. The acceleration is smooth and at a gentle rate.

Figure 5 shows the levels of NOx emissions both with respect to time and to vehicle distance travelled. The measured values are steady before the step increase in pedal. There is a delay due to gas transport time between the pedal increase and the increase in measured NOx emissions. After the initial increase in NOx emissions levels, there is a slight downward trend in the NOx emissions per kilometre as the dynamometer speed increases.

The paths taken by the controller demand and the engine in the torque speed domain is shown in Figure 4. The initial and final operating points on the NOx ideal operating line are clearly marked. The demand starts at the lower of these and moves immediately to the actual level of engine speed since this is higher and the new ideal point is in this direction. The demand then continues to move toward the new ideal point, the torque being increased ahead of the speed. The actual path followed by the engine also follows this trend. This is shown using both the torque calculated from the EPIC fuel demand and the filtered signal from the torque transducer.

Performance of Controller 2 - rig test example

Figure 6 shows the emissions results produced during a test which was similar to that described above. This time however, Controller 2 was used. The vehicle pedal signal change was of the same level of magnitude, the initial level being slightly different in order to achieve a similar initial dynamometer speed. The NOx ideal operating line was used as before.

Steady state conditions were achieved before the pedal increase. As with the above test, the measured torque signals showed moderate cyclic variations, although the speed and fuel signals do not undergo similar changes. There are small steady state errors between the demanded and actual values of engine speed and torque.

There is an immediate response to the increase in pedal position. The torque demand undergoes a large increase to a value greater than the ideal. This is due to the influence of a trimming torque function which is designed to provide sufficient additional torque to drive the powertrain through the transient. The torque demand is held at this new higher level for some ten seconds before being gradually reduced to the new ideal level. The engine speed demand is gradually increased to the new ideal level over a period of about five seconds. The actual speed also increases gradually, and with no overshoot, due to the demand change and the extra torque available. There are final small steady state errors between the demanded and actual values of engine torques and speeds. The vehicle speed increase is gradual and immediate following the pedal signal change. Again the acceleration is smooth and the rate low.

The levels of NOx production measured during this test are shown in Figure 6. The levels with respect to time and vehicle distance both seemed stable prior to the accelerator pedal increase. Following this there is a definite rise in both traces, the level per kilometre decreasing slightly as the vehicle speed increases.

The path followed by the engine during this transient was of a different form from the shape of the path shown in Figure 4. Using Controller 2, the demanded torque was increased more quickly with respect to the rate of engine speed increase. The use of a different controller strategy had a significant effect on the engine operating points visited and the resulting formation of emissions.

Performance of Controller 3 - rig test example

Figure 7 shows emissions data recorded during a test of Controller 3 on the rig. The levels with respect to time and vehicle distance again increased markedly following the pedal increase, before falling gradually to new levels. The increase in NOx production taking into account transport delays correspond with the pedal increase and the peak power delivery. In this strategy, no particular ideal operating line was followed. Instead the controller was tailored more towards vehicle drivability requirements.

4. CONCLUSIONS

In the three controller tests presented in Figures 5 to 7, the characteristic rate of increase and shape of the NOx emissions traces have been examined. Although the dynamometer speeds and accelerations were not identical in each of the three tests, and so therefore the results cannot be used to evaluate numerical emissions advantages of the controller strategies, the tests do show the significant impact of the alternative strategies on the formation of NOx emissions during a CVT powertrain transient. More recent work, still to be published, gives direct comparisons of the three controller performances, in terms of powertrain emissions and economy, over the European legislative emissions drivecycle. In the theoretical and practical supporting work, a transient powertrain test rig has been built and computer models of the powertrain have been validated and used to test the alternative controller strategies.

Further work will include optimisation of controller strategies for emissions reduction, investigation of vehicle drivability under such controllers and could involve the linking of the strategies with other vehicle systems having an impact upon emissions such as the engine cooling and the exhaust catalyst systems.

REFERENCES

(1) Cuypers MH, Metal V-belt and V-chain traction drives. International Symposium on Advanced and Hybrid Vehicles, Strathclyde, UK, 1984, ASME.

(2) Hendriks E, Qualitative and Quantitative Influence of a Fully Electronically Controlled CVT on Fuel Economy and Vehicle Performance. SAE 930668, 1993.

(3) Hendriks E, Heegde ter P, Prooijen van T, Aspects of a metal Pushing V-belt for Automotive Car Application. SAE 881734, 1988.

(4) Dorey RE, Guebeli M, Real time powertrain simulation for dynamic engine testing using a hydrostatic dynamometer. IEE Colloquium on Powertrain Control, 16 May 1990.

(5) Richards CW, Tilley DG, Tomlinson SP, Burrows CR, Bath*fp* - A second generation simulation package for fluid power systems. 9th International Conference on Fluid Power, Cambridge, UK, 1990.

(6) Deacon M, Brace CJ, Guebeli M, Vaughan ND, Burrows CR, Dorey RE, A modular approach to the computer simulation of a passenger car powertrain incorporating a Diesel engine and continuously variable transmission. IEE International Conference on Control, University of Warwick, UK, 21-24 March 1994.

(7) Brace CJ, Deacon M, Vaughan ND, Burrows CR, Horrocks RW, An operating point optimiser for an integrated Diesel/CVT powertrain. IMechE International Seminar on Application of Powertrain and Fuel Technologies to meet Emissions Standards for the 21st Century, IMechE HQ, London, UK, 24-26 June 1996, IMechE.

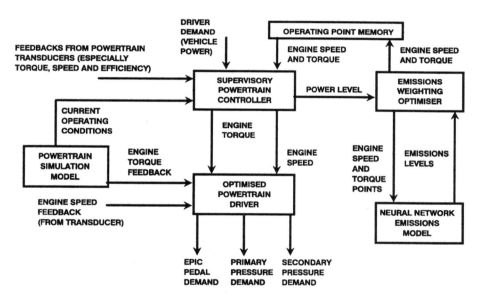

Figure 1 Chosen controller architecture

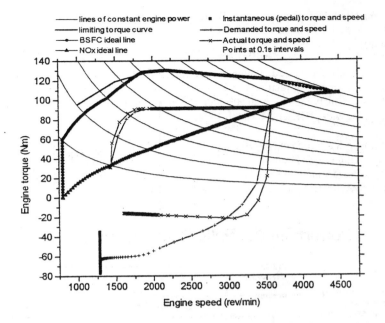

Figure 2 Controller 1 - predicted engine path followed during transient

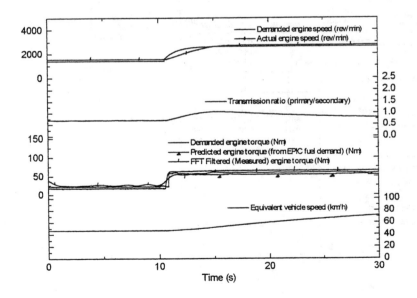

Figure 3 Controller 1 - rig test

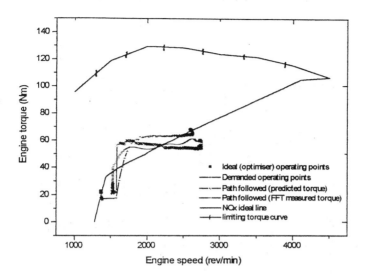

Figure 4 Controller 1 - rig test

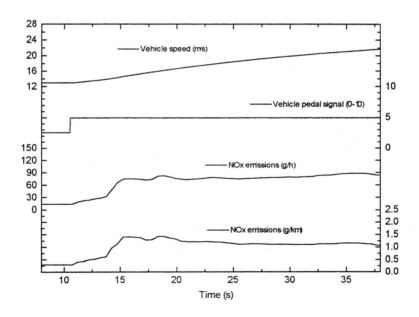

Figure 5 Controller 1 - rig test

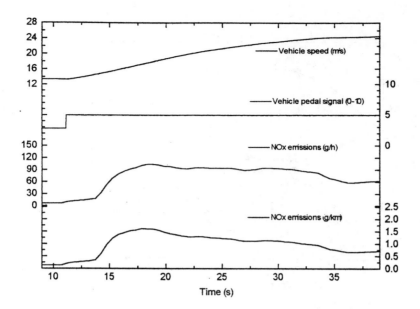

Figure 6 Controller 2 - rig test

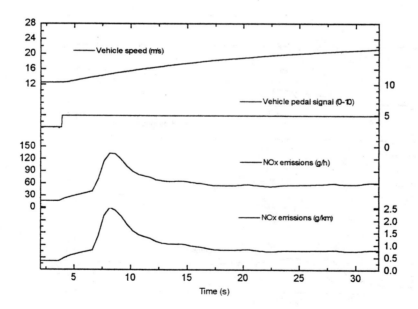

Figure 7 Controller 3 - rig test

C517/035/96

Control technology for future low emissions diesel passenger cars

PORTER BSc, ACGI, **T J ROSS-MARTIN** BSc, PhD, and **A J TRUSCOTT** BSc, PhD
Ricardo Consulting Engineers Limited, West Sussex, UK

Synopsis

This paper describes a Ricardo research project aimed at meeting the European Stage 3 emissions targets on a direct injection diesel engine with competitive responsiveness, driveability and refinement. The objective is to demonstrate on a vehicle the technologies required for future engines integrating advanced design, fuel injection equipment, refinement and control technologies. It illustrates Ricardo's control system approach by referring to a model based transient fuelling control strategy being developed for a demonstrator vehicle. It shows how testbed, vehicle and simulation studies are integrated together for control development purposes. Simulation and vehicle results demonstrating the effectiveness of Ricardo's control methodology are also presented.

Notation

AFR	Air Fuel Ratio	
CAN	Control Area Network	
ECU	Electronic Control Unit	
EGR	Exhaust Gas Recirculation	[%]
EMPS	Engine Management Prototyping System	
FIE	Fuel Injection Equipment	
FSN	Filtered Smoke Number	[0-10]
MAF	Mass Air Flow	[kg/s]
NEng	Engine Speed	[r/min]
PEngI	Intake Manifold Gas Pressure	[Pa]
PEngO	Exhaust Manifold Gas Pressure	[Pa]
PWM	Pulse Width Modulation	

QCmd	Actual fuel mass per injection	[mg/inj]
QFie	Demand fuel mass per injection	[mg/inj]
QLim	Fuel mass per injection limit	[mg/inj]
TCool	Coolant Temperature	[K]
TEngI	Intake Manifold Gas Temperature	[K]
VNT	Variable Nozzle Turbocharger	
XEgr	EGR Valve opening position	[0-1]
XThrot	Intake Throttle closing position	[0-1]
XVnt	VNT opening position	[0-1]

1 INTRODUCTION

Direct injection diesel engines offer an attractive route for achieving good fuel economy in passenger cars, but advances are required to meet emissions targets such as European stage 3 with competitive responsiveness, driveability and refinement. Ricardo is carrying out a major research project to demonstrate on a vehicle the technologies required for future direct injection diesel engines integrating advanced design, fuel injection equipment, refinement and control technologies.

The control system is required to give precise management of the highly interactive air handling, fuel and EGR functions on the engine. A second important requirement is minimization of development and calibration time.

This paper illustrates Ricardo's approach to the control system development by consideration of the transient air fuel ratio and EGR control strategy developed for the demonstrator vehicle. Section 2 describes how the the Ricardo methodology integrates testbed, vehicle and simulation studies for controller design purposes. The third section describes the design of the control strategy. Implementation of the strategy in the vehicle and vehicle test results are presented in Section 4. In the final section conclusions are drawn. The remainder of the introduction is given over to a description of the system to be controlled and Ricardo's control development methodology.

1.1 System Description

The engine to which the control system is applied is a turbocharged intercooled 2.2ℓ four cylinder direct injection diesel Ricardo research engine fitted to a mule vehicle. It features four valves per cylinder, employs vertical centrally mounted injectors with ports designed to give high swirl characteristics and incorporates dual balancer shafts.

A high pressure rotary fuel pump is employed giving full electronic control of fuelling quantity and timing. Two stage lift minisac injectors are used. To provide good turbocharger matching over a wide load and speed range a pneumatically actuated variable nozzle turbocharger (VNT) is fitted. The engine incorporates an Exhaust Gas Recirculation (EGR) system comprising an exhaust manifold mounted vacuum actuated EGR valve, an EGR cooler and a vacuum actuated intake throttle.

1.2 Methodology

Ricardo's control system development methodology has been designed with the objective of minimizing development time. A staged approach is taken making maximum use of data as it becomes available and computer aided engineering tools. This enables concurrent development of the control system alongside the vehicle and powertrain. The control development process begins with the building of a simulation model to characterise system behaviour. Using this, system behaviour is analysed and control strategies and code are developed and tested in simulation prior to implementation. Rapid implementation is achieved by making use of Ricardo's Engine Management Prototyping System (EMPS), described later. The block diagram shown in Figure 1 illustrates the approach.

2 MODELLING

To minimize simulation model development time Ricardo make extensive use of pre-written libraries of validated powertrain component Simulink and MatrixX models. Where necessary for new applications these are supplemented by new component models which are developed and later added to these libraries.

Typically, for control development, quasi-steady state (mean value) lumped-parameter models based on non-steady flow energy assumptions are used. These reproduce the steady-state and transient bulk phenomena of the engine whilst ignoring the individual cylinder combustion events. This approach results in models giving good representations of system behaviour and simulation times which are practical for control development purposes.

2.1 Parameterisation

Many of the values needed to parameterise component models are available directly from design data. (Typical parameters might be engine displacement, manifold volume and compressor characteristic). Where this is not the case data is integrated from other sources.

Some parameters may be estimated, utilizing Ricardo's experience with similar engine/system types. Frequently parameters needed to characterise the effects of unmodelled dynamics can be obtained using the Ricardo unsteady flow simulation program WAVE (1). Typical of these parameters would be those needed to account for the effects of manifold pressure pulsations on turbocharger operation and EGR flow.

Finally, parameters may be derived directly from engine and component test data, when this becomes available. Here use is made of automated data reduction techniques. Generally, Matlab macros are used to compute model parameters from WAVE or testbed data. This approach enables automated parameterisation providing significant time savings when system changes are introduced, for example, a change of turbocharger.

2.2 Validation

Where possible system model behaviour is validated by comparison with both steady-state and transient experimental data, model errors are analyzed and, when necessary, models and their parameters are revised to improve correlation. This process is continuous throughout the development process.

2.3 Model Description

A dynamic simulation model of the research engine was developed using Matlab/Simulink simulation tools. In addition to the engine processes, the system model includes the following component models:

- EGR valve
- Inlet throttle
- VNT
- Charge air cooler
- EGR cooler
- Actuators for EGR valve, inlet throttle and VNT
- Pressure, temperature and actuator position sensors

Simulink block diagrams of the model are given in Figures 2a and 2b.

The EGR valve and inlet throttle are modelled as one-way valves. These control the EGR flow into the inlet manifold where perfect mixing with air is assumed to take place. The flow of this mixture through the engine is assumed to be a steady-state pumping process whereby volumetric efficiency is mapped against engine operating condition. The energy transfer during the combustion process to the crankshaft and exhaust gas is computed using 2D look-up tables also based on engine condition. The exhaust gas flowing through the turbine side of the VNT is modelled using 3D look-up tables based on turbocharger speed, turbine pressure ratio and VNT opening. Compensation factors are applied to the turbine efficiency and the pressure upstream of the EGR valve in order to account for pressure pulsations arising from unmodelled individual combustion events. The corrected turbine efficiency is used to compute the torque required for driving the compressor through which air mass flow and temperature rise are computed using 2D look-up tables. Pressure drops through the air filter, charge air cooler and exhaust pipe are computed based on incompressible flow.

Actuator and sensor responses are modelled using first order transfer functions. For the former, different opening and closure rates are modelled by switching the transfer function time constant between two values accordingly.

Engine torque is obtained from the proportion of combustion energy which is transferred to the crankshaft and smoke density (Filtered Smoke Number - FSN) is obtained from a 1D look-up table against AFR.

2.4 Model Parameterisation
The simulation model was parameterised using both design data and steady-state test data. System volumes, valve areas and turbocharger look-up tables where parameterised using design data whilst look-up tables for engine volumetric efficiency, combustion energy transfer, pulse compensation factors, and smoke density were derived from steady-state testbed data.

3 CONTROL DEVELOPMENT

Control strategies are initially synthesized, developed and tested in simulation. To speed development use is made of a comprehensive modular library of control algorithms which are pre-coded as Simulink S-functions. Examples of these are conventional boost limit and model based fuel control algorithms with variants using different sensor combinations. From this library, individual algorithms can be selected to run within a generic controller block. This approach allows different control strategies to be tested and compared without modification to the simulation model.

3.1 Objective
The objective of the control development process is to produce a strategy that provides excellent emissions performance and driveability whilst remaining robust to system parameter variations and utilizing the current types and numbers of sensors available. A secondary objective is to minimise the calibration effort that the strategy demands. To achieve the latter a model based strategy has been adopted.

In the initial phase of development, reported here, effort has been concentrated on development of transient fuelling and EGR strategies to optimise emissions and engine torque response during load changes.

3.2 Control Strategy

Whilst developing the transient strategy, a simple map following steady state strategy has been adopted in which demanded VNT, EGR valve and intake throttle actuator positions are looked up on the basis of engine speed and load.

The model based transient strategy that has been developed is shown in Figures 3a and 3b. The strategy employs the following sensors: MAF, NEng, XEgr, XThrot, TCool, TEngI, PEngI, PEngO. In addition, demanded and actual fuelling quantities are inputs. Outputs drive the EGR valve, intake throttle and VNT actuators and limit fuelling.

The transient controller acts to reduce EGR flow and limit fuelling on the basis of an estimate of air-fuel ratio. When the estimated AFR approaches a mapped AFR limit the controller reduces the EGR valve opening and fully opens the inlet throttle while simultaneously limiting the fuelling. The air-fuel ratio estimate accounts for both inlet gas CO_2 and manifold dynamics. Limiting the minimum AFR in this way reduces the production of transient smoke and particulates.

Fuelling is limited on the basis of an estimate of the clean airflow into the cylinders so that limiting to a minimum air fuel ratio (AFR) can be achieved. The fuelling limit is calculated from the estimate of mass airflow into the engine and a mapped AFR limit.

The mass airflow estimator is a model of the EGR circuit and intake system providing an estimate of the mass flow rate of oxygen entering the engine cylinders. The main stages in this calculation are shown in Figure 3b.

Volumetric efficiency is estimated from a steady state map with engine speed, inlet manifold pressure and estimated temperature as inputs.

Inlet manifold and exhaust temperature are estimated from the engine speed, fuelling, the estimated pumped flow and the fraction of exhaust gas in the manifold. Steady state derived maps for the fuel energy entering the exhaust etc. enable the temperatures to be estimated without the delays imposed by measurement. The sensed inlet manifold temperature is used as a feedback to modify the inlet temperature state to match the measurement in the steady state.

The EGR concentration estimation consists of a model of the flow through the EGR valve and the dynamics of the concentration of exhaust gas in the inlet manifold. Estimated engine pumped flow, exhaust temperature, and volumetric efficiency are input together with measured engine speed, load, EGR valve position and manifold pressures. Characteristics of the valve flow etc. are contained in maps derived from the steady state mapping. A mapped EGR fraction is used as a feedback to modify the EGR fraction estimate to match the mapped value. On a system with a reliable means of determining steady state EGR rate this could be used in place of the map.

The engine mass flow rate estimator simply uses the measured inlet manifold pressure, engine speed and the estimate of temperature with the estimated volumetric efficiency to calculate the pumped flow of the inlet manifold gas mixture into the engine.

The final stage calculates the mass flow rate of "clean" air, i.e. not including the proportion of the gas which is EGR, entering the cylinders, used for the calculation of the AFR based fuel limit, QLim in Figure 3a.

3.3 Simulation Results

This section presents some of the simulation results depicting the performance of the transient control strategy. These results were compared with those obtained from simulations without model-based fuel limiting and/or EGR valve closure.

At a constant engine speed of 2000rev/min a load step of 10 to 50% was applied. The performance of the control strategy is depicted in Figure 4a. In this example an AFR limit of 22 was imposed and the EGR valve was re-opened at 0.5sec after the point at which the fuelling became unlimited. The small peak observed in the AFR following the step is mainly due to combined phase lag in the measurement and signal conditioning of inlet manifold pressure and EGR valve position sensing. This is also reflected in the EGR estimate immediately after the fuelling step. The rapid closure of the EGR valve and subsequent gradual re-opening is due to the combined actuator model dynamics and a one-way signal filter applied to the transient actuator position demand. The effect of this non-linear filtering is to reduce valve chattering when operating close to the AFR limit. This helps to improve AFR control by reducing fuelling fluctuations and hence reduces smoke.

Various strategy alternatives were simulated to examine the effects of fuel limiting and EGR valve closure. The following cases were compared:

A   No fuel limiting and no EGR valve closure
B   No fuel limiting and EGR valve closure
C   Fuel limiting and no EGR valve closure
D   Fuel limiting and EGR valve closure

The simulation results, given in Figure 4b, clearly illustrate the benefits of the control strategy. In case A, with neither fuel limiting nor valve closure, smoke levels are shown to be excessively high. Furthermore, since the EGR valve is open the flow path around the EGR circuit prevents the rapid build-up in exhaust manifold pressure. This reduces the amount of exhaust energy available to the turbocharger resulting also in a slower build-up in turbo speed and boost pressure.

In case B, closing the EGR valve has shown improvements in the response of the boost pressure and the amount of smoke. Limiting the fuelling without closing the valve (case C) limits the amount of smoke effectively but gives rise to a very slow boost pressure response. Not only is this due to the flow path around the EGR circuit but also because the steady-state EGR valve demand is dependent on current engine load.

From case D it can be seen that the strategy combining fuel limiting and EGR valve closure effectively controls transient smoke emissions without excessive fuel limiting. Simulation results have therefore shown the effectiveness of the model-based controller in providing a trade-off between engine performance and smoke emissions.

4. IMPLEMENTATION

To test the model-based control strategy further it has been implemented in a demonstrator vehicle. To speed development and eliminate duplication of effort, the FIE manufacturer's own Vehicle ECU has been retained to provide primary fuelling rate and timing control together with pump diagnostics. Alongside this the Ricardo diesel AFR and EGR strategy has been implemented using the Ricardo Engine Management Prototyping System (EMPS).

4.1 EMPS System

The EMPS system has been developed to enable the rapid development and testing of new control systems and strategies. The system is based around a PC, on which the control algorithms and the EMPS software are run, and up to three special purpose ECUs which act as input/output modules for the PC. The principal ECU module provides the main interface to the system's sensors and actuators. It provides analogue, digital, speed and frequency sensor inputs together with high current digital, PWM and crank angle synchronous outputs, as summarised in Table 1. Other modules provide respectively, inputs for 8 type-k thermocouples and 6 opto-isolated ratiometric analogue outputs. The ECUs communicate with the PC via a Control Area Network (CAN) operating with 1 Mbit/s bit rate.

Table 1   EMPS ECU Inputs and Outputs

6 channels for 0-5.1V analogue inputs (10 bit, high impedance)
12 channels for 0-5.1V analogue inputs (10 bit, medium impedance)
2 channels for resistive temperature sensor analogue inputs (8 bit, switchable impedance)
3 channels for resistive temperature sensor analogue inputs (8 bit)
6 channels for 0-2.6V analogue inputs (10 bit)
2 channels for 0-26V analogue inputs (10 bit)
4 channels for high resolution digital TTL level frequency inputs (16 bit)
1 channel for high resolution TTL level PWM input duty cycle measurement (10 bit)
1 channel for low resolution digital TTL level frequency input (8 bit)
8 channels for slow digital TTL level inputs
10 channels for high current PWM outputs
6 channels for high current digital outputs.

The EMPS software on the PC provides a user interface and runs the control strategy in three real time synchronous tasks operating typically at 1, 10 and 100Hz execution rates. The user interface provides a configurable real time display and user controlled logging of controller variables. It additionally allows the user to modify controller parameters while running and displays system error status codes.

The EMPS software is designed to facilitate rapid programming of new strategies. The coding process is speeded by a number of features. To simplify the code a library of pre-coded functions is available (providing look-up tables, PID control algorithms, digital filters and pseudo random binary sequence generation). To avoid real time scheduling difficulties, automatic real time task scheduling is pre-coded. To eliminate time consuming low level programming, code is written in a high level language ('C'). Also, full use is made of the PC's processor's ability to run floating point code. This eliminates variable scaling problems so that variables within the code can be scaled in conventional engineering units, simplifying the code and improving readability, as well as avoiding the time consuming and error prone process of conversion to fixed point format. Additionally, to facilitate rapid changes, EMPS automates the compilation and linking of code.

For the Ceres project the EMPS control system is connected to a large number of sensors providing the inputs shown in Table 2. The system also receives analogue voltage inputs from the Vehicle ECU, as show in Table 3. Additional miscellaneous inputs are from driver controlled dashboard switches, the key line and ECU watchdog circuits. Only actual fuelling quantity, inlet manifold temperature and pressure, EGR system differential pressure, charge air cooler outlet temperature and the actuator position measurements are used in the strategy presented here. The outputs from the control system are given in Table 4.

Table 2   Ceres research vehicle EMPS system sensor inputs

vehicle speed
engine speed
accelerator pedal position
intake throttle position
EGR valve position
VNT positions
mass air flow
compressor inlet pressure
compressor outlet pressure
inlet manifold pressure
exhaust manifold pressure
EGR system differential pressure
coolant temperature
oil temperature
compressor inlet temperature
compressor outlet temperature
inlet manifold temperature
exhaust manifold temperature
EGR cooler outlet temperature
charge air cooler outlet temperature
turbine outlet temperature
catalyst outlet temperature

Table 3 Ceres research vehicle EMPS system inputs from Vehicle ECU

demanded fuel per injection
actual fuel per injection
demanded start of injection timing

Table 4 Ceres research vehicle EMPS system outputs

EGR valve vacuum modulating valve PWM drive signal
intake throttle vacuum modulating valve PWM drive signal
VNT actuator pressure modulating valve PWM drive signal
maximum fuel limit analogue output (to Vehicle ECU)
timing trim analogue output (to Vehicle ECU)
relay drives (3 off)
warning lamp drives (3 off)

The actuator position control loops and the AFR and EGR control strategies are all implemented as a task executed at 100Hz. Prior to implementation "C" code modules for the controller were tested in simulation as Simulink S functions. This approach enabled a "right first time" implementation of the quite complex control strategy in the vehicle.

4.2 Calibration Data

Throughout development, the EMPS control system was used on an engine testbed as a flexible control system in order to achieve a high degree of automation in engine testing. This aided rapid testing through the implementation of closed loop control of EGR rate, boost pressure, turbo speed and injection timing controls using feedback from testbed measurement systems. Using the system, data with which to calibrate the controller was logged during steady state engine testing. From this data, all the maps used by the control strategy were generated automatically. This was done using processing macros developed in MATLAB and tested during the System Modelling and Control Development stages of the development process. Subsequently both the engine and the control system were installed in the demonstrator vehicle. This approach ensured consistency in the measurement system and avoided component to component variation as a potential source of error.

4.3 Testing

The vehicle and the control strategy are at an early stage of development, however, the effectiveness of the strategy has been examined both in road tests, to subjectively assess driveability, and in European emissions test cycles driven on the rolling road (ECE15 plus EUDC).

In addition, tests have been carried out to compare the strategy described above with other strategy variants including a conventional empirical boost limited fuelling based strategy. The boost limited fuelling strategy calibration was derived directly from testbed measurements. For these comparisons a simple automated second gear acceleration test has been used, starting from a steady state driving condition at 1000 r/min and accelerating to approximately 2200 r/min. This is a an engine speed range of particular relevance with regard to the performance and emissions. The tests were carried out on the rolling road and emissions responses were logged. The EMPS system was used to drive the accelerator pedal input to the Vehicle ECU over a predefined duty cycle. The high repeatability inherent in this technique has been found to be beneficial in reducing the numbers of comparisons necessary to establish the effects of control parameter changes on emissions responses

For the initial tests, results of which are presented here, no refinement of the control system calibration has been carried out. The calibration data for the model based strategy and for the boost limit based strategy, with which it is compared, were derived purely from steady state testbed data.

4.4 Vehicle Test Results

Figure 5 shows the road speed, Celesco smoke and dilute NOx data recorded during the hot ECE phase of an emissions test. For this test the vehicle was fitted with a close coupled oxidation catalyst sourced from a current production vehicle. It is notable that smoke level remains low throughout and that even the peaks during the accelerations rarely exceed 2% opacity. Similarly the dilute NOx emissions remain well controlled and low even during the accelerations. Table 4 shows the overall NOx and particulate results obtained in comparative tests of the model based and boost limit based strategies and also the latest results. The results for the model based strategy are significantly better than those achieved using the boost limit strategy and are within the current prosed limits for European stage III legislation.

Table 4   Comparison of emissions results for European test cycle

	NOx (g/km)	Particulate (g/km)
EC Stage 3 Proposal	0.37	0.05
Model based strategy (with oxidation catalyst)	0.37	0.036
Boost limit based strategy (with oxidation catalyst)	0.405	0.055
Model based strategy (with lean NOx catalyst)	0.312	0.035

Figure 6 presents a comparison of the effects of different types of fuel limiting strategy on vehicle responses to a step pedal input. Examination of the fueling shows that both fuel limiting strategies restrict fuelling significantly, in comparison with the unlimited case (which effectively represents the demanded fuelling for all cases). The Ricardo AFR strategy initially limits fuelling at a similar level to the boost limit based strategy. However as intake manifold

EGR concentration drops, as a result of the EGR valve closure at the start of the transient, the model based strategy is able to react and increase the fuelling limit above that of the boost limit based strategy. Thus the model based strategy achieves an improved torque response, as can be seen from a comparison of the vehicle speed traces. The peak smoke levels produced during the transients are similar in both fuel limited cases, although the duration of the smoke peak is greater for the model based case, reflecting the increased fuel delivery. For the unlimited fuelling case, the peak smoke opacity was 15%.

These observations correlate well with the subjective driveability assessments of testers driving the vehicle, who report improved responsiveness and acceleration from the model based strategy variant, with no visible increase in smoke.

Figures 7 also shows the impacts of the strategies on dilute NOx emissions. In comparison with the conventional boost limiting strategy the model based strategy produces slightly more NOx during the acceleration. This is to be expected since the fuelling quantity delivered is greater.

As evidenced by the results above the Ricardo model based control strategy provides a good mechanism for achieving a satisfactory trade-off between transient smoke emissions and torque response. In comparison with the conventional boost limit fuelling based strategy it offers significant advantages in several areas:

i) Reduced initial calibration effort. The strategy is calibrated directly from steady state testbed data, without the necessity for expensive transient testbed operation or time consuming vehicle based calibration.

ii) Reduced recalibration effort. Because the strategy is model-based it substantially compensates for changes in other parts of the strategy which effect transient intake manifold conditions. For example, changes to the transient EGR valve closure strategy may be made or a transient VNT control strategy can be introduced without the need for recalibration.

iii) Improved overall performance. While convensional boost limit strategies must be optomized for individual transients because of their implicit assumptions regarding manifold conditions, the model-based strategy is applicable to all transients without individual optimization. Thus a better overall performance is achievable.

5. CONCLUSIONS

Ricardo has applied an integrated control strategy development methodology to the problem of HSDI transient air fuel ratio control. This has resulted in the development through simulation of a model based AFR control strategy. The strategy has been implemented on a Ricardo HSDI research engined demonstrator vehicle using the Ricardo Engine Management Prototyping System. Tests of the vehicle have confirmed the value of a model based strategy in terms of reduced calibration effort and good emissions and driveability performance.

5.1 Further Work

Ricardo are continuing development of the diesel AFR control strategy, including the integration of turbocharger control, hard acceleration and active driveability algorithms as well as refinements to the FIE, combustion system and turbo charger matching and the construction of a second demonstrator vehicle. Improvements are also planned for accelerating the calibration and emissions/driveability/fuel economy process further.

6. REFERENCES
1. C.S.WREN, O.JOHNSON, "Gas Dynamics Simulation for the Design of Intake and Exhaust Systems - Latest Techniques", SAE 951367, 1995.

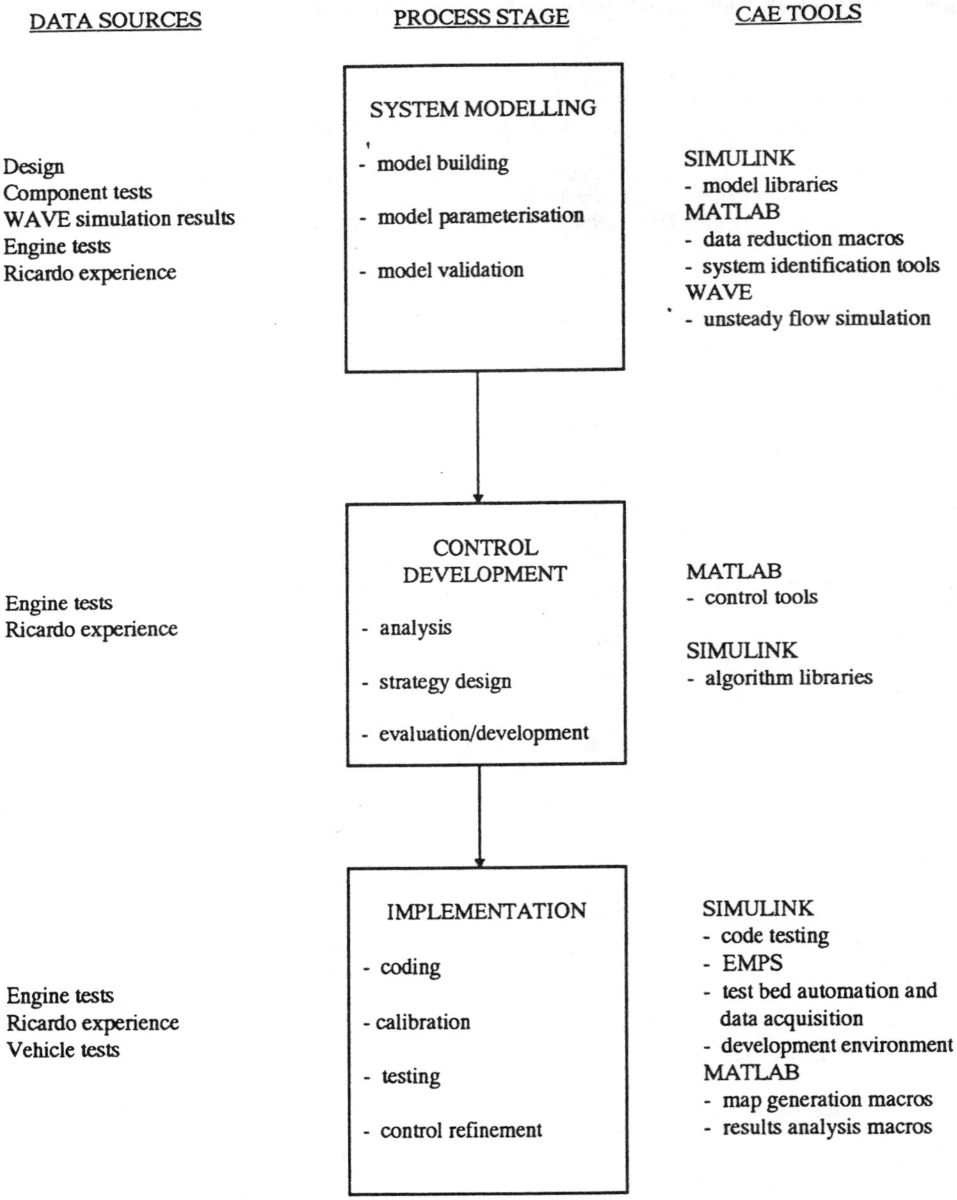

DATA SOURCES	PROCESS STAGE	CAE TOOLS
Design Component tests WAVE simulation results Engine tests Ricardo experience	**SYSTEM MODELLING** - model building - model parameterisation - model validation	SIMULINK - model libraries MATLAB - data reduction macros - system identification tools WAVE - unsteady flow simulation
Engine tests Ricardo experience	**CONTROL DEVELOPMENT** - analysis - strategy design - evaluation/development	MATLAB - control tools SIMULINK - algorithm libraries
Engine tests Ricardo experience Vehicle tests	**IMPLEMENTATION** - coding - calibration - testing - control refinement	SIMULINK - code testing - EMPS - test bed automation and data acquisition - development environment MATLAB - map generation macros - results analysis macros

FIGURE 1   RICARDO'S CONTROL SYSTEM DEVELOPMENT PROCESS

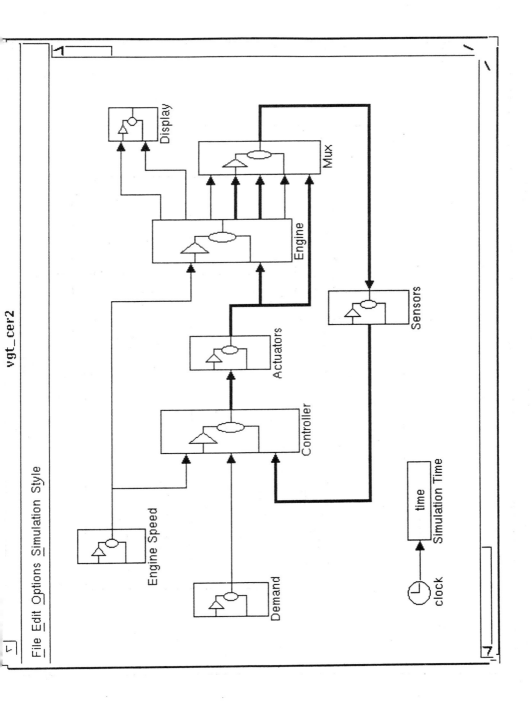

Figure 2a : Simulink Model of Research Engine : Top Level

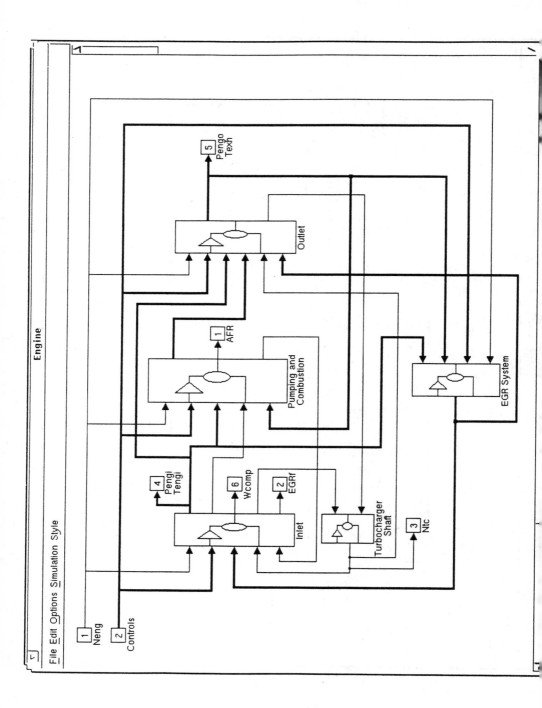

Figure 2b : Simulink Model of Research Engine : Engine Block

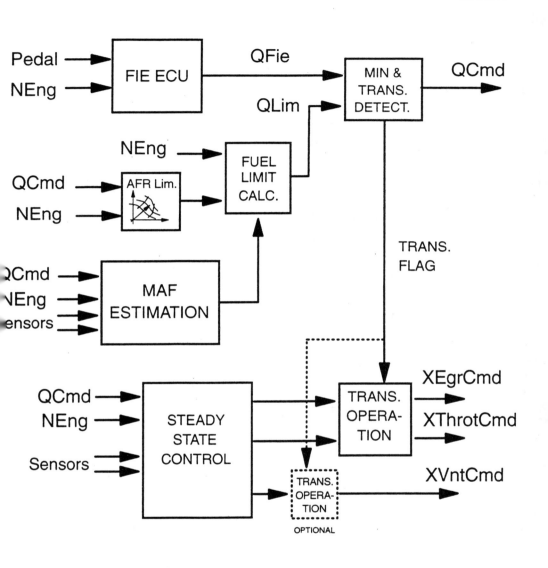

Figure 3a : Controller Overview

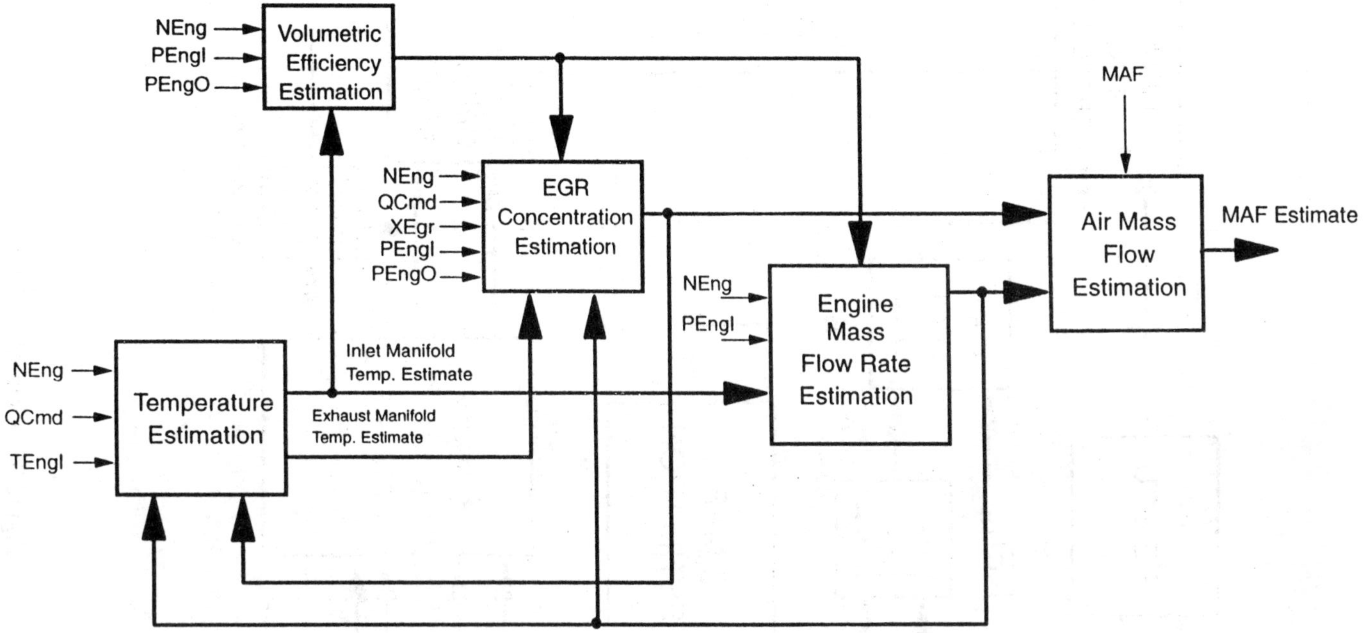

Figure 3b : Mass Air Flow Estimator

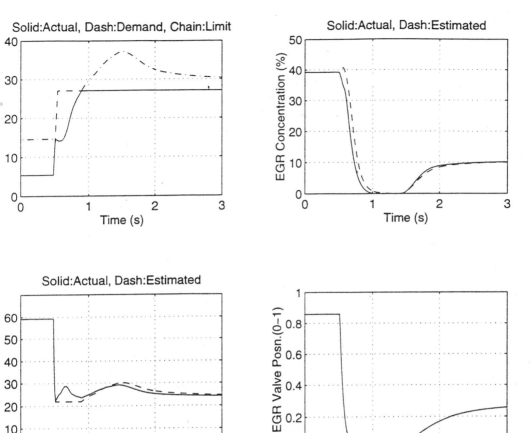

Fig 4a   Model Based Transient Control Strategy results for 10-50% load step at 2000 r/min

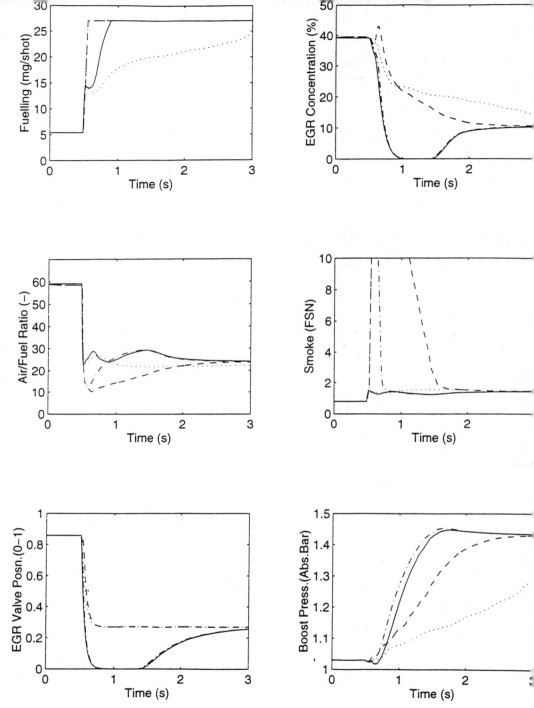

Fig 4b Comparison of Model Based Control Strategy with alternative strategies for 10-50% load step at 2000 r/min. (Solid line: Model based fuel limit with EGR valve closure. Chain line: EGR valve closure but no fuel limiting. Dashed line: No fuel limiting nor EGR valve closure. Dotted line: Model based fuel limit but no EGR valve closure).

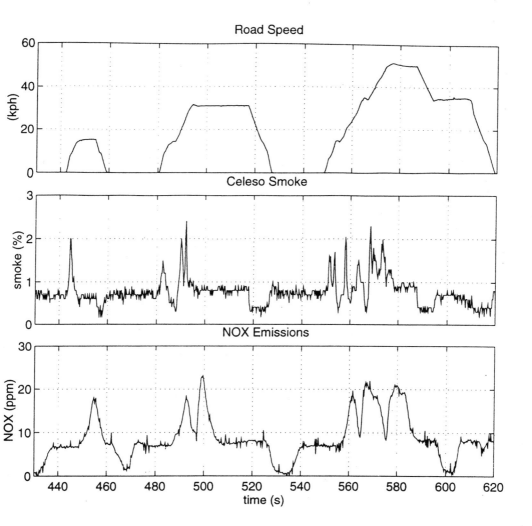

Fig 5   Celesco Smoke opacity and dilute NOx emissions in ECE test.

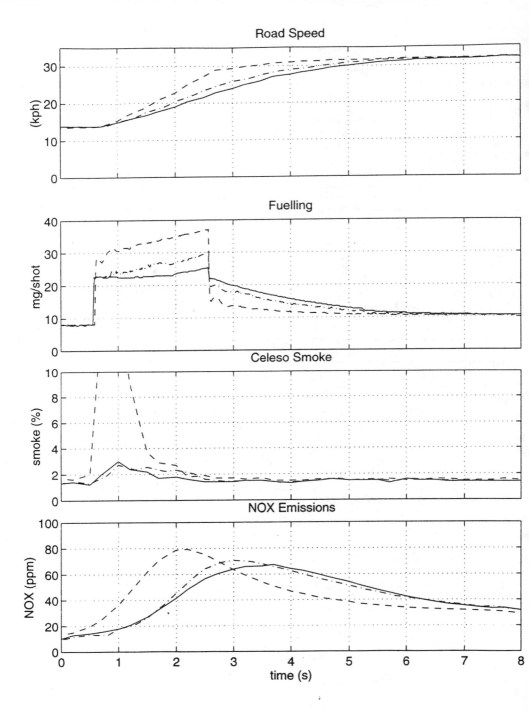

Fig 6 Comparison of effects of fuel limiting strategies on performance and emissions in acceleration test. (Solid line: Boost limit strategy. Chain line: Model based strategy. Dashed line: No fuel limiting).

517/043/96

Operating point optimizer for integrated diesel/CVT powertrain

C J DEACON BEng, MIMechE and **R W HORROCKS**
Ford Motor Company Limited, Essex, UK
C J BRACE BEng, CEng, MIMechE, **N D VAUGHAN** BSc, PhD, MIMechE, and **C R BURROWS**
School of Mechanical Engineering, University of Bath, UK

Synopsis

A technique has been developed to return the optimum emissions and economy from an engine and CVT using integrated electronic control. A supervisory controller is used to place the engine at the optimum operating point for the demanded power. The *Ideal Operating Line* (IOL) is generated automatically using a simple neural network based routine and updated continuously during operation to reflect changing conditions. Results from a simple simulation are presented here which show the effect of the IOL on vehicle performance over the ECE15 + EUDC.

1 Introduction

The drivetrain under consideration consisted of an experimental 1.8TCi DI Diesel engine, an oxidation catalyst and a belt drive CVT. The fuel injection pump and transmission were electronically controlled. The project used experimental and computer simulation studies to develop the necessary control algorithms. Reduced order models were employed to investigate the relative performance of candidate control strategies by simulation. Selected control methods were then to be tested experimentally in the laboratory to determine the emission and fuel economy improvements. The modified system was also installed in a vehicle for the conventional ECE15 + EUDC drive cycle test and an assessment of drivability.

2 Operating Point Prediction

2.1 Definition

One of the fundamental concepts in the Integrated Driveline Control project was that of an Ideal Operating Point (IOP). This is defined as the engine speed and load which delivers the desired power whilst producing the lowest level of undesirable emissions. A locus of IOPs may be drawn across the engine speed/load map and referred to as an Ideal Operating Line (IOL). The undesirable emissions are more wide ranging than the traditional concern relating to CO_2 (directly analogous to fuel consumption). In the Diesel engine they are CO, uHC, NOx and PMs. If the Diesel engine is functioning correctly there should be insignificant production of CO compared to a SI petrol fuelled engine. As such CO was not considered in the optimisation process.

2.2 Effect of Varying Operating Conditions

The emissions performance of the engine will change with operating conditions. Gradual changes in operating conditions, such as changing water temperature, may be referred to as *long term transients*. Such events are important to include in the emissions model as they will have a bearing on steady performance. Boost pressure, injection timing, EGR fraction and exhaust manifold pressure vary comparatively rapidly. Such variables are not useful when designing an IOL generator. Generally the plant will take some time to move to a new operating point. As such the boost pressure and similar variables will track the operating point with sufficient speed to make their inclusion as independent variables in the IOL generator superfluous. Deviations will be inevitable, the variables will, however, return to their nominal values quickly and may be referred to as *short term transients*.

The effects of the catalyst should also be considered when formulating an IOL. The catalyst performance will vary dramatically with temperature. Consideration of HC production is less important if the catalyst can be relied upon to oxidise the major proportion. The engineer may wish to use this fact to allow the optimisation of NOx. The catalyst will only partially remove PM, which prevents full exploitation of this effect as regions which are good for NOx are generally poor for PM. The catalyst will be ineffective below its light-off temperature, requiring attention to be paid to minimising HC.

2.4 Implications

Given the fact that in general the various emissions are lowest at different operating points there cannot be a global optimum operating point. The engineer must arrive at a compromise solution taking into account the relative importance of each pollutant. These rankings may be used to arrive at a weighted sum of the emissions which allows an IOL to be set. Before fixing the relative weightings some idea of their effect is required.

- User defined weights to prioritise the different pollutants
- Limited engine emissions model including slow transients only
- Variable weights to account for varying catalyst performance and other relevant environmental considerations

3 Optimiser Structure

The structure of the IOP generator developed is shown schematically in Figure 1. The task may be broken down into several sub-sections.

3.1 Operating Line Interpolation

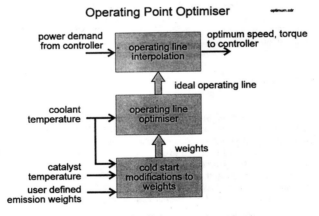

Fig 1 Ideal Operating Point generator - structure

Upon receiving a power demand from the supervisory controller the optimiser interpolates along a pre-determined ideal optimum line (IOL) to return an ideal engine speed and torque (IOP) for the demanded power. The IOL is returned to the supervisory controller which may either move the plant to this condition or may decide to overrule the optimiser to achieve good transient drivability.

3.2 Operating Line Optimiser

The IOL used to set the IOP is determined by a second module, the *operating line optimiser*. This function incorporates a model of the engine which is used to revise the IOL according to changing operating conditions such as coolant temperature.

The line consists of a series of points spaced at 5kW intervals from minimum to maximum engine power. The routine finds the engine speed which gives the lowest predicted weighted sum of emissions for each 5kW power step. Initially a straight 'optimum' line is drawn between the fixed minimum power (at idle speed) and maximum power (at maximum engine speed) points. The engine model is used to predict the emissions at the point where the initial line crosses the 5kW constant power curve. The various emissions are scaled according to pre-determined weightings and summed. This step is repeated at two further points which lie on the same power curve but at lower and higher engine speeds respectively. The point with the lowest weighted sum is chosen as the new 'ideal' point for 5kW. This process is repeated for each power step up to 95% full power. The procedure is repeated at the next controller iteration, which will move the line again in the direction of reduced emissions.

3.3 Modifications to Weights

A weighted value is assigned to each of the various emissions considered to describe their relative importance. These weights need not be fixed and can be updated continuously to reflect the prevailing conditions.

The network outputs are normalised over the range 0-1. This process allows the relative weightings of each pollutant to be set more simply. The weights may be modified as a function of coolant temperature and catalyst temperature. The weights may also be varied with power demand. For example, it may be desirable to optimise for HC at low power outputs but as higher power levels are demanded, the emphasis could be switched to minimise NOx production.

4 The Engine Model

4.1 Structure

The model used to predict the emissions at each step is an empirical model developed using data from a steady state test cell [1]. The data were used to train a neural network of the form shown in Figure 2. The neural network is a convenient way of building a fast running empirical model with some useful smoothing capabilities built into the training process.

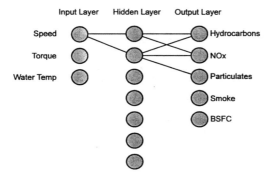

Fig 2 Structure of Neural network

The network was trained using a subset of the data gathered for a transient engine model [2]. In addition to the *on design point* data the maps generated at low and high water temperatures are included.

4.2 Training & Validation

The validation task for this model is relatively straightforward compared to the fully transient engine model. The model only has three inputs, as such the main two inputs (speed and torque) may be varied across their ranges while holding the third (water temperature) fixed. The resulting data may be presented conveniently as a two dimensional map or three dimensional surface. The experimental data may be superimposed and the fit assessed subjectively. A more objective method of evaluation is to inspect the RMS and standard deviation of the error in the network prediction compared to the training data. The RMS figures are shown in Table 1.

Table 1 - Optimiser Network Training Errors (RMS as % full scale)

BSuHC	BSNOx	BSPM	Bosch	BSFC
7.80	6.74	4.64	9.87	2.43

The figures above show that the network represents the engine data presented to it during training with a reasonable degree of accuracy. The largest RMS error is for the Smoke prediction. This reflects the highly non linear nature of the smoke map. The simple network used here cannot represent a sufficiently high order surface to match the experimental data completely. The RMS error of the HC and NOx predictions is better but still significant. However, this is primarily due to noise in the training data. The network will train to a smooth surface with a minimum error. This is useful for the application considered here as a highly convoluted surface would result in a volatile IOL with only minimal improvement in performance.

4.3 Subjective Appraisal of Neural Network Emissions Model

An example of the optimiser network output is shown in Figure 3. Experimental data points for BSNOx are superimposed on a surface mesh generated by the network. It can be seen that the network has generalised quite successfully to represent the real data.

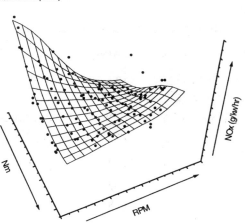

Fig 3 Neural Network representation of NOx data

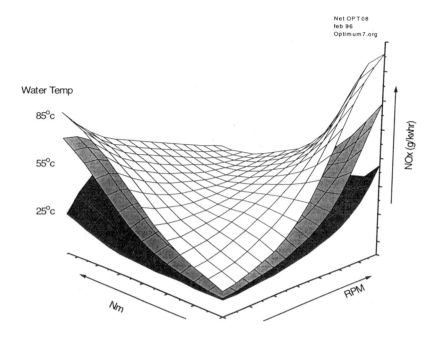

Fig 4 Variation of BSNOx with water temperature

Figure 4 shows how the shape and position of this surface changes with varying coolant temperature. The general characteristic is for less NOx to be produced at lower water temperatures although the shape of the surface does not alter greatly.

5 Results

The optimiser can be used to generate a vast number of subtly different IOLs by varying the weightings. In the first instance it was instructive to investigate the optimum lines for each of the pollutants in isolation. This was easily achieved by setting all the weightings to zero with the exception of the pollutant under investigation. In each case the network reproduced the line which would be considered to represent the minimisation of each species. The HC, and BSFC lines are quite similar but the major exception is NOx.

However, in normal operation the user defined weightings for each of the pollutants will ensure a line which is a compromise between these five extremes. Figure 5 shows a line optimised for NOx and HC. The weighting for NOx varies from 0 at 0kW to 1 at 50kW. The weighting for HC varies from 1 at 0kW to 0 at 50kW.

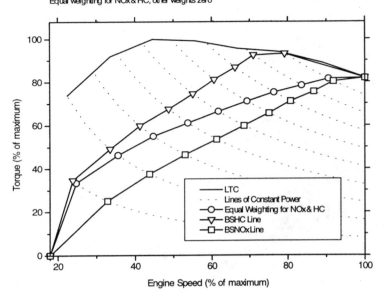

Fig 5 Ideal line with modulated weightings

6 Simulated Drive Cycle Performance

6.1 Model Structure

In order to investigate the effect of different Ideal operating lines on the emissions performance during drive cycles a simulation was carried out. The simulation is a very simple representation of the system dynamics implemented within an *Excel* spreadsheet. The test is split into one second intervals. For each second the power required to achieve the required vehicle speed is calculated. The engine is constrained to operate on the relevant IOL except in regions of clutch slip when starting from rest or where the ratio range of the unit would be exceeded. The crucial task of emissions prediction at each step is performed by a neural network model of the engine. This model is similar to the network used in the IOL generation except that in this case the mass flow rates of the various emissions are predicted directly in order to simplify the analysis. Results presented here are from a simulation which does not include a catalyst. This simulation tool allows many different strategies to be investigated quickly and easily. Experimental results have indicated that the simulation is sufficiently accurate to allow its use in this fashion. Where more detailed predictions of engine or controller performance are required a fully dynamic simulation may be used [3].

6.2 Results

Table 2 presents results from a selection of simulations. Separate simulations were performed with IOLs designed for HC, NOx and fuel consumption. Figure 6 as an example shows the PM mass flows for each run. It can be seen that the HC and BSFC lines are quite similar in their predictions as expected.. This was also true for HC emssion and fuel consumption. The major exception is in their production of NOx. The BSFC line produces significantly more NOx. This is due to the engine speeds selected being generally lower for a given power. This is the trade off for the improved fuel consumption, which was the aim of the line. Similarly the NOx line, although producing less NOx as expected, returns significantly worse PM predictions. On balance the HC line appears the best compromise. By inspecting the results from such lines a selection can be made which means that the penalties of each of the more extreme lines are avoided whilst retaining most of their benefits.

Table 2 - Drive cycle Predictions

Test No.	Ideal Line	HC g/km	NOx g/km	HC+NOx g/km	PM g/km	Fuel L/100km	Excel File
4	HC	0.23	0.50	0.73	0.17	5.86	HC_LINE
8	NOx	0.26	0.46	0.72	0.21	6.16	NOX_LINE
12	BSFC	0.23	0.59	0.82	0.18	5.76	BSFCLINE

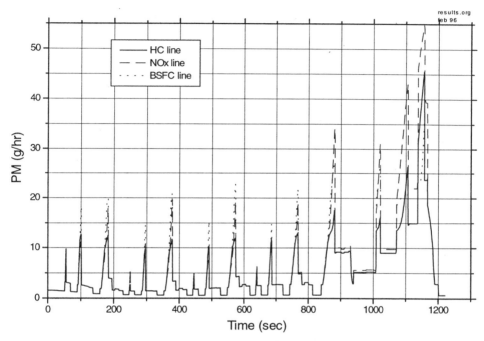

Fig 6 PM predictions

7 Conclusion

An optimising routine has been developed which makes use of a neural network based engine emissions model to set an ideal operating line (IOL) for use with an integrated Diesel engine/CVT driveline controller. This allows the supervisory controller to locate the engine at the optimum operating point for the demanded power. The approach used is flexible allowing easy adjustment and tuning of the balance between emissions. The IOL may be updated during operation to reflect the effect of changing conditions on emissions production.

The simple drive cycle simulation technique presented allows a useful preliminary representation of the effects of choice of the IOL. Results are presented which suggest that it is possible to alter the emissions performance of the vehicle significantly over the ECE15 + EUDC test by varying the IOL. The engineer may choose the pollutant to be minimised although a trade off with other emissions is unavoidable.

References

[1] Charlton, SJ., Cox, A., Somerville, BJ., Watts, MJ., Horrocks, RW. - **An Investigation of the Emissions Characteristics of the Passenger Car IDI Diesel Engine.** *IMechE paper C448/025 1992*

[2] Brace, C.J., Deacon, M., Vaughan, N.D., Charlton, S.J., Burrows, C.R. - **Prediction of Emissions from a Turbocharged Passenger Car Diesel Engine Using a Neural Network.** *IMechE International Conference 'Turbocharging and Turbochargers' 7-9 June 1994.*

[3] Deacon, M., Brace, CJ., Guebeli, M., Vaughan, ND., Burrows, CR., Dorey, RE - **A Modular Approach to the Computer Simulation of a Passenger Car Powertrain Incorporating a Diesel Engine and Continuously Variable Transmission.** *IEE Conference 'Control 94*

EGR for Diesel

The effects of carbon dioxide in EGR on diesel engine emissions

N LADOMMATOS BSc, MSc, PhD, MIMechE, **S M ABDELHALIM** BSc, PhD, MSAE, MInstE, WES, and **H ZHAO**
Department of Mechanical Engineering, Brunel University, Uxbridge, UK

SYNOPSIS

This paper investigates the way that exhaust gas recirculation (EGR) influences diesel engine combustion and emissions. The different effects of carbon dioxide (CO_2), which is a principal constituent of EGR, were analyzed and quantified. The engine tests were carried out while the engine speed, fuelling rate, injection timing, inlet charge total mass rate and inlet charge temperature were all kept constant.

The overall and individual effects of carbon dioxide on combustion and emissions were studied by replacing oxygen or nitrogen in the inlet air with carbon dioxide and/or inert gases. Additional tests were carried out which investigated the effects of oxygen replacement on ignition delay. Also tests were carried out in which the inlet charge temperature was raised gradually in order to quantify the effects of EGR temperature on emissions.

Findings from the tests included: the effect of CO_2 dissociation (chemical effect) on exhaust emissions is small; the high heat absorbing capacity of CO_2 (thermal effect) had only a small effect on exhaust emissions including NO_x. The reduction in the inlet charge oxygen (dilution effect) is the dominant effect on emissions, resulting in very large reductions in exhaust NO_x at the expense of higher particulate and unburnt hydrocarbon emissions and lower engine power output and fuel economy. The higher inlet charge temperature that is normally associated with hot EGR increased NO_x; it also increases the exhaust smoke and particulate emissions, but also reduces unburnt hydrocarbon emissions.

1. INTRODUCTION

One of the most effective techniques for reducing NO_x emissions in internal combustion engines involves the use of EGR. However, the application of EGR incurs penalties on the specific fuel consumption and particulate emissions. In diesel engines the trade-off between NO_x and particulate emissions is highly aggravated by the application of EGR, especially at high loads (1,2). The higher soot emissions also adversely affect the lubricating oil quality and engine durability (3,4,5).

The introduction of EGR mainly involves the replacement of inlet oxygen with CO_2 and water vapour. This alters the combustion process in various ways. The reduction of oxygen in the inlet charge is called the dilution effect. The higher specific heat capacities of CO_2 and water vapour in comparison to that of the replaced oxygen results in what is known as the thermal effect. As these EGR constituents dissociate at the high temperature of combustion to form free radicals, they can also participate in the combustion process; this is called the chemical effect of EGR. EGR can have an additional effect on combustion if the exhaust is recirculated without being cooled, mainly, the increase in inlet charge temperature associated with the use of hot EGR.

1.1 Dilution Effect of EGR

Plee et al varied the inlet charge oxygen in separate investigations and showed that the flame temperature had the major influence on NO_x emissions (6) while particulate and CO emissions were mainly influenced by oxygen mole fraction in the inlet charge (7). Mitchell et al (8) also found that the flame temperature rather than the oxygen availability was the dominant factor for NO_x formation when the inlet charge to a DI diesel engine was diluted with nitrogen or CO_2.

The above work was done whilst the start of combustion was kept constant. Keeping the injection timing fixed, Lida and Sato (9) showed that oxygen enrichment of the inlet charge resulted in a decrease in ignition delay, an increase in NO_x emissions, and in a reduction in particulate emissions. More detailed work on soot formation and oxidation showed that the soot formation was higher for a higher surrounding gas temperature, whereas soot oxidation was higher for a higher flame temperature (10).

The results showed by Ropke et al (11) were also in agreement with other workers' results, where oxygen availability influenced the flame temperature and, thereby, NO_x formation. The reduction in NO_x formation associated with lower flame temperature was suggested to be due to the reduction in dissociation of CO_2 and water to form atomic oxygen that is required for NO_x formation.

1.2 Thermal Effect of EGR

Many investigators (1,12,13) suggested that the main effect of EGR on combustion and emissions was through its high heat absorbing capacity which reduces the combustion temperature. The higher heat absorbing capacity of EGR was attributed to the high specific heat capacity of the recirculated CO_2 and H_2O and also due to the dissociation of these gases. Lida (10) observed a remarkable reduction in the surrounding gas temperature at TDC when CO_2 was added to the inlet charge of a rapid compression machine. The surrounding gas temperature, which also affects the ignition delay period, was shown by the author to be one of the major factors that influence soot formation in diesel combustion.

On the other hand, Wilson et al (14) were among the first to point out the insignificant role on the flame temperature of the higher specific heat components of EGR over that of oxygen. They suggested that the thermal effect of EGR is very low and is offset by the high temperature of the recirculated gases when hot EGR is used.

1.3 Chemical Effect of EGR

The dissociation of CO_2 at high temperatures of combustion was suggested by Ropke et al (11) to provide atomic oxygen that could increase the NO_x formation according to the Zeldovich mechanism. The dissociation of CO_2 was also shown (8,10) to cause a reduction in soot production in diesel combustion. Mitchell et al (8) suggested that this could be due to the increase in premixed burning associated with the longer ignition delay and the enhancement of soot oxidation.

1.4 Inlet Charge Temperature Effect

The recirculation of the uncooled exhaust gas raises the temperature of the inlet charge well above ambient temperature, especially for high EGR fractions.

Torpey et al (1) compared the emissions of hot and cooled EGR and found that the hot EGR resulted in higher NO_x and lower unburnt hydrocarbon emissions in comparison to cold EGR. Similar results were also found by Durnholz et al (13) and Ropke et al (11). However, both have noticed diminishing NO_x with increased EGR which they suggested to be due to the thermal throttling when hot EGR is used. Thermal throttling is known as the reduction in inlet charge mass as a result of increase in inlet temperature, since the intake volume of the engine is fixed.

2. EXPERIMENTAL TECHNIQUE

The engine used in these tests was a four cylinder 2.5 l DI

diesel engine which was naturally aspirated. The engine was modified so as to separate the inlet and exhaust for cylinder number one. A thermostatically controlled electric heater was located upstream of the inlet to the cylinder number one so as to vary the inlet charge temperature of this cylinder independently. Cylinder number one was used for monitoring independently the inlet and exhaust gas composition of this cylinder. The first cylinder was also equipped with a fuel injector needle lift sensor and cylinder piezoelectric pressure transducer. The engine modification involved mechanically locking the injection timing at 10 deg. c.a. before TDC. Tests were conducted at 2000 r/min (constant speed) and constant fuel consumption equivalent to that of about 40 percent load. The fuel consumption and the inlet charge mass were kept constant for all tests and the same batch of diesel fuel was used throughout all tests.

In order to investigate the overall effects of O_2 replacement by CO_2 in the inlet charge, as well as isolate the dilution, thermal, and chemical effects of this substitution, the composition of the inlet charge was modified in a controlled manner as shown in Figure 1. The total effect of CO_2 was analyzed by progressively replacing oxygen in air with CO_2. The engine performance and emissions were then compared to those of the baseline case (when no O_2 was being replaced).

The dilution effect of CO_2 was discerned by replacing oxygen in the inlet charge by a controlled amount of nitrogen and argon. This ensured that the specific heat capacity of the inlet charge remained constant and equal to that of air. Thus, the effects of higher heat capacity and dissociated CO_2 on combustion and emissions were eliminated. In another set of tests, some of the nitrogen in the inlet air was progressively replaced by a carefully controlled mixture of helium and nitrogen. In this way the thermal effect of CO_2 on combustion and emissions was isolated and quantified. The chemical or dissociation effects of CO_2 were isolated by replacing nitrogen in the inlet air with a controlled amount of CO_2 and argon. The specific heat capacities for different inlet charge mixtures over a wide range of temperature are shown in Figure 2.

The injection of different gases in the inlet air altered the length of the ignition delay period. Therefore, the effect of ignition delay with 6 percent dilution was offset by bringing the ignition delay back to what it was before dilution using 2-ethyl hexyl nitrate additive to the fuel as an ignition improver (in concentrations of up to 6000 ppm).

In another test, the temperature of the inlet charge was progressively raised from 40 to 120 °C in order to study the effect on emissions of higher inlet charge temperatures associated with hot EGR. For this test the inlet charge mass, which was kept constant during the test, was initially reduced so as to compensate for thermal throttling that occurs at higher inlet temperatures. The test was carried out with a diluted inlet charge and correction for the ignition delay was performed with the aid of the ignition improver as mentioned above. Table 1 shows the characteristics of the inlet charge mixtures for different tests conducted in the present work.

3. RESULTS AND DISCUSSION

3.1 Effects of CO_2 on Emissions

The NO_x emissions for the overall as well as the isolated effects of CO_2 are shown in Figure 3. This figure clearly shows that the most significant effect of CO_2 on NO_x emissions is due to the reduction of the inlet charge oxygen concentration i.e. the dilution effect. The dissociation of CO_2 also reduces the NO_x emissions but to a lesser extent in comparison to that for the dilution effect. On the other hand, the high specific heat capacity of CO_2 hardly influenced the NO_x emissions, even for up to 7 percent inlet charge replacement level.

The combined dilution and chemical effects of CO_2 were investigated at 6 percent inlet charge replacement level and designated by an asterisk in all figures. Assuming that the overall effect of CO_2 equals the algebraic addition of the dilution, chemical, and thermal effects, the difference between the overall effect of CO_2 and the combined dilution and chemical effects is equivalent to the thermal effect of CO_2. Therefore, the combined dilution and chemical effects were isolated from the overall effect by replacing 6 percent by mass of the inlet charge oxygen with CO_2 (overall effect of CO_2) whilst keeping the inlet charge specific heat capacity constant and equivalent to that of air. The specific heat capacity of the inlet charge was kept constant and equal to that of air by replacing part of the nitrogen in air with argon. This test was performed in order to study the influence on emissions of the dissociation of CO_2 at a reduced inlet charge oxygen level. This was also useful for confirmation of other results. In most cases, the combined dilution and chemical effects were almost equal to the overall effect of CO_2. This indicates that, in contrast to what was traditionally believed, the high specific heat capacity of CO_2 plays an insignificant role in the reduction of NO_x emissions.

Figure 4 shows that the introduction of the CO_2 in the inlet charge results in a prolonged ignition delay. This was mainly due to the reduction in oxygen availability. The chemical effect of CO_2 also contributes to the extension of the ignition delay period, however, this contribution does not exceed 25 percent of the overall effect of CO_2. The thermal effect of CO_2 on ignition delay was almost negligible. This is also obvious from the rest of the ignition delay results where the sum of the isolated dilution and chemical effects (or the two effects combined) were equivalent to the overall effect.

The effect of displacing 6 percent by mass of the inlet charge oxygen on NO_x emissions was separated into ignition delay effect and oxygen availability effect as shown in Figure 5. When the ignition delay of the diluted charge was brought back to what it was with normal air (baseline level) the NO_x emissions slightly increased. This implies that the oxygen availability played the dominant role in reducing the NO_x emissions, whilst the lengthening the delay with CO_2 only accounted for about 15 percent of the NO_x reduction.

The total unburnt hydrocarbons increased with increasing inlet charge CO_2 concentration, as shown in Figure 6. The dilution effect was the most significant of all effects. The chemical effect slightly increased the unburnt hydrocarbons. The thermal effect had virtually no effect on these emissions. However, neither the algebraic addition of the separate dilution and chemical effects nor the two effects combined together equalled the overall effect of CO_2. Thus, the extensive increase in unburnt hydrocarbons due the introduction of CO_2 in the inlet charge remains partly unexplained.

Figure 7 shows the overall and the individual effects of CO_2 on particulate emissions. The dilution of the inlet charge can be seen to be greater than the overall effect of CO_2. This is, obviously, due to the fact that the chemical effect of CO_2 reduces the particulate emissions as shown in Figure 7. The chemical effect of CO_2 in reducing particulate emissions (which could be either through suppression of soot formation or enhancement of the soot oxidation) was found by other workers too (8,10,15).

The carbon emissions followed the same trend as that for total particulates as shown in Figure 7 and 8. This could indicate that carbon emissions dominated the particulate emissions. Figure 9 shows the particulate carbon and volatile fractions for the dilution effect. This indicates that the increase in the particulate emissions with dilution was mainly attributable to the increase in the carbon. The ignition delay had no significant effect on the particulate emissions.

Figures 10 a and b show the trade-offs between NO_x and particulates, and NO_x and unburnt hydrocarbons respectively. Clearly, the application of EGR in practice will not be beneficial beyond a certain limit where no further significant reductions in NO_x are achieved; in contrast, both particulates and unburnt hydrocarbons continue to increase.

3.2 Effect of Inlet charge Temperature

Increasing the temperature of the inlet charge resulted in a decrease in ignition delay. The effect of inlet charge temperature and that of ignition delay on nitric oxide emissions is shown in Figure 11. At least 40 percent of the increase in these emissions was attributable to the reduction in ignition delay which occurred when the inlet charge temperature was increased. Although the fraction of premixed combustion was reduced by the reduction in ignition delay, however, the fact that combustion occurred closer to TDC during the compression stroke resulted in the increase in NO_x emissions.

The particulate emissions also increased with the increase in inlet charge temperature as shown in Figure 12. However, the ignition delay had no effect on the total particulate emissions; thus, the increase in these emissions was mainly due to the increase in inlet temperature.

Figure 13 shows that increasing the inlet charge temperature resulted in a decrease in the unburnt hydrocarbon emissions. Apparently, this was mainly due to the decrease in the ignition delay.

4. CONCLUSIONS

1. The most dominant effect of CO_2 on emissions was through the reduction of oxygen in the inlet charge; the higher specific heat capacity of the CO_2 in comparison to the O_2 it replaced had virtually no effect on emissions.

2. The introduction of CO_2 in the inlet charge resulted in a reduction in NO_x, an increase in the particulate emissions (which was dominated by dry soot), and also resulted in an increase in the unburnt hydrocarbon emissions.

3. The chemical effect showed that a moderate reduction in NO_x and particulates could be achieved through the introduction of CO_2 in the inlet charge.

4. Cooling of the EGR is recommended because increasing the inlet charge temperature resulted in an increase in both NO_x and particulate emissions. However, there was a reduction in the unburnt hydrocarbons mainly due to the reduction in the ignition delay that accompanied the increase in inlet charge temperature.

5. REFERENCES

(1) Torpey, P.M., Whitehead, M.J., and Wright, M., "Experiments in the control of Diesel Emissions", Conference on Air Pollution Control in Transport Engines, 9-11th Nov. (1971).

(2) Narusawa, K., Odaka, M., Koike, N., Tsukamoto, Y., and Yoshida, K., "An EGR Control Method for Heavy-Duty Diesel Engines

under Transient operations." SAE Paper No. 900444 (1990).

(3) Tokura, N., Terasaka, K., and Yasuhara, S., "Process through which Soot Intermixes into Lubricating Oil of a Diesel Engine with Exhaust Gas Recirculation", SAE Paper No. 820082 (1982).

(4) Nagai, I., Endo, H., Nakamura, H., and Yano, H. "Soot and Valve Train Wear in Passenger Car Diesel Engine", SAE Paper No. 831757 (1983).

(5) Cadman, W. and Johnson, J.H., "The Study of the Effect of Exhaust Gas Recirculation on Engine Wear in a Heavy-Duty Diesel Engine Using Analytical Ferrography", SAE Paper No. 860378 (1986).

(6) Plee, S.L., Ahmad, T., Myers, J.P., and Faeth, G.M., "Diesel NO_x Emissions- A Simple Correlation Technique for Intake Air Effects", Nineteenth Symposium (International) on Combustion/ The Combustion Institute, pp.1495-1502 (1982).

(7) Plee, S.L., Ahmad, T., Myers, J.P., and Siegla, D.C., "Effects of Flame Temperature and Air-Fuel Mixing on Emissions of Particulate Carbon from a Divided-Chamber Diesel Engine", in Particulate Carbon-Formation During Combustion, pp.423-487, Plenum press, N.y. (1981).

(8) Mitchell, D.L., Pinson, J.A., and Litzinger, T.A., "The Effects of Simulated EGR Via Intake Air Dilution on Combustion in an Optically Accessible DI Diesel Engine", SAE Paper No. 932798 (1993).

(9) Lida, N. and Sato, G. T., "Temperature and Mixing Effects on NO_x and Particulate", SAE Paper No. 880424 (1988).

(10) Lida, N., "Surrounding Gas Effects on Soot Formation and Extinction- Observation of Diesel Spray Combustion Using a Rapid Compression Machine", SAE Paper No. 930603 (1993).

(11) Ropke, S., Schweimer, G.W., and Strauss, T.S., "NO_x Formation in Diesel Engines for Various Fuels and Intake Gases", SAE Paper No. 950213 (1995).

(12) Ohigashi, S., Kuroda, H., Hayashi, Y., and Sugihara, K., "Heat Capacity Changes Predict Nitrogen Oxides Reduction by Exhaust Gas Recirculation", SAE Paper No.710010 (1971).

(13) Durnholz, M., Eifler, G., and Endres, H., "Exhaust- Gas Recirculation- A Measure to Reduce Exhaust Emissions of DI Diesel Engines", SAE Paper No. 920725 (1992).

(14) Wilson, R.P., Muir, E.B., and Pellicciotti, F.A., "Emissions Study of a Single-Cylinder Diesel Engine", SAE Paper No. 740123 (1974).

(15) Lida, N. and Watanabe, J., "Surrounding Gas Condition Effects on NO_x and Particulates", International Symposium COMODIA 90, Kyoto, pp. 625-632 (1990).

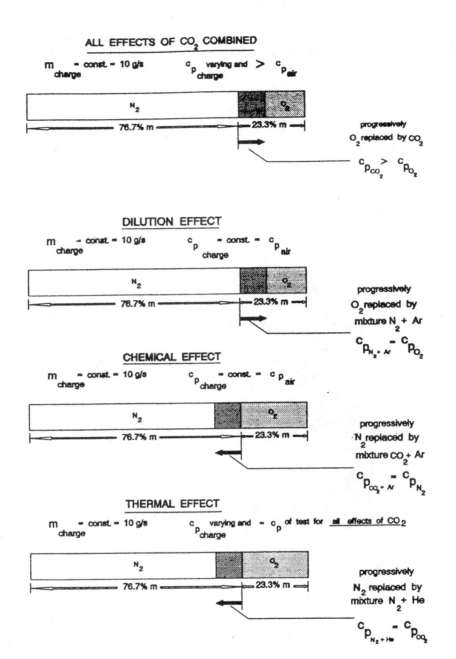

Fig.(1): Summary of the Methodology Used for Separating the Effects of CO2 on Combustion and Emissions

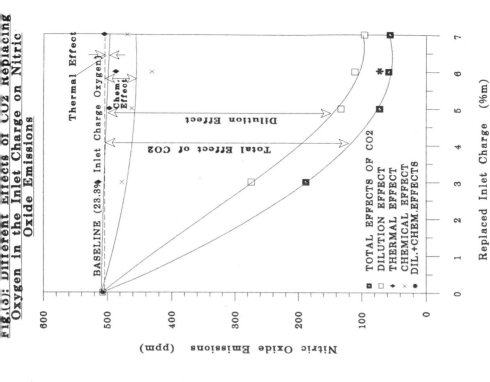

Fig.(6): Different Effects of CO2 Replacing Oxygen in the Inlet Charge on Nitric Oxide Emissions

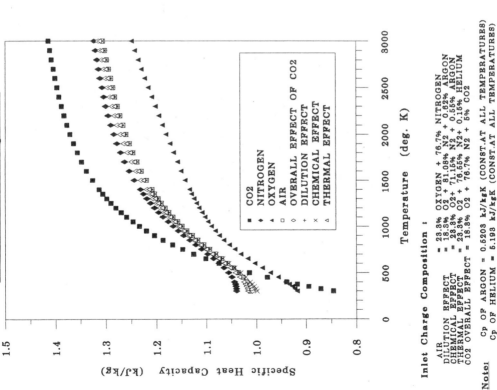

Resembling Different Effects of Five Percent Inlet Charge CO2

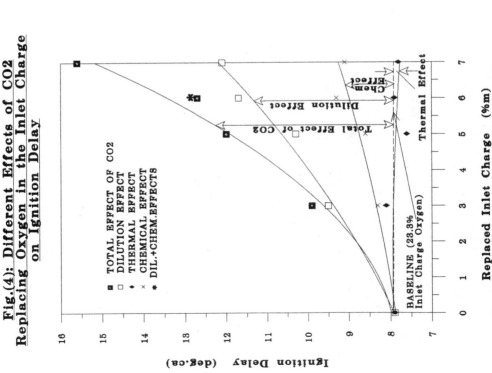

Fig.(4): Different Effects of CO2 Replacing Oxygen in the Inlet Charge on Ignition Delay

Fig.(5): Nitric Oxide Emissions at Constant Inlet Oxygen Concentration (17.3%m) with Ignition Delay Progressively Reduced Using Cetane Improver

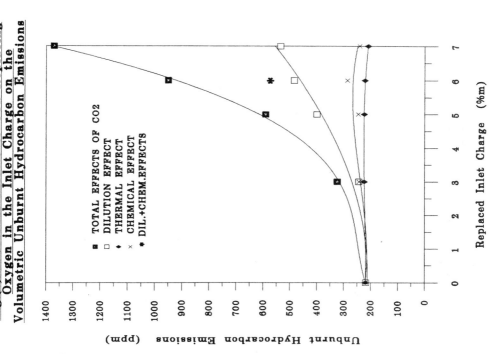

Fig.(7): Different Effects of CO2 Replacing Oxygen in the Inlet Charge on Total Particulate Emissions

Fig.(6): Different Effects of CO2 Replacing Oxygen in the Inlet Charge on the Volumetric Unburnt Hydrocarbon Emissions

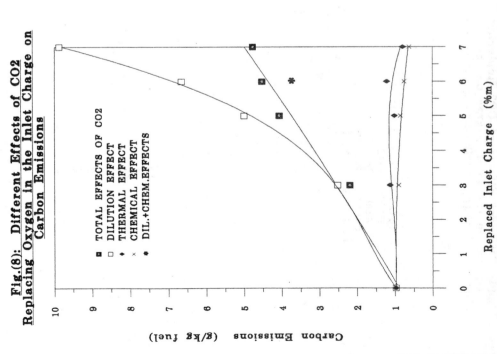

Fig.(8): Different Effects of CO2 Replacing Oxygen in the Inlet Charge on Carbon Emissions

Fig.(9): Variation in Particulate Mass Fractions with Inlet Oxygen Concentration

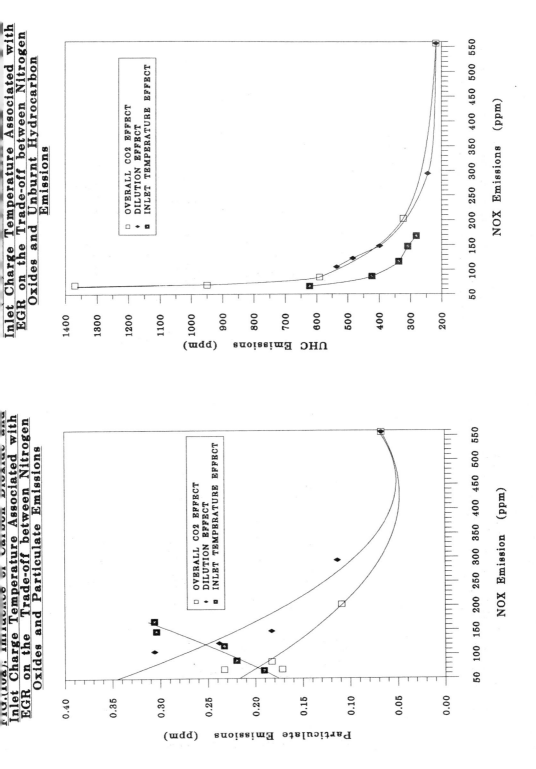

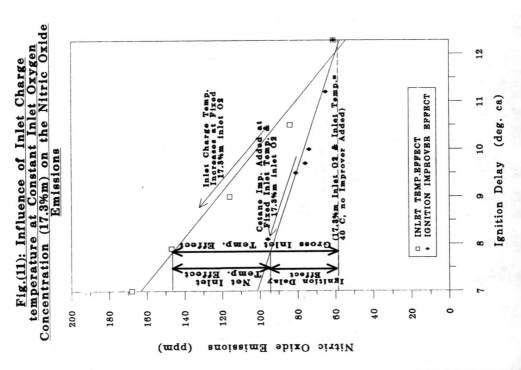

Fig.(11): Influence of Inlet Charge temperature at Constant Inlet Oxygen Concentration (17.3%m) on the Nitric Oxide Emissions

Fig.(12): Influence of Inlet Charge Temperature at Constant Inlet Oxygen Concentration (17.3%m) on the Particulate Emissions as a Function of Fuel Burnt

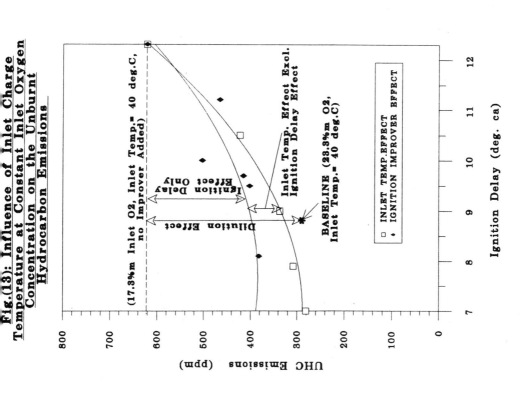

Fig.(13): Influence of Inlet Charge Temperature at Constant Inlet Oxygen Concentration on the Unburnt Hydrocarbon Emissions

TEST	CO2 Overall Effect	Dilution Effect	Chemical Effect	Thermal Effect	Dilution + Chemical Effects	Ignition Delay Effect	In. Ch. Temp. Effect
Fuelling rate/Cy. (g/s)	0.224	0.224	0.224	0.224	0.224	0.224	0.202
In.Ch.rate/cyl inder (g/s)	10	10	10	10	10	10	9
In.Ch.Temp (deg.C)	30	30	30	30	30	30	40 - 120
Oxygen (%m)	23.3-16.3	23.3-16.3	23.3	23.3	17.3	17.3	17.3
Nitrogen (%m)	76.7	76.7-82.8	76.7-68.95	76.7-76.49	75.33	81.95	81.95
CO2 (%m)	0.0 - 7.0	-	0.0 - 7.0	-	6.0	-	-
Argon (%m)	-	0.0 - 0.9	0.0 - 0.75	-	1.37	0.75	0.75
Helium (%m)	-	-	-	0.0 - 0.21	-	-	-

Table (1): Fuelling Rate and Inlet Charge Characteristics for Different Tests

EGR technology for lowest emissions

BAERT, D E BECKMAN BSc, and A W M J VEEN
TNO Road-Vehicles Research Institute, Delft, The Netherlands

SYNOPSIS

An EGR system for turbocharged and aftercooled HD diesel engines has been demonstrated on a 12 litre 315 kW engine with 4 valves per cylinder and a high pressure injection system. In this system exhaust gas is tapped off before the turbine, run through a cooler and mixed with the intake air after the compressor and aftercooler. The EGR system combines a novel, very efficient venturi-mixer unit with a VGT turbocharger. The venturi-mixer is positioned between the aftercooler and the intake manifold and provides a suction power to the EGR gas.

A first report showed that optimization of EGR quantity and injection timing enabled emission levels in the ECE R49 test below 3 g/kWh for NO_x and around 0.10 g/kWh for particulates without a substantial increase in fuel consumption. Since then tests have continued with different fuels and with different EGR hardware. This paper gives further and more detailed information on the venturi-mixer characteristics and on how its interaction with the turbocharger influences engine operation and EGR potential. It is also shown that especially with some oxygenated fuels increase of particulate emissions with EGR could be minimal.

1 INTRODUCTION

Over the last 5 to 10 years emissions legislation in Europe for heavy duty (HD) vehicles has continuously sharpened. Moreover, as the bulk of the HD vehicles in Europe are diesel engine powered, the primary targets of this legislation have been the emissions of nitric oxides (NO_x) and of particulate matter (PM) by these engines. There are no signs that this trend will change. A recent study revealed that, unless special measures are taken, by the year 2005 up to 50% of all nitric oxide emissions in Western Europe could come from heavy duty trucks and buses [1]. This has been recognized by the legislator, as shown in Table 1 where the current and future European emission requirements for HD engines are summarized. The future EURO3 and EURO4 emission levels in this table are only projections. In the final EURO3 test procedure both the selection and the weighting factors of the engine working points will differ from those in the present ECE R49 13 mode procedure (as will the corresponding limits). Furthermore the future EURO4 test procedure will probably be a fully transient one. The EURO3 and EURO4 values mentioned in Table 1 should therefore be considered as target levels for technology development. The challenges imposed by such levels are then supposedly equivalent to those imposed by the actual future test procedures and their limiting values. Similar values have been suggested as a possible long term standard in Europe [14], and comparable long term emissions targets are forecasted in the US [2] and in Japan [3].

Table 1   European emission standards for HD vehicles (> 3.5 tonne GVW; > 85 kW) ECE R49 13-mode test procedure.

Effective date	EURO 1 1992	EURO 2 1995	EURO 3 (*) 1999	EURO 4 2004
NO_x (g/kWh)	8.0	7.0	< 5.0	< 3.0
PM (g/kWh)	0.36	0.15	< 0.10	< 0.10
HC (g/kWh)	1.1	1.1	0.7	0.5
CO (g/kWh)	4.5	4.0	2.5	1.0

(*) with introduction of EURO3 legislation the test procedure will change

In response to this legislative action the manufacturers of heavy duty vehicles are introducing diesel engines with increasingly advanced features. At present EURO2 diesel engines are appearing on the European market [4]. They are the result of the implementation of turbocharging and aftercooling and of the introduction of new FIE allowing higher injection pressures. Results published in the literature indicate that similar efforts aiming at further combustion system optimization will bring the diesel engine close to meeting the EURO3 emission targets [5]. This will however be at the expense of a 3 to 5 % increase in fuel consumption. This confirms what has been recognized already some time ago i.e. that, in order to achieve emission levels that are substantially below EURO3, new technology will have to be introduced [6,7].

In 1993 research activities were started at TNO, aiming at the demonstration of a HD diesel engine with EURO4-like emission levels but with a fuel consumption and transient behaviour comparable to that of its EURO2 predecessors. At the start it was also decided to concentrate research efforts on the evaluation of two of the potential key-technologies for further reduction of NO_x emissions : exhaust gas recirculation (EGR) [8] and catalytic exhaust gas NO_x reduction ($deNO_x$) [9]. The necessary reduction of particulates would then come through the use of low sulphur fuels, through minimizing oil consumption and from further developments in FIE.

2   BACKGROUND : THE TNO EGR PROJECT

In the first stage of the project a multicylinder TCA HD diesel engine with EURO3 potential was procured and on this engine an EGR system with a control system was built and demonstrated to enable emission levels in line with the EURO4 projections in Table 1. Earlier reports focused on EGR hardware development and steady state engine optimization [8] and on control system development and transient engine behaviour optimization [10]. Since then more experiments have been performed with the objective to :
1. further study the potential of the EGR system developed before
2. compare this with other solutions for realizing EGR
3. study the influence of fuel formulation.

As before, the emphasis in this work was on the design and demonstration of a practical EGR system. So far the combustion system has not been optimized for EGR application.

3 EGR STRATEGY FOR LOWEST EMISSIONS WITH HD DIESEL ENGINES

Whether or not an EGR strategy is appropriate for implementation on an engine is determined by :
1. The frequency at which that engine will be used across its operating range; this is supposedly characterized in the emissions test procedure.
2. The target reduction in NO_x that needs to be achieved (when measured according to this test procedure).
3. The effectiveness of EGR in reducing NO_x without increasing PM emissions and fuel consumption.

Compared with the EURO3 levels, the EURO4 levels mentioned in Table 1 require a 50 % decrease in NO_x at constant PM. Currently in Europe overall emission levels are measured according to the ECE R49 13-mode test procedure. This procedure attributes a large weighting factor to the maximum torque and rated power working points (Figure 1a). That is why with a conventional (EURO2-like) TCA HD diesel engine typically around 60 % of the NO_x in the ECE R49 test comes from the full load points. To realize a 50% reduction in overall NO_x emissions the EGR system should therefore enable sufficient exhaust gas recirculation at high load in the intermediate to rated speed range. The quantity of EGR needed was estimated from the results of a pilot study on a 5.7 L 6-cylinder ISUZU diesel engine. In this study a 50 % NO_x reduction was achieved with a 10 % EGR admission at full load in the maximum torque to rated speed range. EGR admission being defined as mass fraction of exhaust gas in the intake manifold flow and calculated from :

$$EGR\ (\%) = \frac{CO_{2,intake} - CO_{2,ambient}}{CO_{2,exhaust}} * 100$$

where $[CO_2]$ represents the volume fraction of CO_2 in the gas stream.
Based on these results it was decided that the EGR system should be able to realize 15 % or more EGR in the full load 13-mode points.

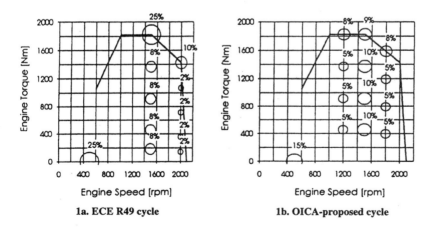

1a. ECE R49 cycle **1b. OICA-proposed cycle**

Figure 1 Different steady-state test cycles [9]

The same pilot study however also confirmed earlier observations that EGR at the same time tends to increase PM emissions [7,11] especially at high load conditions where air/fuel ratio (AFR) is lowest. These investigations also indicated that this PM increase could be minimised by keeping AFR constant. This observation implies that for this study the EGR system lay-out should be such that, under low AFR conditions, it enables "additional" rather than "replacement" EGR. When applying "additional" EGR at full load, the original air mass flow through the engine is maintained at its baseline (no EGR) level. Thus the original torque curve can be maintained and NO_x reduced without excessive smoke penalty.

Of course the question can be raised whether the expected change in emission test procedure will affect the above mentioned EGR strategy. Information in [12] suggests a somewhat modified 13 mode test as shown in Figure 1b. In this test, which is supposedly more representative of the actual HD truck driving pattern, emphasis (weighting) is more evenly distributed across the higher speed range. But even then, for the same EURO2 engine mentioned above, around 45 % of the NO_x will come from the full load points. And the 75 % and 100 % load points taken together account for 70 % of all NO_x emissions (as compared to 80 % in the ECE R49 test). It seems therefore unlikely that the proposed change in test procedure will exclude the demand for a strategy with "additional" EGR at high load conditions.

4 BASIC EGR SYSTEM CONFIGURATION

4.1 Baseline engine/fuel specification

Table 2 gives the main specifications of the 4 valve engine that was used for most of the EGR work. Since it was the intention of this work to demonstrate the potential of EGR to reduce NO_x and PM emissions from EURO3 to EURO4 levels an engine configuration was used that was expected to have EURO3 potential (prior to EGR). In original build this non-production type engine combines a conventional fixed geometry turbocharger with a unit pump fuel injection system. Combustion is swirl-supported and piston bowl geometry is reentrant.

Table 2 *Baseline/original engine specifications prior to EGR.*

	DAF WF 4V ENGINE	
	Original Build	Changes prior to EGR
Bore / Stroke (mm)	130 / 146	
No. cylinders	6	
Displacement (litre)	11.63	
Compression ratio (nom.)	16	15.3
FIE type	Bosch PLD Mk 1	(unit pump)
Turbocharger geometry	Fixed	Garrett VGT
Charge cooling	air-to-air	
Max. power	315 kW	at 2000rpm
Max. torque	1740 Nm	at 1500 rpm

In Europe a maximum limit of 0.05% mass S will be mandatory from October 1996 onwards. Unless specified differently, the results mentioned refer to a conventional diesel fuel conforming with this sulphur limit.

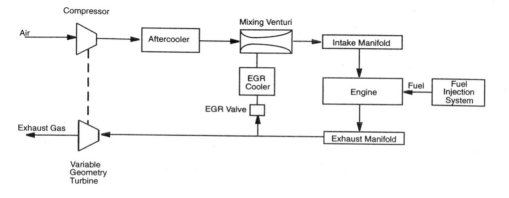

Figure 2   *Turbocharged Diesel Engine with EGR System*

4.2 EGR system lay-out

After analysing the challenges that oppose the use of EGR on a TCA HD diesel engine an EGR system was selected in which exhaust gas is tapped off before the turbine, run through a cooler and mixed with the intake air after the compressor and aftercooler [8]. A schematic of the system is shown in Figure 2. This socalled "short" route was selected because it has the considerable advantage that the compressor and aftercooler are not exposed to particulates, hydrocarbons and sulphate present in the exhaust gases. Thus problems with fouling and corrosion are limited. Also, the mass flows through the turbocharger remain more or less unchanged. Cooling the recirculating exhaust gas reduces the volume of exhaust gas flowing into the intake manifold and therefore limits the boost pressure increase necessary for realising an "additional" EGR approach. Cooling also limits the temperature rise with EGR of the trapped gases, thus assisting NO_x reduction.

As part of the EGR system development a dedicated EGR cooler was designed and constructed. Air was chosen as the coolant medium (in this way EGR temperature could easily be varied). As mentioned in [8] the design target for the pressure loss across the cooler was set at 5 kPa at the maximum EGR flow (in "fouled condition"). A lower pressure loss would have resulted in an excessive EGR cooler volume, and a higher pressure loss would have further increased pre-turbine pressure levels (thus increasing fuel consumption).

The need to recirculate exhaust gas under full load conditions posed an additional challenge: over an important part of the engine operating range the mean exhaust manifold pressure is lower than the inlet manifold pressure. With a "short" route EGR system lay-out this opposes EGR. In the basic EGR system the intake manifold pressure is locally reduced by means of a venturi to aid EGR. This is done in combination with a variable geometry turbine (VGT) turbocharger. As the turbine flow area is reduced (VGT is "closed") both the power taken up by the turbine and the pressure before the turbine increase. As soon as this pre-turbine pressure exceeds the pressure at the venturi throat an EGR flow occurs. An EGR control valve situated between cooler and mixer gives the possibility to modulate air flow and EGR admission independently. It can also be used to shut off EGR in certain parts of the engine operating field.

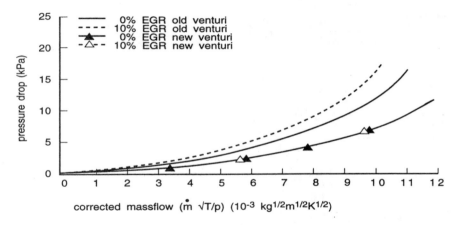

Figure 3   *Flow characteristics of old and new venturi-mixers on flow-rig*

4.3 Venturi-mixer design

The venturi-mixer unit is a key-component of the EGR system. It should combine a sufficient reduction in intake manifold pressure (i.e. suction power to the admitted gas) with a minimum of remaining pressure loss and a uniform EGR gas distribution over the cylinders through fast, homogeneous mixing. TNO has considerable experience with mixing gaseous fuels and EGR to inlet air, because of its long standing experience with gaseous fuelled engines. This technology was developed further, leading to a new venturi design, for which a patent application has been filed.

In this design a good macroscopic EGR gas distribution is obtained by distributing the gas through a cross-shaped insert over the cross sectional area of the venturi. The EGR gas is further mixed with the air flow by creating turbulence. Flow resistance is kept low because of the aerodynamic design of the cross in relation to the venturi and by creating turbulence only there where EGR gas is admitted to the air. Figure 3 compares the flow resistance of the "old" versus the "new" venturi-mixer design as measured on a flow-rig. Flow resistance is characterized by the remaining pressure drop across the mixer. Both venturi-mixers shown have the same contraction ratio (i.e. suction power) $m = (\frac{d}{D})^2$ where d and D represent the equivalent diameters of the inlet pipe and venturi throat respectively. It is clear that the new design has a considerably lower pressure loss for the same suction power. It will be shown later that mixing performance was also good.

4.4 Engine modification prior to EGR

For this work a VGT with adjustable nozzle guide vanes was used. Compressor and turbine were selected to enable the higher pressures needed with an "additional" EGR approach. After fitting the VGT to the engine, the first step taken was to modify the engine combustion system such that it achieves EURO3-like emission levels while at the same time it is ready for implementing EGR. Compression ratio was lowered to allow the required charging to be achieved within the cylinder pressure limit. (It was accepted that advanced FIE and EGR would ultimately need to compensate for effects on cold-starting, noise and light-load HC). Next an estimate was made of the nozzle hole size that would maintain an acceptable fuel air mixing at the increased charge densities. For this, results were used from an experimental study on fuel spray growth at charge densities in the range 10-60 kg/m^3 [13]. Limited engine testing resulted in the selection of engine hardware (nozzle protrusion, nozzle hole size and fuel cam static timing) and boost (VGT guide vane position) and timing settings that were predicted to give EURO3-like emission levels. (These R49 13-mode emission predictions are shown in Table 3). At that point the basic EGR hardware was built onto the engine and EGR testing could start.

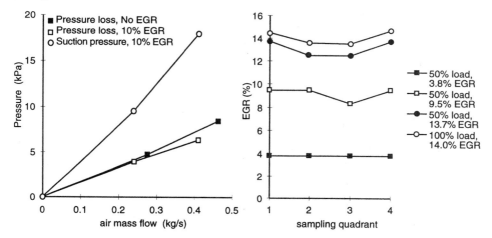

Figure 4 Pressure characteristics of new venturi mixer (1500 rpm; 50 and 100% load)

Figure 5 EGR distribution at 1500 rpm

5   PRELIMINARY RESULTS WITH BASIC EGR SYSTEM

5.1   Venturi assessment

First, tests were conducted to check EGR admission and mixing characteristics with the new venturi-mixer design. Figures 4 and 5 show some of these characteristics at 1500 rpm and 10 % EGR both for 50 % and 100 % load. Figure 4 shows the pressure loss across the venturi and its suction pressure versus mass flow, suction pressure being defined as the difference between pressure at venturi entrance and throat section. The new venturi-mixer realizes a suction pressure of up to 20 kPa while pressure recovery is about 60%. Its mixing performance was assessed by dividing the intake pipe downstream of the venturi outlet into 4 quadrants, thus freezing the EGR distribution over these quadrants. The EGR % was then calculated from the CO_2 content in the gas tapped off each quadrant at the exit of the pipe (before it enters the intake manifold). As shown in Figure 5 the variation of EGR over the quadrants is small.

5.2   Effect of EGR temperature

In all tests the temperature of the EGR gas leaving the cooler was controlled and set at 413 K (140 °C). This temperature is a compromise. NO_x would benefit from an EGR temperature as low as possible. On the other side of the trade-off are however increasing cooler size and fouling and corrosion due to condensation of sulphate and possibly water. In the end, cooler-out temperature was primarily chosen such that condensation of corrosive gases would be prevented.

In order to evaluate the consequence of this decision the effect of EGR cooler-out temperature on the engine performance and emissions was examined at full load, 1500 rev/min conditions. As expected NO_x decreased with EGR temperature. However, as shown in Figure 6, for T > 415 K this effect is small. Effect of EGR temperature (in the range examined) on BSFC and PM was minimal. Although EGR temperatures below 413 K would result in lower NO_x levels, the first temperature was retained.

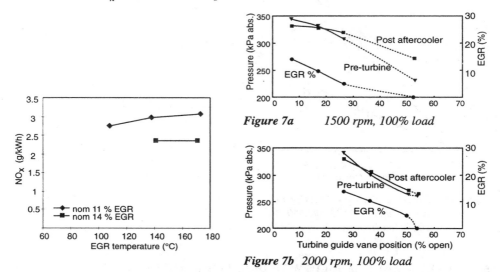

Figure 6 Influence of EGR temperature on NO_x

Figure 7 Change of EGR percentage and exhaust and inlet pressure with turbine guide vane position

5.3. Preliminary tests with the basic EGR system

These tests confirmed that the target of 15 % "additional" EGR could be achieved in most of the engine operating range. When determining engine settings it was found that for given NO_x and PM level, fuel consumption was lowest with the EGR control valve fully open. Thus for the remainder of the tests EGR flow was controlled by modulating the VGT guide vane position only. Figure 7 shows how intake and exhaust manifold pressure and EGR change with VGT guide vane position (VGT position) at full load both at 1500 rpm and 2000 rpm. As the VGT closes, exhaust manifold pressure rises faster than intake manifold pressure. In combination with the suction power of the venturi, at some point this is sufficient to create an EGR flow. The different pressure curves at these two operating points reflect the way the turbocharger is matched across the engine speed range (at full load). At 1500 rpm, even with the assistance of the venturi the turbine must be driven to a relatively low efficiency before EGR flows. Once EGR flows the turbine is further deprived of energy and the compressor delivery falls away. At 2000 rpm however, closing the turbine to achieve increasing EGR flows results in higher turbine efficiencies, such that compressor delivery (and AFR) is maintained or even increased with EGR.

6 EGR SYSTEM ALTERNATIVES

6.1 Description of alternative EGR systems/hardware

Apart from the basic EGR system described before, a number of alternative systems have been studied. They differ from the basic EGR system in the method used to promote a favourable pressure gradient (for EGR flow) between the exhaust and inlet manifolds, and in the way the EGR is introduced into the inlet manifold. The aim of these tests was to compare their potential for "additional" EGR with minimum fuel consumption. The systems are listed below :
- (VGT, new VM1). This is the basic system described above; EGR flow is controlled by modulating the VGT and EGR is added to the intake by use of a venturi-mixer of the "new" design. Throat area of the venturi-mixer is 1450 mm2.
- (VGT, new VM2). As above but with a smaller venturi throat area (1140 mm2); this gives larger suction power to the venturi.
- (VGT, old VM). As above but admission of the exhaust to the inlet manifold utilises a venturi-mixer of a type originally developed for gas engines. Venturi throat area as for the first combination.
- (VGT, no VM). Exhaust pressure is modulated only by VGT (guide-vane setting). Exhaust gas is admitted upstream of the inlet manifold with simple pipe geometry.
- (EBPV, no VM). Setting up of an EGR flow by modulation of an exhaust back pressure valve (EBPV) located after the turbocharger turbine exit. The VGT is not modulated; guide vane position is fixed. Exhaust admission to the inlet manifold as for the (VGT, no VM) situation.

6.2 Steady state results

Each EGR system variation was tested at 4 engine operating points: 3 full-load points (1200, 1500, and 2000 rpm) and 50 % load at 1500 rpm. At each operating point injection timing was kept constant and EGR was varied between zero and maximum. Although an EGR control valve was present in the EGR circuit it was completely open for all tests. Finally, with each system, fuelling was adjusted to maintain constant load at each working point as EGR was modulated. Figures 8 to 10 show some of the NO_x-BSFC and NO_x-PM trade-offs that were obtained. PM-values mentioned are predicted. The prediction utilises measurements of Bosch smoke and gaseous HC and the known sulphur-content of the fuel.

Figure 8 (1200 rpm, 100 % load) is the simplest to view. With the exception of the EBPV, the trade-offs are similar for the different systems. Lower NO_x (i.e. higher EGR flow) is accompanied by increasing PM and BSFC (Figures 8a and 8b). The limited reduction of NO_x with the (VGT, no VM) is due to the onset of compressor surge as the compressor responds to closure of the turbine guide vanes. The smaller new venturi provides the greatest assistance to EGR admission through local reduction of the inlet air pressure and achieves the lowest NO_x.

Using the EBPV seems to have a different effect on the engine behaviour. As shown in Figure 8c closing the EBPV achieves almost twice as much NO_x reduction for a given EGR ratio as the other systems. This reflects depressed air-fuel ratios and higher trapped residuals resulting from poor scavenging and low turbine work. Figure 8d shows the change of fuel consumption with the pressure drop across the engine. Again the EBPV curve deviates from the rest, and again the reason is similar: for more or less constant AFR BSFC shows only moderate changes with pressure drop. However, when AFR drops below a certain limiting value, BSFC responds sharply to AFR. The resulting PM and BSFC increases with the EBPV are unacceptable.

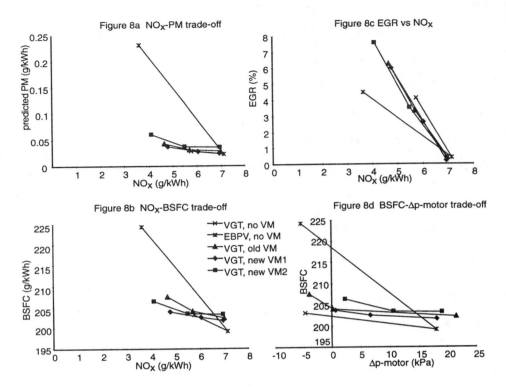

Figure 8 Engine performance and emissions with different EGR systems: 1200 rpm, 100% load

Figure 9 (1500 rpm, 100 % load) shows similar characteristics as Figure 8. EBPV1, EBPV2 and EBPV3 denote modulation of the back pressure valve at different (but constant in each case) positions of the VGT guide vanes. Larger reductions in NO_x at this higher speed (from around 6.5 to 2.5 g/kWh) are due to higher EGR flows before compressor surge is approached. These results indicate that the best trade-off between NO_x and BSFC was achieved using an EGR system supported by a VGT alone (VGT, no VM). Although this system required higher exhaust pressures to enable EGR it also results in the highest charge air pressures resulting in higher air-fuel ratios. The advantage of higher AFR outweighs the loss of positive pumping work due to the increased exhaust pressure - both for BSFC and PM. This emphasizes the initial objective of the system to apply EGR whilst maintaining AFR. Although adding a new venturi-mixer enabled EGR to be applied with very little loss of AFR - due to efficient pressure recovery - this was never sufficient to match the AFRs achieved using the (VGT, no VM) solution.

At maximum power the range of NO_x is more dependent on the system than seen at other points (Figure 10). The lowest NO_x recorded with (VGT, no VM) was 3 g/kWh. This was extended to 2.6 g/kWh using the older type venturi (VGT, old VM) and to 2.2 g/kWh with the new venturi (VGT, new VM1). The smaller new venturi (VGT, new VM2) enabled 2 g/kWh. Unlike the 2 lower speeds, airflow increases simultaneously with increasing EGR. As explained before, this is due to the increase in turbine work that is derived from closing the guide vanes at the higher speed. However, this also means that as the guide vanes are opened, compressor delivery falls. With the venturi systems in particular this results in low AFRs while EGR is still flowing and as a consequence in an upturn in PM and BSFC as EGR decreases. Thus these systems are appropriate when the EGR strategy demands high EGR quantities.

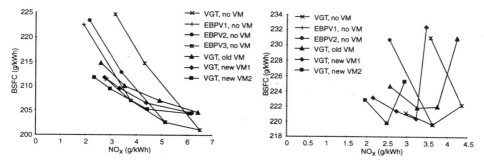

Figure 9 NO_x-BSFC trade-off with different EGR systems: 1500 rpm, 100% load

Figure 10 NO_x-BSFC trade-off with different EGR systems: 2000 rpm, 100% load

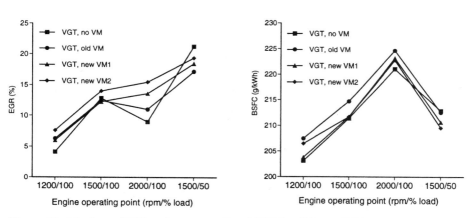

Figure 11 Maximum EGR and corresponding BSFC for different EGR systems and engine operating points

Figure 11 compares the maximum attainable EGR and corresponding BSFC at each of the different engine operating points and for the non-EBPV systems. In terms of engine emissions and fuel consumption, the simplest system (i.e. VGT only) - that is the system offering least resistance to flows - is seen to provide the highest airflows and this proves to be the decisive factor in obtaining the lowest BSFC and PM for a given NO_x emission. However, adding the venturi-mixer has advantages. It enables the turbocharger to operate at lower pressures which could influence durability. At some operating points it also significantly extends the range of EGR and therefore the overall NO_x reduction which can be achieved.

7   13-MODE TEST RESULTS WITH THE BASIC EGR SYSTEM

For the basic EGR system, a set of experiments has been defined and conducted to enable analytical models of the main engine responses to EGR, AFR and injection timing to be constructed. Based on these data and some additional measurements, engine settings were optimized for ECE R49 testing aiming at EURO4 NO_x emission levels of 2.5 g/kWh. 13-mode tests were performed, the results of which are shown in Table 3 "EURO 4 Low NO_x". Both NO_x and BSFC met the test targets but the PM exceeded the target by 7 %. It was not judged practical (in this test programme) to reduce PM by additional optimization of EGR and timing. An additional EGR/timing strategy was tested aimed at relaxing the NO_x target to 3.0 g/kWh in the interests of improved economy (Table 3 "EURO4 Economy").

Table 3   Summary of ECE R49 test results (g/kWh).

	EGR	NO_x	PM (1)	PM (2)	HC	CO	BSFC
Reference EURO2 [8]	No	6.92	100	0.112	0.23	0.68	212.5
Baseline (3)	No	5.73	63	-	-	-	215
EURO 4 Low NO_x	Yes	2.42	82	0.107	0.2	0.6	220
EURO 4 Economy	Yes	2.88	88	-	0.18	0.58	216

(1) predicted PM as percentage of EURO2 Reference engine; pred. for 0.05 % mass S fuel
(2) measured
(3) baseline (prior to EGR); values estimated from R49 test modes 4,6,8 and 10

At this point it is worthwhile to remind the reader of the fact that no combustion system optimization was attempted on this engine. This indicates scope for further reduction of PM (while maintaining the above mentioned NO_x).

8   PARTICULATES CONTROL BY FUEL COMPOSITION

Another possibility for restricting the PM emission when using EGR is a change of fuel composition. Various fuels have been tested on the engine when operating with EGR. These fuels include commercial blends as well as "exotic" fuels designed to investigate the effect of specific properties of the fuel. As the results from these tests are currently being evaluated only some preliminary results are presented here.

Tests were done under different combinations of EGR and AFR at 1500 rpm and 75% load. A throttle was used before the turbocharger compressor intake to extend the range of AFR possible using only the VGT and EGR valves. At this test point the most encouraging results were obtained with a blend of Tall-oil Methyl Ester (TME) with Finnish city diesel fuel. TME is a waste bio-product from wood pulping which is esterificized to give it diesel-like properties. The Finnish fuel (50 ppm/m S and 20 %/v aromatics max.) is similar to Swedish city diesel class 2. A 20 %/v blend of TME and the Finnish fuel was used in the tests. This imparts a 2.4 %/m oxygen content to the blend.

Figure 12 shows the NO_x and predicted PM determined for a conventional fuel and for the TME-blend. The effect of EGR is shown at 3 AFRs. The response of NO_x to EGR and AFR is similar for both fuels. Also, the reduction of NO_x with EGR is similar for each of the AFRs. For the conventional fuel the combination of EGR at restricted AFRs leads to high predicted PM as discussed before. For the TME-blend the predicted PM is much less sensitive to AFR and EGR.

These results reflect the insensitivity of the Bosch smoke number which is used in the PM prediction principally to estimate the carbon fraction. Although the prediction works well with conventional fuels a tendency to under-predict the PM has been noticed when using the TME-blend. Nevertheless, chemical analysis of PM have confirmed substantial reductions of the carbon fraction when operating on the TME-blend in place of conventional fuel. Providing the smoke measurements recorded with EGR are confirmed as a valid indicator of the carbon fraction with the TME-blend we would anticipate a combination of very low NO_x and PM using this fuel.

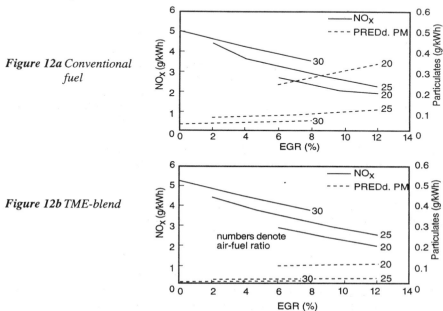

Figure 12a Conventional fuel

Figure 12b TME-blend

Figure 12 Influence of air-fuel ratio and EGR on emissions: 1500 rpm 75% load

9 CONCLUSIONS

1. For a reduction of emissions to EURO4 like levels it seems likely that, even with a change in test procedure, a strategy of between 5 to 10 % EGR or more across the engine working range will have to be realised.
2. To minimize BSFC penalty the EGR system lay-out should be such that it realises these EGR ratios while maintaining AFR with "additional" EGR. Tests with several methods of admitting EGR showed that the best trade-off between NOx and PM and BSFC were achieved with the system which enabled the highest air-fuel ratios. However, these tests also indicated that it seems unlikely that it will be possible to realize such EGR ratio's across the entire engine working range without the help of a venturi-mixer.
3. Use of a venturi-mixer after the turbocharger compressor to encourage EGR flow resulted in an air-fuel ratio and BFSC penalty, already at zero EGR. However, the further change

of BFSC with increasing EGR was smaller than with VGT only. Furthermore, the venturi suction enables EGR to be achieved with lower turbocharger pressures (compressor and turbine) than when using the VGT only. This provides larger margins from compressor surge (at low engine speeds) and turbocharger over-speed (at high engine speeds). Lower operating pressures would also be an advantage in turbocharger durability.
4. Use of an exhaust back pressure valve after the turbine resulted in high PM and BSFC due to the low air-fuel ratios which resulted from the low expansion ratio across the turbine.
5. The PM increase with EGR could be reduced by running the engine with an oxygenated fuel.

REFERENCES

[1] Lenz, H.P., Kohoutek, P., "Global air pollution and the influence of vehicle emissions.", 25th Fisita Congress, Beijing (China), Paper 945115 (1994).

[2] Wilson, R., "Diesel engine prospects boosted by EPA-CARB-EMA accord on emissions", Diesel Progress (Engines & Drives), pp. 26, August 1995.

[3] Joko, I., Aoyagi, Y., "21st Century Key Technologies for Fuel Economy and Emission Needs of Heavy Duty Diesel Engines.", Paper F-02, 21. CIMAC Congress, Interlaken, May 15-18 (1995).

[4] Schittler, M., Fränkle G., "Entwicklung der Mercedes-Benz-Nutzfahrzeugmotoren zur Erfullung verschärfter Abgasgrenzwerte. Mercedes-Benz Commercial Vehicle Engines to Meet the Emission Limits EURO II and EPA 94.", Motortechnische Zeitschrift, Vol. 55, No. 10, pp. 576- 585 (1994).

[5] Pütz, M., and Kotauschek, W. "Die neuen wassergekühlten Deutz-Dieselmotoren BFM 1015 - Verbrennungs- und Einspritzsystem. New Watercooled Deutz Diesel Engines BFM 1015.", Motortechnische Zeitschrift, Vol. 55, No. 9, pp. 492-501 (1994).

[6] Needham, J.R., Such, C.H., Nicol, A.J., "Fuel efficient and green - the future heavy duty diesel.", 2nd ImechE Seminar "Worldwide Engine Emission Standards and How to Meet Them", London, May 25-26, pp. 197-207 (1993).

[7] Herzog, P.L., Buergler, L., Winklhofer E., Zelenka, P., Cartellieri, W., "NO_x Reduction Strategies for DI Diesel Engines." SAE paper 920470 (1992).

[8] Baert, R.S.G., Beckman, D.E., Verbeek, R.P., "New EGR Technology Retains HD Diesel Economy with 21st Century Emissions.", SAE paper 960848 (1996).

[9] Havenith, C., Verbeek, R.P., Heaton, D.M., van Sloten, P., "Development of a Urea $DeNO_x$ Catalyst Concept for European Ultra-Low Emission Heavy-Duty Diesel Engines.", SAE paper 952652 (1995).

[10] Dekker, H.J., Sturm, W.L., "Simulations and Control of a HD Diesel Engine Equipped with New EGR Technology.", SAE paper 960871 (1996).

[11] Yu, R.C., and Shahed, S.M.,"Effects of Injection Timing and Exhaust Gas Recirculation on Emissions from a D.I. Diesel Engine.", SAE paper 811234 (1981).

[12] Fränkle, G., Havenith, C., Chmela, F., "Zur Entwicklung des Prüfzyklus EURO 3 für Motoren zum Antrieb von Fahrzeugen über 3,5 t Gesamtgewicht. (On the development of the EURO 3 Test cycle for Engines of Vehicles with over 3,5 t Total Weight.)", 5. Aachener Kolloquium Fahrzeug und Moorentechnik, October 4-6, pp. 223-248 (1995).

[13] Baert, R., Vermeulen, E., "Experimental study of charge density effects on fuel spray penetration.", Proc. of ImechE Seminar on Measurement and Observation Analysis of Combustion in Engines, March 22, pp. 27 - 37 (1994).

[14] Havenith, C., Edwards, S.P., Such, C.H., Needham, J.R., "Development of a Heavy-Duty Diesel Engine towards EURO 3 using Exhaust Gas Recirculation.", Paper SIA/9506/A16, 5th EAEC Int. Congress, Strasbourg, 21-23 June (1995).

Experimental characterization of turbocharging and EGR systems in an automotive diesel engine

C APOBIANCO MSAE, ATA, **A GAMBAROTTA** PhD, ASME, MSAE, ATA, AIMETA, and **G ZAMBONI**
Istituto di Macchine e Sistemi Energetici, University of Genoa, Italy

SYNOPSIS

A broad experimental investigation is being developed at the Istituto di Macchine e Sistemi Energetici (IMSE) of the University of Genoa on a DI automotive Diesel engine, fitted with a variable geometry turbocharger turbine (VGT) and an exhaust gas recirculation system (EGR). Two experimental facilities allow to develop engine dynamometer tests and define the flow characteristics of the main intake and exhaust circuit components. Several experimental results are discussed in the paper: engine performance and emissions in the baseline arrangement without EGR are first considered, focusing on the matching of the VGT turbocharger. The interactions between the turbocharger and the EGR system are then analysed for different operating conditions and EGR rates.

NOMENCLATURE

Notations

bsfc	brake specific fuel consumption
f	mass flow fraction
m	specific mass emission
n	rotational speed
p	pressure
t	temperature
A	variable geometry system opening degree
A/F	air/fuel ratio
FSN	Filter Smoke Number
P	power
X	volumetric concentration
η	efficiency
Δ	difference

Subscripts

c	compressor
i	intake
m	mechanical
t	turbine
tc	turbocharger
EGR	exhaust gas recirculation
2	compressor exit
3	turbine inlet

1 - INTRODUCTION

In recent years, the application of small high speed direct injection (DI) Diesel engines to cars is becoming more and more attractive since they comply with very low fuel consumption (1, 2). In order to get higher specific power, these engines are usually coupled to an exhaust turbocharger, fitted with an appropriate turbine regulating system (such as a waste gate valve or a variable geometry device), which allows an acceptable torque curve and car drivability.

However, DI Diesel engines main problems with respect to exhaust emissions are nitrogen oxides (NO_x) and soot (1, 3). Present EC regulations (stage 1 limits) yet require the use of specific systems (such as exhaust gas recirculation, EGR) and/or of external aftertreatment devices (catalysts and particulate filters), which will become a must in order to comply with the stricter emission limits which will be incorporated in future legislation (EC stage 3 limits, with effect from October 1999). The broad research work to be developed on this subject should take account of particular exhaust boundary conditions, which make not possible the use of conventional three-way catalytic converters because of the high oxygen content in the exhaust gases. Besides, the application of NO_x catalysts shows several difficulties, due to low temperatures and to the absence of reducing components.

Feasible techniques for filtering the particulates out of Diesel engines have been set up, but regeneration procedures are still not reliable in all vehicle operating conditions, when the exhaust temperature is not adequate for ignition of soot deposits.

A wide research work is then to be developed in order to optimise the engine intake and exhaust system, with reference to both its layout and components matching. Within this study, the behaviour of selected elements has to be deepened, such as the turbocharger, the EGR system and the aftertreatment devices (catalyst and particulate filter), taking into account the effect of the relevant regulating systems and the influence of pulsating flow operation. Starting from the knowledge of the different components performance, proper control strategies can be developed and integrated in the engine management system in order to allow lower exhaust emissions with better engine performance and fuel consumption.

An experimental and theoretical investigation on this subject is being carried out by the authors at the Istituto di Macchine e Sistemi Energetici (IMSE) of the University of Genoa. Two dedicated facilities are available, which allow tests on the whole engine as well as on single components or subassemblies of the intake and exhaust systems. The study is developed on an automotive DI Diesel engine, fitted with a variable geometry turbocharger turbine (VGT) and an exhaust gas recirculation (EGR) circuit. In a first step the research activity was focused on the behaviour and the interaction of these components. Other exhaust devices, such as catalytic converters and particulate filters, will be considered in a following stage, together with different turbocharger regulating systems (i.e., waste gate valves or alternative variable geometry devices).

The goal of the study is the development of optimised control strategies for the components of the intake and exhaust circuit, taking into account their reciprocal influence. These strategies will be referred both to engine normal operation (with particular attention to exhaust boundary conditions for the aftertreatment systems) and to specific operating states (cold transient, particulate filter regeneration, etc.).

As a theoretical support of the experimental activity, a matching model was developed (4, 5) to allow parametric studies to be used in the design of control strategies. Within this model the behaviour of several components (such as turbocharger compressor and turbine, EGR valve, etc.) is described through their steady flow characteristics, measured on IMSE components test bench.

In the paper the results of the experimental programme are presented: the engine and turbocharger performance was first investigated in absence of exhaust gas recirculation; the influence of different EGR rates and the transient induced by EGR activation on both engine performance and emissions were then analysed.

2 - EXPERIMENTAL APPARATUS

Two dedicated test rigs are available at IMSE, by which measurements on the whole engine as well as on the main components or complete intake and exhaust circuits can be performed. Fig.1 shows a schematic of both facilities.

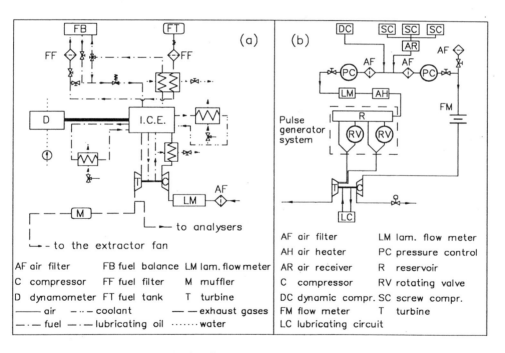

Fig.1 - *Schematic of IMSE engine and components test facilities.*

The engine test bed (fig. 1a) (4) is equipped with an eddy current dynamometer D, governed by an electronic control unit which allows to set independently the engine speed and load within the operating range. The coolant and lubricant temperatures at the engine inlet are controlled through dedicated thermostatic circuits.

An automotive 1.9 l DI Diesel engine (Fiat M711.KT) was installed on the test rig. The engine is intercooled and equipped with a Garrett VNT25 turbocharger, fitted with a variable nozzle turbine controlled by a pneumatic spring-diaphragm actuator, driven by the pressure level at the compressor impeller exit. An uncooled EGR circuit, managed by a dedicated electronic unit, is provided.

Engine air flow rate and fuel consumption are respectively evaluated through a laminar flow meter LM and a fuel balance FB. The latter is inserted in a dedicated circuit fitted with an heat exchanger which allows to keep constant the fuel temperature in the measuring line.

Pressures and temperatures are evaluated in several sections related to the engine, the turbocharger and the EGR system. Pressures are measured through strain gauge transducers, while platinum resistance thermometers and K type thermocouples are used for temperatures. Turbocharger rotational speed is evaluated through an inductive probe.

An exhaust gas analysis system is also available on the test rig, fitted with two non-dispersive infrared analysers (NDIR) for measuring CO and CO_2 concentrations, a heated chemiluminescent analyser (HCLA) for NO_x and an amperometric sensor analyser for O_2. A variable sampling smoke meter is used to evaluate the soot content in the exhaust gases.

The IMSE components test bed (fig. 1b) (6, 7) allows to define steady and unsteady flow behaviour of single devices or complete intake and exhaust systems of automotive engines. The testing circuit is fed with dry compressed air supplied by two different stations (screw and dynamic compressors). Two separate lines are provided, fitted with pressure governors and mass flow rate measuring devices.

Tests on exhaust turbochargers are therefore possible, using the turbocharger compressor C as a dynamometer: in this case the control of the compressor inlet pressure and the replacement of the original volute with a modified casing fitted with injection nozzles fed with compressed air (6) allow to extend the investigation of turbine curves in a wide range of expansion ratio without using a dedicated high speed dynamometer.

Compressed air through the main line may be heated up to about 400 K (to avoid water condensation) and then is sent to the reservoir R which acts as flow distributor to the successive pulse generator system. This allows to develop tests on single and two-entry components both in steady flow (with independent control of thermodynamic parameters at each entry (8, 9)) and in pulsating flow conditions. Pressure pulse characteristics at each entry of tested components may be controlled in order to simulate real on-engine conditions (6, 8). Flow unsteadiness is generated by two rotating valves RV, which allow to vary in a wide range the frequency of pressure pulses (between 10 and over 200 Hz) and their phase angle (9).

Instantaneous and mean wall static pressures are measured at the inlet and outlet of tested devices through strain gauge transducers. Average temperature levels in the same sections are evaluated by means of platinum resistance thermometers. Mean air mass flow rates are measured through a laminar flow meter and a sharp edged orifice. Turbocharger and rotating valves rotational speed is evaluated with inductive probes.

Both engine and components test rigs are equipped with an automatic data acquisition system. Dedicated measuring instruments (multi channel scanners, high speed A/D converters, oscilloscopes, counters, etc.) are connected on a IEEE bus and controlled by a personal computer using a purpose-developed software. Measured and calculated data are then stored and plotted.

3 - ENGINE TESTS RESULTS

3.1 - Baseline Configuration Performance

Baseline engine performance and emissions were measured without EGR, with the original set of the turbocharger VGT regulating system (configuration no. 0): the relevant experimental data will be used as a reference in the development of the study.

Starting from constant speed engine curves (defined between 1500 and 4200 r/min), steady state maps of the main operating parameters of both the engine and the turbocharger were determined. In this phase the pneumatic actuator of the EGR system was disconnected, but the relevant electronic control unit worked: it was therefore possible to measure in each operating condition the EGR governing pressure signal p_{EGR} determined by the control unit.

Fig.2 shows the maps of the boost pressure p_2, measured at the compressor exit (fig.2a), and of the difference between the turbine inlet pressure p_3 and p_2 (fig.2b). At high engine rotational speed (over 2500 r/min), full load boost pressure level results nearly constant, depending on the set point of the VG pneumatic control device. On the contrary, p_2 values aren't fully satisfactory in terms of available engine torque and car drivability at low rotational speed. An optimised turbocharger matching, together with an enhanced control of the turbine nozzle opening, could probably improve boost pressure level at low engine speed, taking full advantage of the turbocharger VG system.

Turbine inlet pressure p_3 resulted always higher than the corresponding compressor delivery level p_2. This result is typical of slightly overboosted automotive engines, particularly in the case of Diesel engines for their lower exhaust temperature. This is a drawback in respect of cylinder scavenge, but makes possible the application of EGR technique. In the case of tested engine, pressure difference across the recirculation valve, which may be considered comparable with (p_3-p_2), proved to be significant (generally above 0.5 bar), especially within the EGR system operating range (see fig.7), thus allowing for considerable recirculation flow rates.

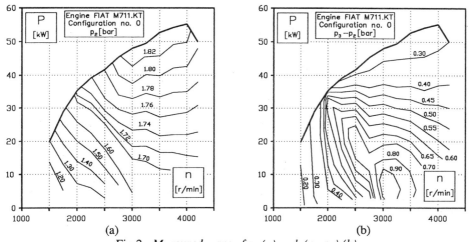

Fig.2 - *Measured maps of p_2 (a) and (p_3-p_2) (b)*

Fig.3 shows the map of the turbine VG system opening degree A, defined as a percentage of the total displacement of nozzle ring push rod (ranging from 0 to 100 per cent when varying the flow area from the minimum to the maximum value). The VG system opening level measured on the engine was strictly related to the relevant pneumatic driving signal (pressure at the compressor impeller exit). In the case of the tested engine, the VG device maximum opening was about 40 per cent. The availability of a large residual displacement is probably intended to prevent turbocharger overspeed: however, a smaller turbine size, resulting in a wider regulating range of the VG system when operating on the engine, might probably improve engine torque at low rotational speed. Turbine matching proved to be optimised with respect to efficiency: in fact, measurements on the component test bed (5) showed that turbine peak efficiency is achieved for a setting of the variable geometry nozzle ring of about 30 per cent, probably corresponding to the optimum velocity distribution in the turbine scroll.

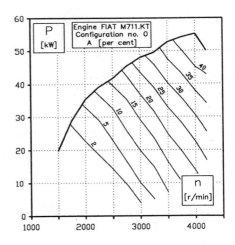

Fig.3 - *Turbine VG system opening degree.*

Turbocharger compressor and turbine efficiency (fig.4) was calculated on the basis of measurements of average thermodynamic parameters. Compressor isentropic efficiency map (η_c, fig.4a) shows small changes of calculated values over the whole engine operating range: however, the highest efficiency levels ($\eta_c \equiv 0.70$) proved to be significantly lower than maximum values measured on the component test rig ($\eta_c \equiv 0.78$) (5).

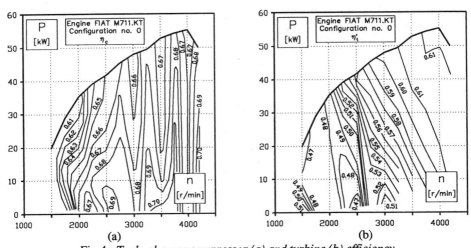

Fig.4 - *Turbocharger compressor (a) and turbine (b) efficiency.*

Turbine efficiency (fig.4b) is reported in terms of η'_t (turbine isentropic efficiency η_t multiplied by turbocharger mechanical efficiency η_m). Engine values often resulted higher than the corresponding levels (at the same turbine non-dimensional speed, expansion ratio and VG setting) measured on the components test rig in steady flow conditions (5): this result may be related to flow unsteadiness through the turbine when it operates on the engine, which can induce errors in efficiency if evaluated on the basis of average pressure and temperature levels at the turbine entry (7).

Fig.5 shows the map of nitrogen oxides specific emission (m_{NOx}), calculated from measured NO_x concentrations in the baseline engine configuration without EGR. Plotted data outline high NO_x emissions. In the evaluation of engine bench results, the EC 1996 m_{NOx} limit (7 g/kWh) for heavy-duty Diesel engines can be assumed as an indirect reference (EC regulation 542/91, ECE R49 dynamometer steady state procedure). The utilisation of an EGR system may be helpful to meet emission standards, particularly in the case of Diesel engines (like the tested unit) for which a different test procedure and a combined emission limit of nitrogen oxides and unburnt hydrocarbons (UHC) is presently set (EC regulation 441/91, ECE+EUDC driving cycle). However, it is interesting to observe that engine NO_x emissions proved to be substantial also outside the EGR operating range (see fig.7): it seems advisable that future legislation will require a control of NO_x (and, more generally, of pollutant emissions) over the whole engine operating range.

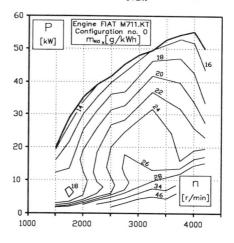

Fig.5 - *Specific NO_x emission map.*

Measured smoke emissions, expressed in terms of the Filter Smoke Number (FSN, ranging between 0 and 10), are reported in fig.6. In the baseline engine configuration (without EGR operation) very low smoke emissions were measured in every operating condition. The trend of FSN contours resulted mainly related to changes of engine air-fuel ratio.

The analysis of baseline engine performance and emissions without EGR confirmed that the optimisation of turbocharger matching and the development of proper control strategies of the VGT can be very helpful to get better engine performance and more suitable torque curves for automotive applications. Besides, an integrated management of turbocharging and EGR systems can help to meet future stricter emission limits through

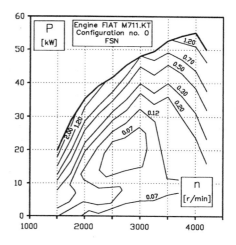

Fig.6 - *Smoke emission map (FSN).*

the improvement of the boundary conditions for applicable aftertreatment devices (NO_x catalyst and particulate filter).

3.2 - EGR Effects On Engine Performance And Emissions

The EGR technique is a well proven method to reduce effectively NO_x emissions, especially for car Diesel engines (10). However, the introduction of exhaust gases in the engine intake manifold (usually from "before turbine" to "after intercooler") affects significantly engine performance and other pollutant emissions, in proportion to the mass of recirculated gases (1). This parameter is usually controlled by a poppet valve which, in the considered engine, is governed by the ECU through a pressure signal p_{EGR} (lower than atmospheric) which drives a spring-diaphragm actuator.

The influence of EGR technique on engine performance and emissions was experimentally investigated. The operating conditions of EGR system were defined with reference to the governing pressure p_{EGR}, expressed as the difference Δp_{EGR} between atmospheric pressure p_{atm} and the pressure signal p_{EGR}. A first set of measurements was developed varying the EGR governing signal Δp_{EGR}, while keeping constant other operating parameters (i.e., engine speed and fuel pump rack position). Referring to five different values of Δp_{EGR}, to a rotational speed of 2500 r/min and to a rack position of 0.614 (defined as its actual displacement divided by the total displacement), engine steady state performance and emissions were measured.

The considered values of Δp_{EGR} were defined within the operating range imposed by the ECU, evaluated in the engine baseline configuration (fig.7). The relationship between Δp_{EGR} and the EGR rate f_{EGR} (i.e., the mass flow of recirculated gas divided by the total intake mass flow) was determined on the basis of the characteristic curves of the EGR valve (measured in steady flow on the IMSE components test bed) and is reported in fig.8a. It is apparent that there isn't any effect for a pressure signal Δp_{EGR} up to 0.15 bar. Therefore the real EGR operating area seems to be narrower than that reported in fig.7: higher EGR rates (near to 20 per cent) are reached at low engine load and at low-medium engine speed (where the pressure difference across the recirculation valve is more favourable, see fig.2b). This control strategy is typical of EGR applications on small automotive Diesel engines, since at higher load this technique has unacceptable effects on engine power and smoke emission (1, 10).

The effects of EGR, evaluated through the experimental results, were analysed taking account of its influence on the combustion process. The addition of exhaust gases to the intake air causes of course an increase in the gas temperature in the intake manifold t_i (as shown in fig.8a) and a decrease in the air mass introduced in the cylinder, due to the mixing with combustion products. Since the rack position was kept constant during the tests, fuel mass flow rate varied slightly and therefore the air/fuel ratio A/F decreased significantly, as it is apparent in fig.8b, with the consequence of a lower oxygen concentration in the exhaust gases.

Notwithstanding the increase in intake temperature and the small variation in fuel

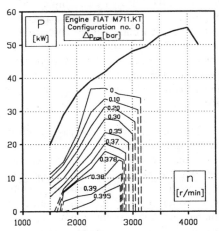

Fig.7 - *EGR governing signal map.*

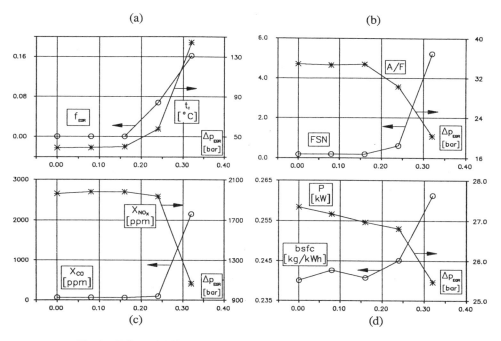

Fig.8 - *Effect of different EGR rates on engine operating parameters.*

mass flow rate, the combustion temperature should decrease, due to the higher heat capacity of CO_2 and H_2O and to changes induced by the EGR in the flame structure (11). This is apparent regarding the decrease in NO_x emissions (of about 50 per cent, with reference to fig.8c), which are extremely sensitive to temperature and relatively unaffected by oxygen concentration (11). On the other hand particulate and CO emissions usually increase, probably due to lower values of combustion temperature and oxygen concentration (10, 11, 12): this trend was clearly observed measuring the smoke and CO content in the exhaust gases (figs.8b and 8c).

It is interesting to observe that the NO_x reduction might be insufficient if a specific limit for NO_x emission should be met, as in the case of EC 542/91 regulation for heavy-duty Diesel engines. However, in the case of car engines, present EC standards (regulation 441/91) are referred to a combined emission limit of NO_x and UHC; this allows the adoption of low EGR rates.

With reference to the thermodynamic cycle, the EGR should cause a decrease in heat rejected through exhaust gases (1, 13). However, power output of the tested Diesel engine was negatively affected by EGR rate, as it is apparent in fig.8d. Since fuel mass flow rate didn't change significantly, this resulted in a definite increase in specific fuel consumption (fig.8d), that means a lower thermodynamic efficiency. This drawback may be due to changes in the structure of the flame caused by the EGR which may affect ignition delay and combustion duration (11), therefore deteriorating the indicated pressure diagram. In this direction measurements of in-cylinder pressure will be very helpful in order to outline the influence of EGR on thermodynamic efficiency and the advantages which may be achieved through a proper control of injection timing. Anyway, the higher bsfc may partly cancel the advantages in NOx emissions.

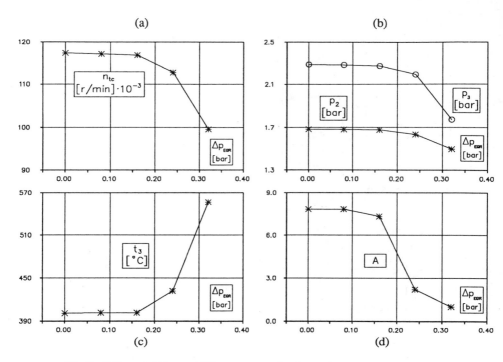

Fig.9 - *Effect of different EGR rates on turbocharger operating parameters.*

On the basis of the above considerations, it may be interesting to investigate the effects of EGR cooling, both on the trade-off between NOx and smoke (10), and on thermodynamic efficiency.

As regards the turbocharger behaviour, increasing EGR rate caused a decrease in rotational speed n_{tc} (fig.9a), which may be due to the lower mass flow rate through the turbine. Moreover, pressure at the turbine inlet p_3 was significantly reduced (fig.9b), while the corresponding temperature t_3 increased probably because of the higher intake temperature and of the loss in thermodynamic efficiency (fig.9c). As a consequence the boost pressure drop was lower, as apparent in fig.9b, due also to the effect of the turbine control system (fig.9d) which reduced the opening degree A of the VG device (7). With reference to compressor and turbine efficiency, a slight decrease was generally observed with EGR, which may be related also to alterations induced in the flow characteristics at the turbine inlet. Measurements of pressure diagrams at the compressor exit and at the turbine inlet will be very useful to investigate on this behaviour, taking into account the substantial changes imposed by EGR on flow unsteadiness in the intake and exhaust manifolds.

The experimental investigation allowed to observe that the activation of the EGR valve induced a transient in engine operating conditions, while steady state was reached after a period of time which was relatively long. With reference to an engine rotational speed of 2750 r/min and to a rack position of 0.617, the EGR governing signal Δp_{EGR} was suddenly varied from 0 to 0.30 bar (corresponding in these conditions to an EGR rate f_{EGR} of about 8 per cent), and engine operating parameters were measured until steady state conditions were reached. After a

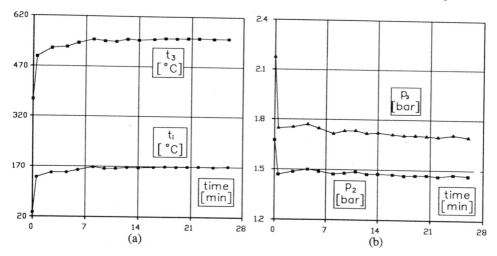

Fig.10 - *Effect of EGR activation on engine and turbocharger parameters.*

first rapid transient due to EGR activation, several measured parameters kept on varying during a longer period of time. Temperature values in the intake manifold t_i and at the turbine inlet t_3 revealed significant deviations, as shown in fig.10a, while variations in measured pressures at compressor outlet p_2 and at turbine inlet p_3 are given in fig.10b. Whereas changes in other operating parameters were less substantial, it is apparent that a long transient follows the activation of the EGR system. To explain this behaviour, which was not observed by other Authors (10), further investigations are needed, with particular reference to the effects of EGR on combustion process and on the possible built up of soot deposits in the combustion chamber and in the EGR circuit components.

The effects of EGR on engine performance and emissions, analysed through the reported experimental data, highlighted interesting advantages which may be achieved by this technique especially in terms of NO_x emissions. However, reported results showed that EGR operation significantly affects many operating parameters of both the engine and the turbocharger. For this reason, it is apparent that the EGR control strategy should be properly integrated into the engine management system, which at present has to cope with a large row of governing parameters concerning also fuel injection equipment and turbocharger control system. With particular reference to the tested engine, whose configuration is typical of small automotive applications, promising results should be obtained through an integrated control system of turbocharger turbine, EGR components and injection timing (which may be optimised with regard to EGR rate).

Further investigations are needed on this subject, taking also account of the utilisation in the exhaust system of particulate traps or catalysts (which presently require definite operating conditions). This area will be explored by the authors, through the definition of catalysts and particulate traps behaviour, with the aim to develop an optimised control strategy for the intake and exhaust systems which may be integrated in the engine management system.

4 - CONCLUSIONS

In order to meet future legislation requirements, small Diesel engines have to cope with pollutant emissions standards, with particular reference to NO_x and soot. A variety of techniques have been proposed and applied to modern automotive Diesel engines to improve performance and to reduce emissions, i.e., variable geometry turbochargers, EGR, catalysts and particulate traps. The engine thus becomes a complex power plant whose intake and exhaust system offers interesting possibilities to control its operating conditions. However, the involved components (turbocharger, EGR, catalyst and particulate trap) have to be properly matched, while an integrated control system is needed to operate them in the most efficient way. Of course this requires adequate information on the behaviour of these components in the real on-engine operating conditions.

The presented investigation has been focused on the interaction between the turbocharger (fitted with a VG turbine) and the EGR system, and on their influence on engine performance and emissions. The experimental study was developed on the test facilities set up at IMSE, which allow measurements both on the engine and on intake and exhaust components and subassemblies.

Performance and emissions of a small automotive Diesel engine were defined in steady state with reference to its baseline configuration without EGR. Experimental data allowed to highlight the role that an optimised matching of the turbocharger plays in the improvement of engine performance. Especially in the case of a variable geometry device, the turbine regulating field should be completely utilised. Higher turbocharger efficiencies could also be achieved through a proper matching, thus increasing boost pressure at low rotational speed. However, it should be noted that the application of EGR requires an exhaust manifold pressure higher than intake one, while on the other hand these operating conditions should be avoided when the EGR valve is closed to get a better cylinder scavenge.

The effects of EGR technique were then analysed. The addition of exhaust gases to the intake air reduces the oxygen concentration and probably lowers the combustion temperature, since NO_x emission, which is extremely sensitive to temperature, decreases significantly. On the other hand, for the same reasons particulate and CO emissions usually increase. With reference to EGR effects on thermodynamic efficiency, experimental data showed an increase in bsfc and a decrease in power output. This may be due to alterations in the flame structure and to changes in ignition delay and combustion duration, which may lead to a worse indicated pressure diagram: measurements of in-cylinder pressure will be developed to study these phenomena and to analyse the opportunity of the definition of a proper control strategy for injection timing when EGR is used.

Tests in the baseline configuration showed high NO_x specific emissions also outside the EGR operating range. However, on the basis of the observed increase in bsfc and smoke emissions, it seems fairly difficult to extend EGR at higher engine loads and speeds. As a consequence, NO_x reduction in these operating conditions will be reached more adequately by using specific aftertreatment devices, such as catalysts.

With reference to the turbocharger, EGR causes a decrease in turbine inlet pressure, while the drop in boost pressure is partly opposed by the turbine control system and by the higher turbine inlet temperature. Compressor and turbine efficiency proved to be slightly lower with EGR: as regards this aspect, pressure diagrams at the compressor exit and at the turbine inlet will be measured to analyse the effects of EGR on flow unsteadiness in the intake and exhaust manifolds.

The presented results highlighted that a proper matching between intake and exhaust components and the engine has a great potential in improving engine performance and emissions. At the same time, this target requires the development of a suitable governing methodology to control the intake and exhaust system, which have to be integrated and tuned with the engine management system. To this aim the investigation will proceed, taking advantage of the experimental facilities set up at IMSE and of a theoretical model of the engine which has been developed by the authors (4, 5).

REFERENCES

(1) HEYWOOD, J.B., *Internal Combustion Engine Fundamentals*, 1988, McGraw-Hill, New York.
(2) TAYLOR, C.F., *The Internal Combustion Engine in Theory and Practice*, Volume 1, 1985, The M.I.T. Press, Cambridge, Massachusetts.
(3) BENSON, R.S. and WHITEHOUSE, N.D., *Internal Combustion Engines*, 1979, Pergamon International Library, Oxford.
(4) ACTON, O., CAPOBIANCO, M., GAMBAROTTA, A. and ZAMBONI G., *Optimum control of variable geometry turbine in automotive turbocharged Diesel engines*, XXV FISITA Congress, Beijing, 1994, paper 945012.
(5) CAPOBIANCO, M., GAMBAROTTA, A., SILVESTRI, P. and ZAMBONI, G., *Intake and exhaust systems of automotive turbocharged Diesel engines: components characterisation and modeling*, XXVI FISITA Congress, Prague, 1996.
(6) CAPOBIANCO, M. and GAMBAROTTA, A., *Unsteady flow performance of turbocharger radial turbines*, 4th International Conference on Turbocharging and Turbochargers, Inst.Mech.Engrs., London, 1990, paper C405/17, pages 123-132.
(7) CAPOBIANCO, M. and GAMBAROTTA, A., *Variable geometry and waste-gated automotive turbochargers: measurements and comparison of turbine performance*, ASME Transactions, Journal of Engineering for Gas Turbines and Power, vol.114, 7/1992, pages 553-560.
(8) CAPOBIANCO, M. and GAMBAROTTA, A., *Performance of a twin-entry automotive turbocharger turbine*, ASME Energy-Sources Technology Conference and Exhibition, Houston, 1993, paper 93-ICE-2.
(9) CAPOBIANCO, M., GAMBAROTTA, A. and SILVESTRI, P., *Effects of volute configuration and flow control system on the performance of automotive turbocharger turbines*, ASME Internal Combustion Engine Division Spring Technical Conference, Marietta, OH, 1995, paper 95-ICE-11.
(10) HERZOG P.L., BURGLER L., WINKLHOFER E., ZELENKA P. and CARTELLIERI W.P., *NO$_x$ reduction strategies for DI Diesel engines*, SAE International Congress and Exposition, 1992, paper 920470.
(11) PLEE S.L., AHMAD T. and MYERS J.P., *Flame Temperature correlation for the effects of Exhaust Gas Recirculation on Diesel particulate and NOx emissions*, SAE Fuels and Lubricants Meeting, 1981, paper 811195.
(12) KAWAKAMI M., TANABE H., MATSUDA J. and SATO K., *Environmental control of advanced medium-speed engines*, 20th International CIMAC Congress on Combustion Engines, London, 1993.
(13) GHEZZI U. and ORTOLANI C., *Combustione ed Inquinamento*, 1974, Tamburini Editore, Milan.

Lubricity of Diesel Fuel Injection Equipment

The lubricity of hydrotreated diesel fuels

R DAVENPORT
Shell Research Limited, Shell Research and Technology Centre, Chester, UK
F LUEBBERS
Deutsche Shell AG, Hamburg, Germany
R GRIESHABER, H SIMON, and **K MEYER**
Robert Bosch GmbH, Stuttgart, Germany

SYNOPSIS :

It is generally agreed that reduced levels of sulphur in automotive gas oil (AGO) lead to lower tailpipe emissions of diesel particulates and sulphur oxides. Therefore, in line with concerns over the environmental impact of road transport, many countries have already, or are planning to reduce the maximum permitted level of sulphur in AGO. In order to reduce the level of sulphur it is necessary to use some form of hydro-processing. Deep hydrogenation - as that required to produce 'environmentally adapted' AGO such as Swedish Class I (10ppm max. sulphur) - is known to lead to a significant reduction in the natural lubricity characteristics of AGO. The purpose of the work described in this paper was to consider the lubricity characteristics of AGO produced by milder hydrotreatment processes - such as those likely to be used in the production of AGO to meet 500ppm (mg/kg) max. sulphur specifications.

Road trials, pump durability tests and bench test results confirm that *some* AGOs manufactured to meet lower sulphur specifications can lead to increased wear in sensitive types of diesel injection equipment and that - where necessary - additives can be successfully used to improve the lubricity characteristics of low sulphur AGOs. The results of this work illustrate the need for technically sound and commercially viable specifications to define minimum levels of AGO lubricity - both to protect sensitive equipment already in the market and to ensure that new equipment is designed to be robust with respect to operation on AGO of defined minimum lubricity.

INTRODUCTION :

The previous 100 years or so of relatively calm, mutual evolution of the diesel engine, the fuel and the fuel injection system were disrupted in the early 1990s by a step change in fuel properties, resulting from a radically different fuel specification which was introduced in Sweden in order to reduce urban air pollution. The specification called for a reduction of the sulphur level by three orders of magnitude, and could be met only by applying severe hydroprocessing techniques. The fuel made in this way had a distillation range which was more like kerosine than a traditional automotive gasoil (AGO), and it is not surprising that there were several areas in which its compatibility with the existing diesel equipment had changed. Lower density means lower energy per unit volume, and lower polarity caused by reduced aromatic content means a change in the behaviour of elastomeric seals, but most importantly the lubricity of the fuel had been drastically reduced (1).

Little is known about the species which historically gave diesel fuel its high lubricity, except that they are polar materials, and that they are removed by hydroprocessing. The distributor type injection pumps which are currently used in a large proportion of passenger cars, vans and small trucks rely on the fuel to provide sufficient boundary lubrication for the moving contacts to be protected These pumps show a range of sensitivity to fuel lubricity; for the most sensitive pumps, the severely treated, environmentally adapted fuels could not meet this need unless treated with lubricity additives.

Sweden represents an extreme case; as shown in Table 1, the maximum allowable sulphur content in many other diesel markets either has, or soon will be, reduced from a typical 0.2 - 0.3 % by mass to a level of 0.05 %.

Table 1 - some representative fuel requirements

Property	Swedish Class I	Swedish Class II	Swiss (winter)	CEN EN590	ASTM D975LS	JIS No. 2 (Japan)
Sulphur, ppm	< 10	< 50	< 500	<2000 (#)	< 500	< 2000 (*)
Aromatics, %v	< 5	< 20	-		< 35	-
Polyaromatics, %v	< 0.02	< 0.1	-		-	-
T90, °C	-	-	-		< 338	< 350
T95, °C	< 285	< 295	< 340	< 370	-	-
Density, kg/m3	800 - 820	800 - 820	800 - 845	820 - 850	-	-
Viscosity, cS (40 °C)			1.5 - 4.0	2.0 - 4.5		> 2.5 (30 °C)
Cetane no.(CN) / Index (CI)	> 50 (CI)	> 47 (CI)	> 47 (CN)	>49 (CN)	> 40 (CN)	> 45 (CI)

(#) <500 ppm by 1/10/96   (*) <500 ppm in 1997

The purpose of this paper is to describe a work programme which has been undertaken by Shell to develop and / or verify reliable lubricity assessment techniques. These have then been used to examine the range of lubricity of 0.05 % sulphur fuels, and to provide sound technical data which will allow rational decisions to be made on performance limits.

BACKGROUND :

There are several ways in which the fuel sulphur can be reduced to meet the 500 ppm limit which will soon apply in Europe, North America and Japan. Modern diesel fuel is blended from a large number of components, and changes in product specifications often involve changing the component balance. The options for 0.05 % sulphur AGO are to apply severe hydrotreatment or hydrocracking to produce very low sulphur gasoils or kerosines as blend components, or to use a lower sulphur crude feedstock, with mild hydrotreatment. These techniques could each give the required sulphur level, but will result in fuels of differing lubricity. In general the sulphur content and lubricity decrease as the processing severity increases, but the relationship between sulphur content and lubricity is weak, and there is evidence that some sulphur-containing species are actually pro-wear (2).

If it is not possible to predict the lubricity of a reduced sulphur gasoil on the basis of its composition, we have to use techniques to measure this property. There are basically two ways of doing this; the first is to carry out tests which involve fuel pumps and perhaps vehicles, such as field trials or pump endurance tests. These provide unambiguous primary reference data, but take too long and cost too much to be of use in quality assurance - a pump endurance test takes between 2 and 8 weeks, while a field trial takes months. The second type of test is one which is carried out in a bench instrument. With appropriate design and test conditions it is possible to accelerate the wear processes which occur in vehicle fuel injection system. By their very nature, bench tests do not give primary data, but instead are used to give an assessment of how the fuel and injection system would perform in practice. With enough data to support a correlation with 'real life', the bench tests can form a useful part of a fuel specification.

One such bench test is the TAFLE - the Thornton Aviation Fuel Lubricity Evaluator. This was originally developed by Shell and the British Ministry of Defence (3) to mimic the conditions inside an aviation fuel pump, and it has been successfully used over a long period to assess the lubricity of aviation kerosines. Work on Swedish Class I and II base and additivated fuels (1) demonstrated that the TAFLE test gave results for AGOs which correlated well with fuel pump wear data from field trials. However, there are two major drawbacks to the TAFLE - a duplicate test takes two days, and there is only one such instrument, so while it has been invaluable in providing high quality research data, it is of little use in a specification.

The lubricity changes in Sweden highlighted the need for a widely available bench test method, and several groups were formed to investigate ways forward. In 1994 a joint round robin programme carried out by the ISO TC22/SC7/WG6 group and the CEC PF006 group examined the performance of 4 bench instruments which might discriminate between fuels on the basis of their lubricity and which could be made widely available (4). Bench test data and pump wear data were generated for a range of fuels, and from these results the HFRR (High Frequency Reciprocating Rig) test was selected as the instrument of choice. This test offered the type of rapid lubricity assessment which would be needed by fuel suppliers to ensure that their fuel was fit for purpose, but there was still a considerable amount of work to be done, as it was not known how well the test would perform with fuels of intermediate lubricity. Further round robin work to generate precision data for the modified test is currently in progress.

THE NEED FOR A RESEARCH PROGRAMME :

In 1993 it was clear that more information would be needed if the widely discussed proposals for reductions in fuel sulphur were to be handled properly. The majority of diesel fuel had shown no historic lubricity problems, yet the environmentally adapted fuels in Sweden could only be used in the more sensitive injection equipment if they contained lubricity additives. In principle it was known that 0.05 % sulphur fuels could be produced by a variety of methods, and several refinery solutions had been found, but no fuels had been produced to that specification. It was thought that 0.05 % sulphur fuels would show a range of lubricities, yet it was not known how wide the range would be, or whether additive solutions existed for those fuels which were not fit for purpose. Perhaps most importantly, it was clear that there would have to be lubricity specifications, but there was no test method which was sufficiently widely available to be used in routine quality assurance, and no information as to where a performance limit would lie.

OBJECTIVES :

With these issues in mind, a programme of work was drawn up within Shell, with the intention of addressing the following aims in advance of the widespread introduction of reduced sulphur fuels :
- Determine the range of lubricity performance.
- Assess the implications for sensitive equipment already in the market place
- Identify options for ensuring 'fitness for purpose' of the new fuels
- Continue the development of a readily available bench test
- Provide a rational basis for technically sound and commercially viable limits.

APPROACH :

Because it was necessary to obtain the maximum information in a short time, an in-house, more rapid alternative to the field trial was needed as a source of in-service data, together with a bench test with the potential for wide use outside the research environment. However, with so many uncertainties it was necessary to adopt a stepwise approach, testing the available data and looking for lack of correlation along the way. The first stage of this process was to create a set of reference fuels of known field performance. These could then be used to verify the performance of a pump endurance test, and to test the predictive ability of a suitable bench test. Having established that the bench test assessment correlated with field performance, it could then be used to examine the lubricity of a wider range of fuels.

INJECTION PUMP WEAR ASSESSMENT :

Measuring the wear of the injection equipment is at the heart of any assessment of fuel lubricity. However, when the work programme started there had been little need for fuel suppliers and pump manufacturers to discuss the wear rating process. Earlier work had shown that the rotary distributor type of pump would discriminate between fuels on the basis of their lubricity, and to remove another unwanted variable from the programme all the pumps tested were produced by Bosch. It was therefore logical for Bosch to carry out the wear rating assessments on our behalf at their Stuttgart headquarters.

The Bosch wear rating technique is fully described elsewhere (5), and involves visual and mechanical measurements of over 30 of the pump contacts. Using the assumption of constant rate of wear, and knowing the service life of the pump, the measurements are scaled to a common 'mileage' of 100, 000 km, and each part of the pump is assigned a wear rating. These ratings are then reduced to a single figure which represents the overall pump condition. The rating scale is from 1 to 10, where 1 indicates a perfect pump, values less than 4 indicate acceptable pump wear, and 10 is a seizure.

GENERATING THE FIELD PERFORMANCE DATA :

Since no fuels meeting an 0.05 % sulphur specification were available, it was necessary to produce them specifically for this work programme. With limited time and resources for field trial testing, two key base fuels were targeted; one which had been severely hydrotreated to give poor lubricity, and one which had undergone milder processing, so as to be close to borderline performance. TAFLE tests on a range of potential fuels were carried out in order to select the two which most closely met these requirements. A third test fuel was produced by treating the one of poor lubricity with additives and measuring the TAFLE response, again with the intention of achieving borderline performance. Finally, two more fuels were needed to represent the worst and best lubricities available; Swedish Class I and a typical mid-European fuel meeting an 0.2% sulphur specification were chosen. The properties of these base fuels are given in Table 2.

Table 2 Properties of the field trial base fuels

Fuel property	Standard AGO (typical)	High lubricity base fuel	Low lubricity base fuel	Class I
Density at 15 °C, kg/m^3	836.5	833.1	831.3	808.5
Viscosity at 40 °C, cS	2.90	2.46	2.00	1.83
Sulphur content, mg/kg	1000	300	200	5
Cetane number	53.3	52.5	53.7	57.9
Total aromatics, %m	24.8	22.3	21.0	4.04
T10, °C	204.0	209.5	192.0	206.5
T90, °C	341.0	317.0	317.5	266.5

Before testing the fuels in field trials it was necessary to check that the pumps in the field trial vehicles would be sufficiently sensitive to fuel lubricity. To do this, three vehicles were run on the Class I fuel. Two of the cars were fitted with brand new pumps; these failed after 30 km and 17 km respectively. The third car, which had been run-in for 1000 km using regular AGO, was driven for a further 10 000 km on Class I fuel. However, it showed unacceptable driveability, with hesitation and an unusually high idle speed, and high fuel consumption. This fuel pump was found to have severe wear at the camplate and rollerbolts.

It was concluded from this short series of tests that the pump type was suitable, and two 100,000 km field trials were carried out, one in Germany, the other in Switzerland. The German trial involved seven cars fitted with new injection pumps; three cars ran on the high lubricity base fuel, three on the low lubricity base fuel and a reference car ran on the standard AGO. The cars were driven under high load and high speed conditions on a 720 km test route. Flow measurements of the fuel pumps were carried out at several points, before during and after the trial, and a number of smoke measurements were made. The Swiss road trial involved three vehicles in mixed driving cycles, accumulating approximately 600 km/day. Each car used the low lubricity fuel with additive, and again there were several flow and smoke measurements during the course of the test. All of the vehicles in the two trials were fitted with tachographs to check that the required driving patterns were being achieved.

FIELD TRIAL RESULTS :

There were no catastrophic pump failures during the field trials, and all 10 vehicles could still be driven after 100,000 km on the test fuels. However, the vehicles which had used the low lubricity fuel had clear driveability problems, in particular a high idling speed and hesitation on accelerating. Further evidence that the poor lubricity fuel had adversely affected the fuel injection equipment came from the exhaust smoke measurements.

Following the field trials, the injection pumps were rated by Bosch. The results are shown in Figure 1, as a function of the TAFLE scuffing load. It will be seen that the correlation between the two measurements is very good, confirming the earlier evidence that the TAFLE is an appropriate technique for AGOs. It can also be seen that the result for the additivated fuel is predicted as well by the TAFLE as those for the base fuels.

Figure 1   Comparison of TAFLE load with Bosch wear rating

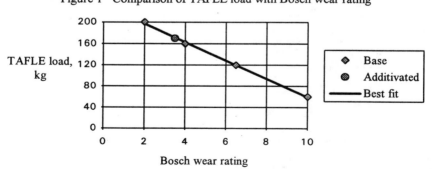

DEVELOPMENT OF A RAPID PUMP ENDURANCE TEST :

The need to examine the performance of a large number of fuels in a very limited time meant that there had to be a faster, reliable alternative to the field trial. All of the fuel injection equipment manufacturers use pump endurance rigs in their design development work and for testing pumps from the field, and this type of equipment seemed to be a rational choice. However, it was clear that selection of the test cycle, and examining the correlation between wear in an endurance test and wear in the field were critically important.

Following discussions with the pump manufacturers, a Lucas Hartridge HA760 pump endurance rig was installed at Thornton. In this equipment the fuel injection pump is driven by an electric motor in a precisely controlled speed cycle. A low pressure transfer pump is used to deliver the fuel from a thermostatically controlled reservoir to the test pump, and out through injector nozzles. The injected fuel is returned to the fuel tank. While this rig had been designed to test injection pumps using a standard calibration fluid, what was required for this work was to use a standard pump to compare different fuels. Few modifications were needed to adopt this change of role, apart from an exterior fuel tank.

The cycle which was chosen was based on one used by Lucas. It lasts for 15 seconds, and as shown in Figure 2, it comprises low, medium and high speed regimes. A standard test consists of 70 000 such cycles, with a fuel change half way through. A pump inlet temperature of 40 °C and an injector lift pressure of 130 bar are used. Pump flow data are acquired by a Bosch dealer before and after each test, and the pump is then sent to Bosch, Stuttgart for wear rating. The overall running time of this test is about 290 hours, compared with 1000 hours which is carried out by Bosch in Stuttgart, and only one pump is usually used, whereas Bosch use two, fed from a shared fuel tank. The advantage of the shorter test is that it allows fuels to be tested more quickly. However, it must be accepted that it produces less wear than a conventional test, and therefore it is more difficult to rate pumps which have operated on fuels of high lubricity.

Figure 2 Pump rig operating conditions

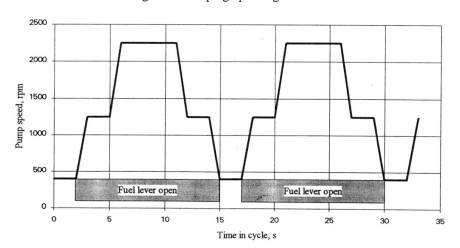

In order to validate the performance of the pump endurance rig, the first tests were performed using the Class I fuel and the high and low lubricity base fuels from the German field trial. The Class I fuel was tested twice, and gave seizure almost immediately - indeed, in one test the pump failed before the first cycle had been completed. The other fuels allowed the pumps to run satisfactorily to the end of test, and when rated by Bosch gave results which agreed very well with those in the field trials. There was not enough of the additivated fuel used in the Swiss trial, or the standard AGOs used as a reference in the German trial to use in this exercise, and fuels with very similar properties from the same refineries (and with the same lubricity additive) had to be used instead. The comparison of pump wear data from the field trials and from the endurance rig is shown in Table 3.

Table 3   Comparison of field trial with endurance rig pump wear ratings

Fuel	Bosch wear rating (Field trial)	Bosch wear rating (endurance rig)
Class I	10	10
Low lubricity 0.05 %S	6 - 7	5 - 7
High lubricity 0.05 %S	4	3 - 4
Low lubricity + additive	3.5	3 - 4 (similar fuel)
High lubricity 0.2 %S	2	2.5 (similar fuel)

Having shown that the field trial and pump endurance rig data were in good agreement, the set of reference fuels was then used to investigate the performance of the high frequency reciprocating rig (HFRR) bench test. This test was selected at the end of 1994 by the ISO TC22/SC7/WG6 and CEC PF006 diesel fuel lubricity working groups as the best of the widely available lubricity test methods, on the basis of an extensive round robin programme. However, there was still a considerable amount of development work to carry out, as the round robin fuels matrix had not included fuels of intermediate lubricity. It had been shown (4) that the HFRR discriminated well between a fuel of low lubricity and one of high lubricity, but it was not known how well the technique would perform for intermediate fuels.

The initial HFRR tests carried out by Shell on the field trial fuels gave promising results, with reasonable discrimination and precision, and the same ranking which the field trial pump wear ratings had shown. However, further work indicated problems, as the results appeared to be different each time the set of fuels was tested. After much test work it was found that the HFRR method which had been used by the ISO and CEC groups gave results which depended on atmospheric humidity. Each of the low sulphur fuels behaved differently, and the ranking of the fuels varied with the moisture level in the air. There was no time to pursue the reason for this behaviour, but a relatively easy solution was identified, and shown to be robust. The solution was to increase the HFRR sample volume from 1 to 2 ml, which suggests that the contact was not fully flooded, and that atmospheric moisture was affecting the wear processes. The importance of a fully flooded condition in this test had been highlighted earlier (6).

With the increased sample volume there is still some effect of humidity, with the HFRR wear scar increasing in line with the moisture content of the air. However, it appears that this effect is directionally the same for a very wide range of base fuels, additivated fuels and blend components, with the important result that the ranking of a set of fuels is independent of humidity with the modified method. In order to reduce repeatability errors, the HFRR at Thornton is operated inside an environmental cabinet set at 25 °C, 60 %RH.

An extensive series of tests with the field trial fuels showed conclusively that the modified method has considerably better discrimination for tests at 60 °C than for those at 25 °C. This is illustrated in Figure 3, which shows the mean wear scar diameter (MWSD) at both fluid temperatures, as measured under the 'standard' ambient conditions. On the basis of these results, all further HFRR tests were performed with a fluid temperature of 60 °C.

Figure 3   Effect of HFRR test temperature on discrimination

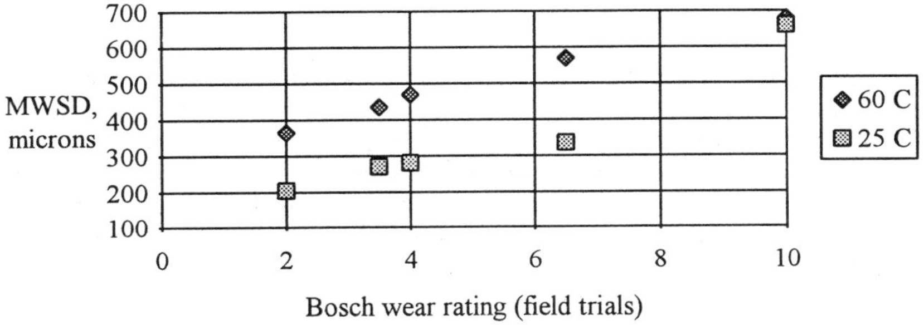

Having shown that the rapid rig endurance test gave results which agreed well with those from field trials, and having modified the HFRR method, we were then in a position to examine a wider range of fuels. A large number of fuels, both with and without lubricity additives, have now been tested by HFRR and in the pump endurance rig, and as shown in Figure 4, the correlation between wear in the HFRR and pump wear ratings from the endurance rig is very good. There is some scatter for the high lubricity fuels (i.e. low wear ratings); as explained earlier, the rapid pump test was designed to produce maximum data in a short time, and it is difficult to carry out precise pump ratings when the actual wear is low.

Figure 4   Comparison between pump wear and HFRR data

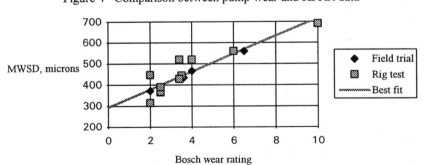

LUBRICITY ADDITIVES :

In some cases, for example fuels made from very low sulphur crudes, where the hydroprocessing severity has been low, 0.05 %S base fuels have been found to exhibit very high lubricity, and there is no need for additive treatment. However, many other 0.05 %S base fuels have been found to have inadequate lubricity - indeed, many are worse than the fuel which was specially produced for the field trial, and which was thought at the time to represent an extreme case. The effect of additive treatment on these fuels has been examined in some detail. The results show that each fuel / additive combination is different, and that prediction of the additive response of a particular fuel is almost impossible - some 'hard' fuels are very responsive to additive treatment, while the lubricity of some 'soft' ones is only improved by high additive treat rates.

It is clear that 'blanket' additivation of all 0.05 %S fuels will not work, as the differing lubricity of the base fuels and the varying response to additives inevitably means that some fuels will show inadequate lubricity, while others will contain unwanted excess additive. Over-treatment is not a good idea, as it will not improve the fuel in any way - indeed, problems such as increased water pickup or interaction with other performance additives may occur, and there may be an increased risk of other unwanted side-effects from excess lubricity additive (7). The conclusion is that each fuel and potential additive needs to be checked for lubricity performance, and the results of this work suggest that the HFRR is a suitable instrument for this type of routine quality assurance.

NEXT STEPS :

From the work carried out in this programme, it is clear that routine monitoring of diesel fuels will be essential to ensure satisfactory lubricity performance. The improvements made to the HFRR as a result of this work have given a test which appears to correlate well with pump rig performance, and the improvements have been incorporated in draft methods (ASTM and ISO DIS12156) for the measurement of lubricity by HFRR. The effect of humidity on the HFRR results needs further investigation, and longer duration pump endurance tests have to be considered for fuels of high lubricity.

There is still a need within Europe to assess the precision of the modified method, without which a meaningful specification cannot be drafted, and to investigate the correlation with a wider range of injection pumps. These items are the focus of further round robin programmes run by the CEC and ISO lubricity working groups, which are currently under way.

CONCLUSIONS :

1. Road trials have shown that reduced sulphur AGOs show a range of performance - some are acceptable, while others are likely to lead to increased wear in sensitive types of equipment.
2. The very rapid, catastrophic failure of fuel injection pumps which has been seen with unadditivated Swedish Class I fuel has not been observed with the less severely processed

fuels, even in the absence of lubricity additive. Instead, the increased wear resulting from fuels of inadequate performance is likely to lead to a gradual loss of performance (driveability, fuel economy) and increased emissions (regulated and smoke).
3. The range of lubricity performance of these fuels means that a lubricity specification is required to ensure 'fitness for purpose'. The minimum acceptable level of AGO lubricity needs to be defined to protect the existing equipment in the market.
4. For specifications to be workable, both for fuel suppliers and customers, there must be a robust, widely available, internationally agreed test method. The work described here resulted in considerable improvements to the HFRR method to meet these needs. The HFRR now appears to have an acceptable level of correlation with 'real life', and can discriminate between fuels of good, intermediate and poor lubricity, both with and without lubricity additives.
5. Future fuel specifications should not mandate the use of lubricity additives, as they are not necessary in some fuels, and much of the diesel parc is insensitive to fuel lubricity.
6. 'Blanket' additivation of fuels without a rational specification will not work, because the differing lubricity of the base fuels and the varying response to additives inevitably means that some treated fuels would have inadequate lubricity, while others would contain unwanted excess additive, which could increase the risk of side effects
7. Further industry-wide collaborative work on the HFRR test is currently under way. The results of these round robins will provide valuable data on the correlation with a wide range of injection equipment, and will be essential in setting lubricity specification limits.

ACKNOWLEDGEMENTS

The authors would like to gratefully acknowledge the contribution of the many people who have participated in this work. Special thanks are due to Mr D.J. Evans (Shell) and Mr A Unterberg (Bosch).

REFERENCES

1. TUCKER, R.F., STRADLING, R.J., WOLVERIDGE, P.E., RIVERS, K.J. and UBBENS, A., "The lubricity of deeply hydrogenated diesel fuels - the Swedish experience", SAE paper 942016, October 1994.
2. WEI, D. and SPIKES, H.A., "The lubricity of diesel fuels", **Wear**, 1986, 111, pp217 - 235
3. HADLEY, J.W., "The measurement of the boundary lubricating properties of aviation turbine fuels", I.Mech.E. Conference on Combustion Engines - Reduction of Friction and Wear, London, March 1985.
4. NIKANJAM, M., CROSBY, C., HENDERSON, P., GRAY, C., MEYER, K. and DAVENPORT, N., "ISO diesel fuel lubricity round robin program", SAE paper 952372, October 1995.
5. MEYER, K. and UNTERBERG, A., "Evaluation of pump endurance tests", Internal Publication, Robert Bosch GmbH, September 1995.
6. BOVINGTON, C., CAPROTTI, R., MEYER, K., and SPIKES, H.A., "Development of a laboratory test to predict lubricity properties of diesel fuels and its application to the

development of highly refined diesel fuels.", 9th International Colloquium, Ecological and economic aspects of tribology, Esslingen, Germany 1994.
7. MIKKONEN, S. and TENHUNEN, E., "Deposits in diesel fuel injection pumps caused by incompatibility in fuel and oil additives", SAE paper 872119, 1987.

© Shell Internationale Research Maatschappij BV

Vehicle tests for LSADO

FARRELL BEng, MEng, AIMechE, MSAE and **N G ELLIOTT** BSc, MSAE
Esso Petroleum Company Limited, Oxfordshire, UK

SYNOPSIS

During 1995/1996 vehicle tests were conducted at the Esso Research Centre, Abingdon (ERCA), to assess the effect of Low Sulphur Automotive Diesel Oil (LSADO = 0.05% mass sulphur) on some types of fuel injection pump. Performance criteria included regulated emissions, black smoke emissions, fuel delivery (assessed on a test rig), pump component weight loss, and rating of the pump components to a procedure defined by the fuel injection pump manufacturer.

Low sulphur base fuels were run with and without lubricity additives, and performance criteria were checked periodically throughout the test duration of 100 000 km. The test concluded successfully with a better understanding of how bench lubricity tests correlate with field tests, and confidence that additive strategies can be employed to restore fuel lubricity performance back to the levels of 0.2% sulphur fuels.

ACKNOWLEDGEMENTS

The authors would like to acknowledge the contributions of their colleagues involved in running this Vehicle Test, in particular those in ERCA's Vehicle Testing Group and Metrology Laboratory.

1 INTRODUCTION

October 1 1996 will see the mandatory reduction of the sulphur content of diesel fuel from 0.2% mass to 0.05% mass throughout Europe. The reduction will help to reduce particulate emissions, which are partly made up from sulphate compounds. The most common way to achieve this sulphur reduction is to increase the severity of the fuel hydrotreatment, although the lower sulphur levels can also be the result of careful crude oil selection by refineries. This increased hydrotreatment can also remove other compounds, which are thought to contribute to the lubricating properties of the fuel. There are therefore concerns about the ability of the fuel to protect some types of fuel injection pump from excessive wear.

Some pump types appear to be more susceptible to increased wear from poor lubricity fuels than others. In-line injection pumps, traditionally used on heavy duty vehicles, are less susceptible, as the critical moving parts - the cam and follower assembly - are lubricated by engine oil from the engine's main lubricating system. The moving parts in rotary type pumps are lubricated entirely by the fuel, and are more likely to wear excessively if the lubricity of the fuel is poor. More than one type of rotary fuel pump design exists, and different designs could be expected to respond differently to poor fuel lubricity. Reports from some Scandinavian countries first highlighted the potential problems, although the fuels causing the majority of injection pump failures were very low sulphur (0.01% mass or less) and low viscosity.

Several bench or rig tests are available to assess fuel lubricity, and these have been reviewed in other published literature (1). However, concerns remain about how these tests correlate with 'real' situations - i.e. fuel injection pumps on vehicles.

Tests have been conducted at ERCA to assess the effects of LSADO on one type of rotary fuel injection pump. The High Frequency Reciprocating Rig (HFRR) test was the rig test used to assess fuel lubricity before the test, and vehicle tests were conducted to verify the rig test results.

2 VEHICLES, FUELS AND ADDITIVES

1.9 litre IDI type passenger cars were used for the test; the vehicles had an identical specification which included a rotary fuel injection pump, EGR and an oxidation catalyst. All vehicles were naturally aspirated. Four vehicles completed a total test distance of 100 000 km.

Table 1 shows details of the vehicle designations, the fuels and additives used, and the sulphur levels in the base fuels.

Table 1 Vehicles, fuels and additives used

VEHICLE	BASE FUEL TYPE	ADDITIVES	SULPHUR CONTENT % mass (nominal)
CAR A	Central European	Performance package	0.2
CAR B	Central European	None	0.05
CAR C	Central European	Performance package + high lubricity additive treat	0.05
CAR D	Central European	Performance package + low lubricity additive treat	0.05

The same base fuel (LSADO) was used for Cars B, C and D. The same performance package was used for Cars A, C and D. This fully formulated package incorporated a detergent and a cetane improver, along with other performance enhancing components.

The LSADO was selected on the basis of its high wear scar diameter in the HFRR test. This was the result of the more severe hydrotreatment that has previously been discussed. The lubricity additive treat rates were chosen to restore the HFRR performance levels typical of 0.2% sulphur fuels.

3 TEST PROTOCOL

3.1 Vehicle preparation

All vehicles were run-in for 5000 km using Esso Diesel 2000 standard forecourt fuel. This was to ensure that all vehicle catalysts were adequately aged, and that combustion chamber and injector deposits were stabilised. This also ensured that none of the fuel injection pumps were adversely affected by the quality of the fuel during the running-in process.

3.2 Injection pump assessment

Fuel injection pumps were removed from the vehicles and stripped after the 5000 km run-in, at 50 000 km and at 100 000 km. The pump components were weighed at each point, and the components rated at 50 000 km and 100 000 km, using a method devised by the injection pump Original Equipment Manufacturer (OEM). The pumps were also tested on a rig to check their performance against the manufacturers recommended settings. These values were recorded, and on rebuilding the pumps were set to the values recorded before the strip (not back to recommended settings).

3.3 Emissions tests

The start of test for each vehicle was defined as the point at which the run-in had been completed, and the injection pump had been inspected, rebuilt and refitted to the vehicle. Emissions tests were carried out at the following intervals: start of test, 10 000 km, 50 000 km (before the pump strip), 50 000 km (after the pump strip), 84 000 km and 100 000 km.

The vehicles were prepared for the emissions tests in accordance with the European passenger car emissions test cycle, the ECE15+EUDC test procedure. This procedure specifies that emissions sampling begins 40 seconds after the vehicle has been started. Each emissions measurement point consisted of three days testing. On each day an ECE15+EUDC test was run, in accordance with the test procedure, followed immediately by two more ECE15+EUDC tests. For three days testing, therefore, a total of three 'cold' tests and six 'hot' tests were used in the data analysis.

Emissions measured included: particulates, NOx, HC and CO. At each measurement point, free acceleration smoke tests were also carried out.

In between emissions tests, the vehicles were driven on the road to accumulate test kilometres. Each vehicle was driven through a prescribed test route, which incorporated some town driving, some motorway driving, and phases which gave a mixture of both types of duty.

4 TEST RESULTS

4.1 HFRR data

Figure 1 shows the HFRR results for the test fuels used in each vehicle. The line drawn at 450 µm denotes the wear scar diameter maximum limit proposed by ISO, one of the bodies active in developing an HFRR test method (2).

Fig 1 HFRR data for test fuels at 60°C

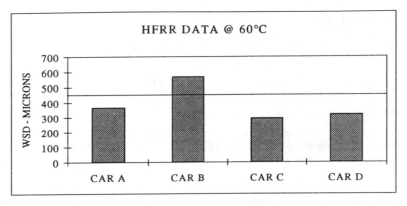

Figure 1 shows that against the proposed maximum limit set by ISO, the base LSADO with no additives (Car B) would fail the HFRR test, while all of the other fuels tested would pass.

4.2 Injection pump rating (using the injection pump OEM rating method)

When the fuel injection pumps were stripped at 50 000 km and 100 000 km, a rating method, devised by the injection pump OEM, was used to assess their condition. The method is intended for use on pumps which have had a life of 100 000 km, and so any interim rating is a forecast to predict the likely condition of the pump at 100 000 km.

The rating method is a subjective assessment of several key components of the pump. The rating is split into two components - sliding wear and vibration wear - to take into account the different types of wear regime that exist in the injection pumps. Figure 2 shows the rating of the injection pumps at 100 000 km. The line on the graph shows the limit identified by the OEM as acceptable performance.

Fig 2 OEM rating method

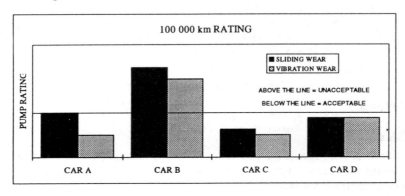

Figure 2 shows that for Car A, running on nominal 0.2% sulphur fuel, the vibration wear pump rating was at a level that would be acceptable to the injection pump OEM. The sliding wear rating was just at the acceptable level. Cars C and D (low sulphur fuel with two levels of lubricity additive treat rate) also exhibited acceptable performance. Car B, which ran on unadditised LSADO was the only one of the four vehicles not to show acceptable performance, in terms of this rating method.

4.3 Injection pump weight loss

The main pump components were weighed at the start of test, at 50 000 km and at the end of the test (100 000 km). The components included the cam plate, and the roller followers. Figure 3 shows the weight loss data for the four test cars over the 100 000 km test. While the numbers are small in absolute terms, the comparison of Car B (unadditised LSADO) with both the 0.2% sulphur vehicle (Car A) and the low sulphur vehicles with additives (Cars C and D) shows the excessive wear that is possible with poor lubricity fuels. The weight loss for Car B was measured at nine times greater than for Car A, the vehicle with the next highest weight loss figures.

Fig 3 Fuel injection pump weight loss data

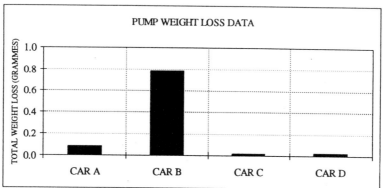

4.4 Injection pump maximum fuelling data

Each time the fuel injection pumps were stripped, checks were made against the manufacturer's specification to assess whether or not the pump was performing as intended by the manufacturer. The values of each check point were recorded and, on rebuilding, the pump was set to its 'as found' operation, not to the manufacturer's recommended settings.

Figures 4 to 6 show the maximum fuelling data for the pumps, at three pump speeds. The manufacturer's specification for fuel delivery at each speed is shown in the graph title, and as lines on the graph.

Fig 4 Injection pump maximum fuel delivery at 1250 rpm pump speed

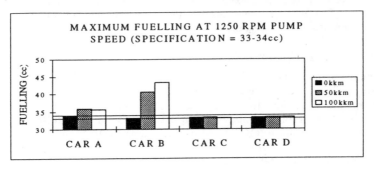

Fig 5 Injection pump maximum fuel delivery at 750 rpm pump speed

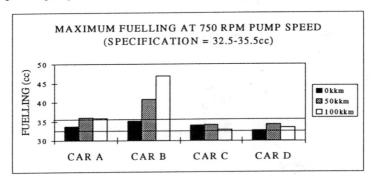

Fig 6 Injection pump maximum fuel delivery at 400 rpm pump speed

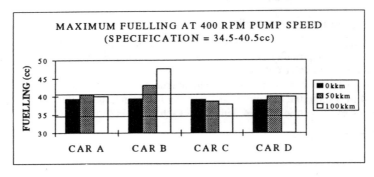

Figures 4 - 6 show that while Cars A, C and D remain within, or very close to, the manufacturer's recommended maximum fuelling rates, Car B showed an increasing maximum fuel delivery as the test progressed. By the end of test, the injection pump from Car B was grossly overfuelling at its maximum governor setting, over the speed range of the pump.

4.5 Emissions tests

4.5.1 Data analysis

The emissions data were analysed in two sets - 'cold' test data (the first ECE+EUDC test on each of the three test days) and 'hot' test data (the second two tests on each of the three test days). Each set was analysed using one-way (single factor) analysis of variance to compare emissions data at the start and end of test. This was used to test whether or not the vehicle emissions increased significantly across the duration of the test. The free acceleration smoke test data were analysed in the same way.

4.5.2 ECE+EUDC Emissions

A total of 3 'cold' tests and 6 'hot' tests were run at each measurement point. The two sets of emissions data were then analysed to check for differences between emissions at the start and end of test. Tables 2 and 3 show which of the emissions differences were significant, and the percentage change of the emissions over the 100 000 km test.

Table 2 Comparison of 'cold' emissions from start (0kkm) to end (100kkm) of test

		COMPARE 0kkm AND 100kkm COLD EMISSIONS		
		Significant at 95%?	Δ g/km	% Increase (+) or Decrease (-)
CAR A	Pm	non significant	0.00347	5.92
CAR B	Pm	non significant	0.00383	7.39
CAR C	Pm	SIGNIFICANT	-0.00630	-11.87
CAR D	Pm	SIGNIFICANT	-0.00787	-15.58
CAR A	NOx	non significant	-0.03793	-6.81
CAR B	NOx	non significant	0.03737	7.11
CAR C	NOx	SIGNIFICANT	-0.06347	-12.82
CAR D	NOx	non significant	-0.03633	-7.94
CAR A	HC	non significant	0.00773	26.70
CAR B	HC	non significant	0.01237	40.77
CAR C	HC	SIGNIFICANT	0.02523	63.40
CAR D	HC	non significant	0.00053	1.27
CAR A	CO	SIGNIFICANT	0.07660	39.99
CAR B	CO	non significant	0.02437	17.14
CAR C	CO	SIGNIFICANT	0.07520	31.36
CAR D	CO	SIGNIFICANT	0.05467	23.30

The 1996 emissions limits for new IDI light duty vehicles are detailed in (3). In summary, the limits are:

CO: 1.0 g/km,   HC+NOx: 0.7 g/km,   Pm: 0.08 g/km

Table 3 Comparison of 'hot' emissions from start (0kkm) to end (100kkm) of test

COMPARE 0kkm AND 100kkm HOT EMISSIONS			
	Significant at 95%?	Δ g/km	% Increase (+) or Decrease (-)
CAR A　Pm	non significant	-0.00075	-1.24
CAR B　Pm	non significant	-0.00288	-5.50
CAR C　Pm	SIGNIFICANT	-0.00488	-9.40
CAR D　Pm	SIGNIFICANT	-0.01153	-21.40
CAR A　NOx	non significant	0.00353	0.63
CAR B　NOx	non significant	0.01380	2.67
CAR C　NOx	SIGNIFICANT	-0.04810	-9.60
CAR D　NOx	SIGNIFICANT	-0.06610	-13.64
CAR A　HC	SIGNIFICANT	0.01343	94.27
CAR B　HC	SIGNIFICANT	0.00943	54.32
CAR C　HC	SIGNIFICANT	0.01752	69.46
CAR D　HC	SIGNIFICANT	0.01558	106.49
CAR A　CO	SIGNIFICANT	0.06925	52.49
CAR B　CO	non significant	0.01903	21.14
CAR C　CO	SIGNIFICANT	0.07863	48.79
CAR D　CO	SIGNIFICANT	0.10385	75.35

The 1996 emissions limits for new IDI light duty vehicles are detailed in (3). In summary, the limits are:

CO: 1.0 g/km,　　HC+NOx: 0.7 g/km,　　Pm: 0.08 g/km

It should be noted that, relative to the legislated emissions limits for new cars, the absolute levels of CO were around one third of the required levels. HC levels were also extremely low. This is why some of the percentage changes for CO and HC shown in Tables 2 and 3 are large.

4.6 Free acceleration smoke test emissions

Black smoke is the visible portion of particulate emissions, and can be measured by assessing how much light the smoke obscures when passed between a light source and a detector. Smoke tests were carried out using a Bosch T100 smoke meter, of the type used in Ministry of Transport (MoT) tests. This automatically determines the number of accelerations to be made, depending on whether an MoT-type 'pass' or 'fail' is registered. The last four accelerations were used to produce a mean black smoke number. This test was then repeated to produce two mean smoke numbers. These two numbers were then averaged to produce the data points shown in Figure 7. The two values of smoke shown at 50 000 km represent measurements taken before the pump strip (50kkm (1)) and after the pump strip (50kkm (2)).

Fig 7 Free acceleration smoke test data

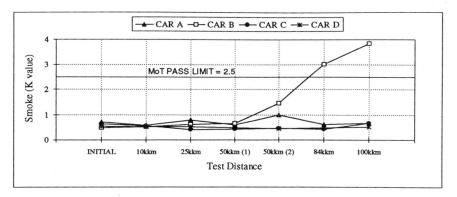

Figure 7 shows that for Car B - the vehicle using LSADO with no additives - the free acceleration smoke increased over the test period. Statistical analysis showed the difference between the smoke data at the start of test and at 100 000 km was significant with 95% confidence. Significant changes in the smoke levels for Cars A, C and D were not seen.

5 DISCUSSION

Data were generated from several sources during the vehicle test, in order to build up a total picture of the effects of LSADO, and to try and understand what was causing the effects. The data can be summarised on a car-by-car basis.

Car A - 0.2% sulphur base fuel with detergent package but no lubricity additive. The data for this vehicle show what can be surmised from the fact that for many years, 0.2% sulphur fuel has been used throughout Europe, with no adverse effects being seen due to fuel lubricity. The HFRR test results show that, according to the ISO recommended limit, the lubricity of the fuel is acceptable. The pump rating using the OEM method also shows acceptable wear at 100 000 km. The weight loss from the pump was at a low level, and the maximum fuelling data shows that, over 100 000 km, the fuelling rates can increase to just above the manufacturer's recommended levels. Most of the emissions gave no significant changes; black smoke changes were also non significant.

Car B - LSADO with no additives. The LSADO failed the HFRR test, according to the ISO recommended maximum limit for wear scar diameter. The pump rating to the OEM method also showed that excessive wear had occurred in the pump, which would lead to a reduced pump life. The weight loss from the pump components was nine times greater than for Car A, and the maximum fuelling was out of specification by the 50 000 km point in the test, and was even worse by the end of test. There were no significant changes in most of the regulated emissions, but black smoke increased significantly.

Cars C and D - LSADO with detergent additive package plus lubricity additive (Car C = high treat rate of lubricity additive, Car D = low treat rate of lubricity additive). Both vehicles had acceptable HFRR wear scar diameters, and good pump ratings. The rating of the pumps ranked the vehicles in a logical order - with the higher treat rate of lubricity additive showing less wear than the lower treat rate. The maximum fuelling data show that the pumps stayed within the manufacturer's specification for the duration of the test. Both vehicles showed significant decreases in some of the emissions, but black smoke did not change significantly.

These data demonstrate that the use of a LSADO with poor lubricity, as determined by the HFRR test, can cause damage to fuel injection pumps. By the end of the test, the drivers of the Car B (base LSADO) were reporting that the levels of smoke under loaded accelerations, for example driving up a hill in fourth gear, were noticeably increased.

One of the pump components which was particularly worn in the pump from Car B was the governor butting ring, which helps to control the maximum fuelling of the injection pump. The maximum fuelling data on the pump rig reinforces the theory that the wear in this pump manifested itself at maximum rack settings in the pump, i.e., under 'foot to the floor' operation. This is further supported by the free acceleration smoke test data which are the result of maximum rack travel under idling conditions.

While the ECE+EUDC emissions tests showed some significant changes between the start and end of test, greater changes were expected, given the levels of smoke seen from Car B. The ECE+EUDC emissions test cycle is relatively lightly loaded and does not simulate full load operation. As a result of this, a full load acceleration test has been developed for use in future vehicle tests, to further quantify emissions changes as a result of excessive pump wear.

One of the objectives of the test was to establish whether or not the HFRR test could predict the performance of a fuel in a vehicle. The test data show that the HFRR can discriminate between a good lubricity fuel and a poor lubricity fuel, although the discrimination between the three fuels shown as having good lubricity in the HFRR is not so distinct. Figures 8 and 9 below show how the HFRR correlates with the two components of the OEM rating - sliding wear and vibration wear. The dotted lines show the 95% confidence band for the correlation. The graphs were the result of simple linear regression analysis of the HFRR data and the pump rating data.

Fig 8 Correlation between HFRR and sliding wear rating

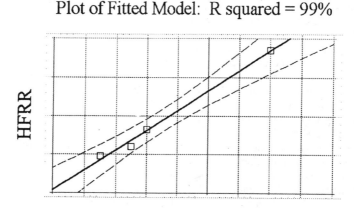

Fig 9 Correlation between HFRR and vibration wear rating

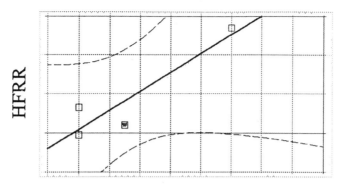

The correlation for vibration wear is not as good as for sliding wear, and this can most likely be explained by the fact that the critical components in this type of injection pump are the cam plate and roller followers. The roller followers roll and slide over the cam plate, with less vibrational movement. It is therefore logical that there is a better correlation between the HFRR results and sliding wear.

HFRR tests have also been carried out on nominal 0.2% sulphur fuels. This was done to ensure that the extent of any lubricity problems was fully understood. Figure 10 shows some examples of the HFRR results for 0.2% sulphur fuels, sampled from across Europe.

Fig 10 Examples of HFRR results for 0.2% sulphur diesel fuels

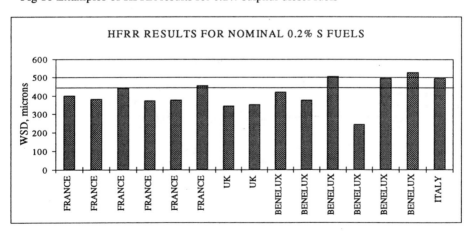

Figure 10 shows that, in some areas of Europe, 0.2% sulphur fuels exceed the proposed ISO limit of 450µm for LSADO. There have been no reports of injection pump failures attributed to the poor lubricity of these fuels. The majority of the fuels are, however, below the 500µm level, and this highlights the need to understand fully the issues associated with LSADO, before acceptable limits are finalised and included in a specification. Further work is needed to identify the effect of a wide range of base LSADOs in vehicles, as well as in rig tests.

Further vehicle tests are being carried out at ERCA, to ensure that fuels from a wide range of refineries can be treated, as necessary, to ensure acceptable lubricity properties.

6 CONCLUSIONS

The following conclusions are drawn from the vehicle test reported here:

- LSADO with failing HFRR wear scars can cause excessive wear in some types of rotary passenger car fuel injection pumps.
- Additive solutions are available to maintain the lubricity at levels seen with 0.2% sulphur fuels. The vehicles in this test which used additised LSADO showed acceptable performance in the critical tests, including the subjective pump rating and black smoke emissions.
- The HFRR test predicts diesel fuel lubricity and has been shown to correlate with the rating method used by the fuel injection equipment OEM to assess the condition of pumps used in vehicles.
- Excessive wear in this type of fuel injection pump governor mechanism has been shown to result in increased maximum fuel delivery rates. This manifests itself primarily by increased black smoke emissions under high load operation.

7 REFERENCES

1. 'ISO Diesel Fuel Lubricity Round Robin Program', SAE 952372, Manuch Nikanjam et al.

2. ISO/TC22/SC7 DIS 12156 Diesel Engines - Diesel fuel lubricity - part 1: Test Method, part 2: Performance Requirements (proposed ISO limit contained in part 2).

3. 'Motor Vehicle Emission Regulations and Fuel Specifications in Europe and the United States, 1995 Update'. CONCAWE report 5/95, prepared for the CONCAWE Automotive Emissions Management Group by its Special Task Force, AE/STF-3, page 50. Table 3.3.

Fuel Technology for Diesel Engines

517/013/96

Feasibility study towards the use of copper fuel additives for particulate reduction from diesel engines

P A NEEFT, M MAKKEE, and J A MOULIJN
Delft University of Technology, Delft, The Netherlands
A SAILE CEng, MIMechE, MSAE
Lubrizol International Laboratories, Derby, UK
A WALKER MA, MIMechE
Consultant, Peterborough, UK

SYNOPSIS

The use of copper fuel additives results in a higher catalytic activity for soot oxidation compared with the use of powdered copper oxide catalysts as a model for catalytic coatings on particulate filters. Diesel engine exhaust tests revealed that with the use of copper fuel additives, constant temperatures of 365 - 435°C are sufficient to establish an equilibrium between soot collection and soot oxidation. When a copper oxide coating on a filter is used, significantly higher temperatures are needed to burn the accumulated soot from the filter.

This study shows that copper fuel additives have a high enough activity and are in principle suitable for application in practice. Explanations for this high activity are given.

1 INTRODUCTION

Metal based fuel additives have been extensively studied and used for the reduction of carbonaceous emissions from combustion processes. The earliest applications dealt with additives that are thought to prevent the formation of soot. Calcium and barium have been reported to be efficient additives in this respect (1,2). It has been found that also other fuel additives decrease soot emissions (1). Although the soot oxidation rate is thus increased by a large number of fuel additives, it is only moderate and consequently calls for extended residence times in the exhaust pipe. Moreover, due to the heat production during oxidation the rates are increased with increasing mass of soot present. The use of filters in combination with such fuel additives is, therefore, a logical choice.

Catalysts for the oxidation of diesel soot can be applied in a number of ways. Dissolution of a precursor material in the diesel fuel (the fuel additive) is one way. Another way is the use of a catalytic coating on the filter, so that the soot, collected on the layer of catalyst, is oxidized at a higher rate compared with soot, collected on a filter without a catalytic coating.

An important difference between catalytic coatings and catalytic fuel additives is that the fuel additive has to be supplied continuously. This has a number of consequences. In the first place, the fuel additive must be added to the fuel, so that for practical applications the logistics of additive distribution must be taken care of. Secondly, the fuel additive must not deteriorate the functioning of the diesel engine, *e.g.* by changing the fuel properties or by formation of inorganic deposits. In the third place, it can only be used in a low concentration as, after combustion of the soot, the metals of the additive form ash which accumulates in the filter. Nevertheless, the oxidation activity of soot, generated from fuel containing such low additive concentrations, has been found to be satisfactory at average exhaust gas temperatures (3-5).

A large number of fuel additives have been studied. Organometallic compounds of Ce, Cu, Fe, Mn, and Pb have been found to be very active (6). The exact reasons for the high activity of these additives is, however, not known. It is self-evident that the contact between the two solids, soot and catalyst, is very important. This contact between catalyst (formed by combustion of the fuel additive) and soot, and the morphology of the catalyst has only been investigated in a limited number of studies (7,8).

The main objectives of this study were to explore the activity of a copper diesel fuel additive, and to make a comparison between the activity of this additive and a catalytic copper coating. Oxidic copper catalysts were selected as (*i*) copper is intrinsically active (9); (*ii*) copper oxide has been previously explored in detail for use as a catalytic coating (6,10); and (*iii*) copper based fuel additives have been studied relatively often, including the assessment of their performance in full-scale engine tests (3,4,11). A further objective of this study was to compare the performance of two different types of wall-flow monolith (which act as filters, *e.g.*, (6)).

2 EXPERIMENTAL

2.1 DIESEL ENGINE AND COLLECTION OF DIESEL PARTICULATES

A one-cylinder, directly injected, naturally aspired 4 kW Yanmar L90E diesel generator set was used (bore 84 mm, stroke 70 mm, swept volume 0.39 l). The diesel fuel was a reference fuel (CEC-RF-03A-84) with a cetane number of 49. It contained 0.27 wt% sulphur and 27 vol% aromatics. The lubricating oil was a commercially available oil (SAE 10W-40). A copper additive was mixed with the reference diesel fuel to obtain a concentration of 50, 100 or 200 ppm copper on a weight basis.

A soot collection set-up was mounted to a side-stream of the diesel engine exhaust pipe line. The set-up is depicted in Figure 1. It consisted of a heated transfer line, an oven in which a segment of wall-flow monolith could be installed, a filter to check for leaks in the wall-flow monolith or its adhesion to the quartz tube on to which the wall-flow monolith was mounted, a flow meter, a pressure-insensitive needle valve and a pump. Figure 2 shows how the segment

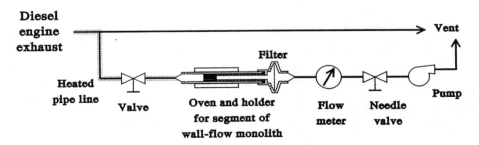

Figure 1 *Flow scheme of experimental set-up. Details of the oven plus holder are shown in Figure 2*

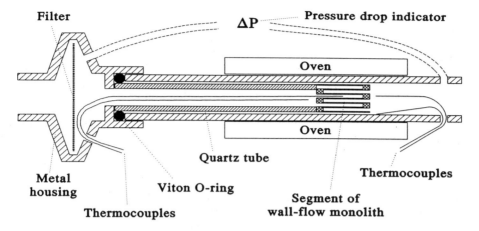

Figure 2 *Detail of experimental set-up, showing the oven plus holder. Collected data are the temperature and pressure drop signals*

of wall-flow monolith was installed in the 24 mm (internal diameter) stainless steel oven. The segments were adhered onto a quartz tube with a ceramic, alumina based glue (Ceramabond 569). The quartz tube was sealed to the cold end of the stainless steel tube by a Viton O-ring. Thermocouples were installed both upstream and downstream of the trap. One thermocouple was placed in the central channel opening at the downstream end of the segment, at half the length of the segment. Thin thermocouples (0.5 mm diameter) were used to assure a fast temperature response. A pressure difference cell (range 475 mbar, accuracy 0.5 mbar) was installed to measure the pressure drop over the segment.

Prior to each experiment, the diesel engine was run for at least an hour to warm up the engine and all exhaust pipe lines. Soot loading experiments were performed under isothermal conditions. At the end of the experiment the oven temperature was raised linearly in order to provoke ignition of the accumulated soot.

Pressure drops that are presented in this paper are all corrected for baseline pressure drop, which is the pressure drop of the clean monolith at the collection temperature when pumping 12 litres of air (without soot) per minute through the trap. In this way, the net pressure drop due to accumulated particulates is obtained. The *equilibrium pressure drop* (P_{eq}) is the steady state pressure drop (corrected for baseline) that is measured in the course of an isothermal experiment and was used as a measure of the activity of the additive and coating.

2.2. SEGMENTS OF WALL-FLOW MONOLITH

The segments of wall-flow monolith had dimensions of 40x20 mm (length x diameter). Two types of wall-flow monolith were used: cordierite ($2\ MgO - 2\ Al_2O_3 - 5\ SiO_2$) and silicon carbide (SiC). The cordierite segments were cut from large wall-flow monoliths. The alternate channels were closed using the ceramic glue (Ceramabond 569). Silicon carbide traps were obtained directly from Stobbe Engineering.

Segments of wall-flow monolith were coated with copper oxide. Two methods of impregnation were used, and the copper concentration was also varied, yielding three different copper coatings A, B, and C. Details on this impregnation are given elsewhere (12).

2.3. OTHER EQUIPMENT

An STA 1500H thermobalance was used for combined thermogravimetrical analysis (TGA) and differential scanning calorimetry (DSC). Experiments were performed with diesel soot obtained with 50 ppm (wt/wt) Cu in the fuel ('diesel soot+additive') and with a commercial model soot Printex-U in tight contact with CuO ('CuO/Printex-U'). Details about catalyst preparation, tight contact, collection of diesel particulates, TGA/DSC sample preparation, and experimental conditions for TGA/DSC can be found elsewhere (9,12).

Transmission electron microscopy (TEM) was performed on a Philips CM-30 at 300 kV. Energy dispersive analysis of X-rays (EDX) was used to determine the elemental composition of soot particulates. For this purpose, an aluminum grid was used instead of a copper grid.

X-ray diffraction (XRD) was performed on an FR 552 Guinier camera. Cu $K\alpha_1$ radiation was used. Diffractograms were recorded at room temperature using long exposure times to obtain a high sensitivity.

X-ray fluorescence (XRF) was performed on a Philips PW 1400 (wavelength dispersive) spectrometer. As no calibration samples were used, matrix effects could not be corrected for, nor could account be taken of them, and the results must be considered to be semi-quantitative.

2.4. KISSINGER PLOTS

The oxidation activity of diesel soot+additive was compared with that of CuO/Printex-U. This comparison was made by measuring the DSC peak temperatures (T_{max}) as a function of heating rate β. The DSC rather than the DTG (Differential Thermogravimetrical Analysis) peak maximum was used as the DTG signal contained more noise and the DTG maximum was difficult to determine, particularly at a low heating rate. By plotting $\ln(\beta/T^2_{max})$ against $1/T_{max}$, so-called Kissinger plots are obtained, in which the slope of the curve can be predicted to be E_a/R (13). It is assumed that (i) the temperature dependency of the reaction rate follows the Arrhenius equation; and (ii) at peak maximum the fraction of converted soot is constant.

3 RESULTS

3.1 ACTIVITY OF Cu-ADDITIVES IN COMPARISON WITH CuO CATALYSTS IN TIGHT CONTACT MODE

In Figure 3, a comparison is shown between combustion experiments of diesel soot+additive and CuO/Printex-U at heating rates of 2.5 and 40°C min^{-1}. The activity of the diesel soot+additive is comparable to that of the CuO/Printex-U in tight contact. The peaks shift to higher temperatures at higher heating rates. The peak shape of the combustion profiles of the CuO/Printex-U sample was found to be very broad at high heating rates (≥ 30°C min^{-1}). At low heating rates the DSC peak shapes are similar for CuO/Printex-U and diesel soot+additive.

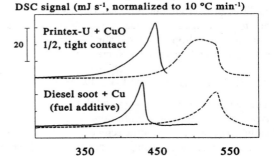

Figure 3 DSC patterns of CuO/Printex-U (2/1 tight contact) and soot+additive (50 ppmw in fuel) (———): 2.5°C min^{-1}; (- - - -): 40°C min^{-1}

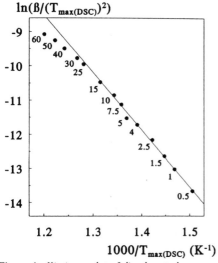
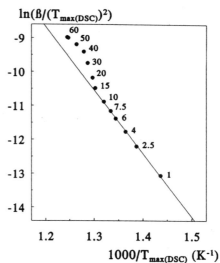

Figure 4 *Kissinger plot of diesel soot plus copper additive at heating rates of 0.5 - 60 K min⁻¹ (heating rates are shown). The line is a fitted curve*

Figure 5 *Kissinger plot of CuO/Printex-U (2/1, tight contact) at heating rates of 1 - 60 K min⁻¹ (heating rates are shown). The line is a fitted curve*

The Kissinger plots of diesel soot+additive and CuO/Printex-U are shown in Figures 4 and 5, respectively. At the low temperature end, the points could be fitted with a straight line, which is also shown in the figures. The activation energies that were determined from these fits were 140 ($\pm$ 5) kJ mol^{-1} for the diesel soot+additive, and 156 ($\pm$ 3) kJ mol^{-1} for the CuO/Printex-U tight contact mixture. The 95% confidence intervals of these two activation energies overlap. The Kissinger plot of diesel soot+additive has a lower slope at higher heating rates. The slope in the Kissinger plot for CuO/Printex-U increases at intermediate heating rates, goes through a maximum and then decreases at the highest heating rates.

3.2 ESTABLISHMENT OF EQUILIBRIUM PRESSURE DROPS IN ENGINE TESTS

3.2.1 Baseline pressure drops and temperatures

The baseline pressure drops varied from segment to segment, and were higher for the cordierite segments (25 - 50 mbar at 365°C) compared with the silicon carbide segments (15 - 25 mbar at 365°C). Some of the copper coatings were found to increase the baseline pressure drop significantly (to 60 - 85 mbar at 365°C for cordierite segments). Baseline pressure drops were found to increase as a function of temperature.

Temperatures reported in this paper are the temperatures measured within the segment under isothermal conditions, unless otherwise mentioned. These temperatures were found to be close to the temperatures upstream and downstream of the segment, but lower than the temperatures of the oven wall.

3.2.2 Wall-flow monolith material and design

The type of the wall-flow monolith segments was found to have some influence on the curves of pressure drop versus time. In Figure 6 the pressure drop curves are shown for silicon carbide and cordierite segments and fuel copper levels of 0 and 100 ppm at a temperature of 365°C. The use of fuel without copper additive resulted in pressure drop curves that did not reach an equilibrium. Instead, the pressure drop increased up to the maximum value of the pressure difference cell without levelling off. The increase in pressure drop was faster for the cordierite segment than for the silicon carbide segment, which was mainly caused by a steep increase in the initial half hour for the cordierite segment. When diesel fuel was used which contained 100 ppm copper, the pressure drops became constant after a few hours. P_{eq} seemed to be somewhat higher for the cordierite segment compared with the silicon carbide segment, though it is not certain if this difference was significant.

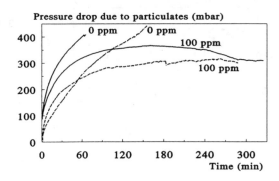

Figure 6 *Soot loading curves at 0 and 100 ppm fuel copper level. Cordierite (———) versus silicon carbide (- - - -) traps at 365°C*

3.2.3 Additive fuel content

The influence of the concentration of copper in the fuel on the pressure drop curves of cordierite segments is illustrated in Figure 7. These experiments were also performed at 365°C. The curve at 0 ppm and one of the two curves at 100 ppm copper are the same as those shown in Figure 6. At a copper fuel level of 100 ppm a P_{eq} of about 300 - 350 mbar was found. An increase in copper fuel level to 200 ppm decreased this P_{eq} to about 120 mbar. The pressure drop was found to go through a maximum of 200 mbar to decrease afterwards. Such a maximum was also observed for one of the experiments at 100 ppm. This maximum was, however, less pronounced compared with the maximum at 200 ppm.

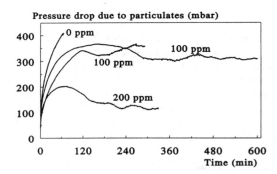

Figure 7 *Influence of fuel copper content on the soot loading curves. Cordierite trap, 365°C*

A number of runs were performed when running the engine on additive-free diesel fuel. It was found that it took a very long time to clear the additive from the exhaust system. Soot, collected one day after changing the fuel from 100 to 0 ppm copper, was found to be still considerably more active than soot which was collected after three days of operation on fuel without the additive. Combustion temperatures, which were determined in the thermobalance, confirmed these observations. The soot which was collected after three days of operation had a combustion temperature similar to that of soot, collected at a time the engine had not yet been run on fuel containing the copper fuel additive. The experiments at 0 ppm copper in the fuel, which are shown in the Figures 6 and 7, were performed after these three days.

3.2.4 Temperature

The influence of temperature on P_{eq} is illustrated in Figure 8 for silicon carbide segments of wall-flow monolith. At increasing temperatures P_{eq} decreases and is reached earlier, and the soot loading curves show a more pronounced maximum prior to reaching P_{eq}. In this figure the different forms of experimental scatter that were found are also illustrated. The pressure drop was often found to fluctuate strongly (curves at 405 and 425°C), the pressure drop during equilibrium often varied some 30 to 40 mbar (405, 425°C), the equilibrium pressure drops could vary up to the same 30 to 40 mbar when experiments were repeated (365°C), and the initial form of the curves was found to fluctuate very strongly when experiments were repeated (365°C). It should be noted that the total mass flow through the segments is the same at each temperature. As a result, higher temperatures result in higher pressure drops for identical segments and particulate loadings.

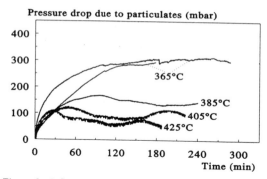

Figure 8 *Influence of temperature on the soot loading curves. Silicon carbide trap, 100 ppm Cu*

For cordierite traps a similar decrease in P_{eq} was found when increasing the temperature. At a temperature of 425°C, however, another observation was made which is shown in Figure 9. The pressure drop reached a maximum and then showed a more pronounced decrease compared with experiments performed at lower temperatures. After reaching a constant value, the temperature in the trap suddenly increased while the pressure drop decreased: a spontaneous regeneration of the trap occurred. This pattern of soot loading and auto-regeneration was found to repeat. The occurrence of a regeneration seemed to be a rather random process as the periods between successive regenerations varied widely.

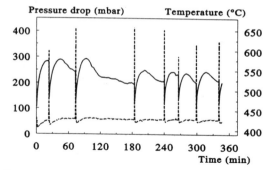

Figure 9 *Soot loading curve for a cordierite segment at 425°C and 100 ppm Cu. (——): Pressure drop over segment; (- - - -): Temperature within segment*

3.2.5 Presence of a catalytic coating

Three Cu-coatings were employed on cordierite segments to study their influence on the soot loading curves. Results are shown in Figure 10. The two curves without a catalytic coating had similar P_{eq}'s. All coatings on the cordierite trap decreased P_{eq}, and also significant differences in P_{eq} between the three coatings were found.

The use of Cu-coatings in combination with fuel without the copper fuel additive resulted in pressure drop curves without an equilibrium, the pressure drop rose to its maximum of 475 mbar at 365°C. This pressure drop curve resembled the blank curve (0 ppm copper additive, cordierite) shown in Figures 6 and 7.

3.3 REGENERATIONS OF PARTICULATE LOADED TRAPS

Runs in which P_{eq} was measured were normally ended by a temperature ramp, during which the accumulated soot ignited. Two typical examples of such regenerations of segments are

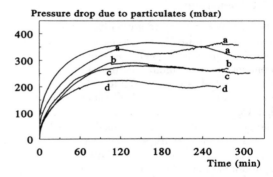

Figure 10 *Effect of Cu-coatings on soot loading curves. Cordierite, T = 365°C, 100 ppm Cu in the diesel fuel. a: no coating; b: Cu-coating A; c: Cu-coating B; d: Cu-coating C*

shown in Figures 11 and 12 for a non-coated cordierite segment and a non-coated silicon carbide segment, respectively. The temperature of onset of ignition was found to be independent of the presence of a catalytic coating, and was consistently at 410 ± 10°C. The segment material did not influence this onset temperature. Some scatter was found due to the high heating rates and the noise on the pressure drop signal. Regenerations of cordierite segments were found to go always hand in hand with very high temperatures within the segment. Temperatures as high as 1300°C were occasionally measured. The temperatures downstream of the segments were always lower, and did not exceed 750°C. During regenerations of the silicon carbide segments the temperatures were lower than in the cordierite segments, the maximum temperatures inside and downstream of the segments were 625 and 600°C, respectively. No correlations could be found between a number of parameters, such as temperatures in or downstream of the segments during regenerations, the pressure drop before the regeneration (the amount of soot present), or the presence of a catalytic coating.

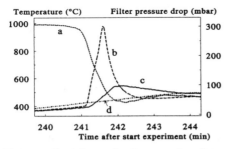

Figure 11 *Typical example of regeneration of a cordierite trap (100 ppm Cu) a: Pressure drop; b: T in trap; c: T after trap; d: T oven*

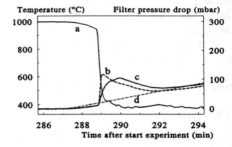

Figure 12 *Typical example of regeneration of a silicon carbide trap (100 ppm Cu) a: Pressure drop; b: T in trap; c: T after trap; d: T oven*

3.4 CHARACTERIZATION OF SOOT PLUS ADDITIVE

The physical appearance of the soot which contained copper from the fuel additive (100 ppm) was studied with TEM. A typical image of the soot is shown in Figure 13. The appearance of the soot is very similar to soot without additive (micrographs shown elsewhere (12)): it consists of elementary spheres with diameters of 36 (± 9) nm, which are clustered together to form agglomerates with dimensions in the order of several hundreds of nanometres.

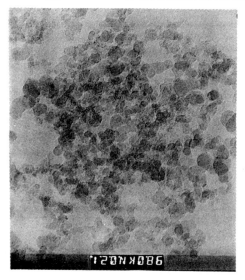

———— : 100 nm   ———— : 100 nm

Figure 13 *TEM micrograph of a soot particulate formed from diesel fuel containing 50 ppm Cu. Typical of an example of particulate showing hardly any contrast*

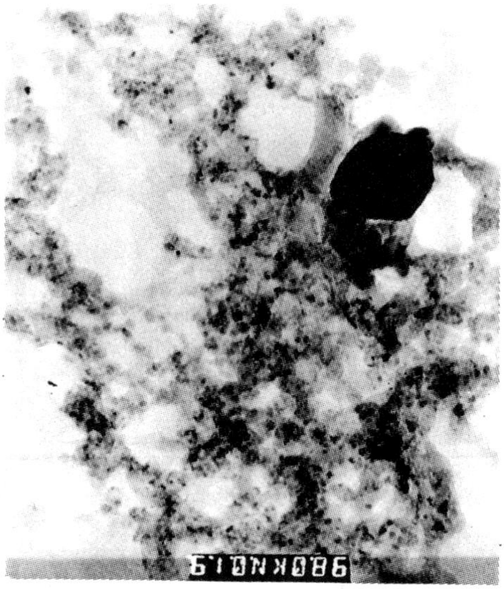

Figure 14 *TEM micrograph of a soot particulate formed from diesel fuel containing 50 ppm Cu. Typical of an example of particulate in which much contrast was observed*

With EDX, the elementary composition of such agglomerates was studied. It was found that the soot particulates as shown in Figure 13 contain copper, as the EDX signal was much higher than the EDX signal from the grid plus carbon film on a location where no particulates were present. The electron diffraction patterns of the sample at a location of a soot particulate did not show spots due to interference of diffracting electrons from crystalline material.

Among the large number of soot particulates that were observed, a few were found that contained material with a high contrast. An example is shown in Figure 14. With EDX, high copper concentrations were found in these particulates. The diffraction patterns also indicated the presence of crystalline material, probably of copper oxide.

With XRD no diffraction pattern was detected for the same soot sample.

3.5 COPPER CONTENT OF THE COATED SEGMENTS OF WALL-FLOW MONOLITH

The amount of copper on segments of wall-flow monolith was determined by XRF, results are shown in Table I. No matrix effects were taken into account, therefore, the data must be considered as semi-quantitative. The relative concentrations among the samples is determined more accurately (the reproducibility is estimated to be ± 10%). The copper coatings contain much more copper than blank segment which was used for particulate collection at a 200 ppm Cu fuel level for 7 hours (4.1 m^3 of exhaust gases). Using data on exhaust gas flow rate and fuel consumption of the Yanmar diesel engine (operating at a load of 3 kW), the amount of copper from the additive present in this 4.1 m^3 of exhaust gas can be calculated to be 35 mg Cu. This amount is higher than the 0.1 wt% in the segments (typical weight 7 g, thus containing about 7 mg Cu) which was found by XRF.

Table I   XRF analysis of blank and copper coated segments of wall-flow monolith

Element (wt%)	Cu	Mg	Al	Si
Blank	bdl[1]	9.2	48.9	38.9
after 4.1 m^3 exhaust gas, 200 ppm Cu in fuel[2]	0.1	7.6	52.4	35.3
copper coating A	17.3	8.5	42.4	29.0
copper coating B	12.1	8.2	45.3	30.6
copper coating C	5.3	7.6	50.2	33.1

[1]: below detection limit
[2]: The segment was regenerated prior to analysis

4 DISCUSSION

4.1 COMPARISON WITH RESULTS REPORTED FOR FULL-SCALE ENGINE TESTS

It is not straightforward to compare results from this study with results reported in literature dealing with engine studies in which the same copper fuel additive was used (3,4). The reason is not only that in these engine studies particulate trap pressure drops are measured under transient conditions, but also that a number of dimensions and conditions of operation of engine and particulate trap differ compared with this study. These differences are: (*i*) the Yanmar diesel engine that was used has much higher particulate emissions than modern heavy-duty engines (about a factor of 5 to 6 at the air-to-fuel ratios used (14)); (*ii*) as a result of the higher particulate emission, the copper fuel content is chosen to be 100 ppm. This is higher than the 10-50 ppm used in practice (3,4); (*iii*) the size of the segments of wall-flow monolith is relatively small in this study. It is estimated that the ratio of size of the segment to exhaust gas flow through the segment is a factor 1.5 to 3 smaller compared with traps under practical conditions (12).

These differences imply that the pressure drop increase in time is faster and P_{eq} is higher compared with practical conditions. The ratio is estimated to be a factor 3 to 10. These higher particulate collection rates are practical from an experimental point-of-view because most experiments can be performed within a time span of one working day.

Although it thus cannot be assessed from this study how copper fuel additives perform under full-scale conditions, the activity found in this study is sufficiently high that it is concluded that copper fuel additives have a high feasibility for application in practice. In service experience (3,4) confirms the use of the copper fuel additive to be effective in line with these studies.

4.2 TIGHT CONTACT ACTIVITY VERSUS ACTIVITY OF FUEL ADDITIVES

Under the conditions of this study, the activity of the soot containing the additive is very high. The activity is even somewhat higher than the activity of copper oxide in tight contact with model soot (which oxidizes at similar temperatures as diesel soot (15)). The concentration of copper in the two soots largely differs: 200 wt% CuO in Printex-U against 2.9 wt% CuO in diesel particulates (calculated from fuel consumption and particulate emission (14)). Therefore, normalized to the amount of catalyst the soot+additive is orders of magnitude more active than the CuO/Printex-U.

4.3 INTERPRETATION OF KISSINGER PLOTS

From the Kissinger plots it is concluded that the activation energies for the soot+additive and the CuO/Printex-U were found to be the same. Besides, these activation energies are also similar to the activation energies for tight and loose contact, CuO catalyzed oxidation of model soot (12).

The points in the Kissinger plots in Figures 4 and 5 deviate from the linear plot at high heating rates. The main cause for this deviation is likely to have been a depletion of oxygen in the TGA sample cup. For CuO/Printex-U, also reduction of CuO to Cu_2O or Cu by the Printex-U is thought to play a role, as is discussed in more detail elsewhere (12).

4.4 EFFECT OF FILTER MATERIAL AND FILTER DESIGN

The filter characteristics of cordierite and silicon carbide filters were found to be different. The back pressure of the cordierite traps rose steeply in the first minutes of soot collection. This phenomenon could be explained from the texture of the cordierite which contains a small number of larger pores (16). It can be imagined that these pores will get clogged up rapidly after the start of particulate collection, resulting in a fast increase in back pressure. After this initial build-up in the filter material, soot deposition will probably proceed in layers, and the pressure rise will be slower. The pores in the SiC trap are more uniform (16,17), and particulate collection can be imagined to occur uniformly on the surface, also at the start of the experiment. These results make the design of the SiC wall-flow monoliths superior in terms of back pressure, compared with the design of the cordierite wall-flow monoliths. Similar results have been reported in literature (16-18). The regeneration behaviour of SiC is also superior compared with that of cordierite, because temperature differences inside the trap are smaller, thus leading to less thermal stress, which has also been corroborated by other studies (18,19). The poor heat dissipation of cordierite, relative to SiC, explains the occurrence of spontaneous regenerations of cordierite traps at 425°C (Figure 9), whereas these ignitions did not occur for the SiC traps. On the basis of these results it is concluded that SiC is an attractive trap material, especially in view of the problems which have been encountered using cordierite in practice (6). However, SiC is a very new material which may have its own problems, for example the formation of small cracks has been reported, which diminish the trapping efficiency (16,18). It is, therefore, too early to express a preference for either of the two filter materials. More study is needed with larger traps as the heat dissipation from the relatively small traps that were used in this study cannot be compared with the heat dissipation in full-size, insulated and canned traps in practice.

4.5 ACTIVITY OF CATALYTIC COATINGS

The use of a catalytic coating, in addition to the use of copper fuel additives, was found to have a positive effect on the equilibrium pressure drop. A catalytic coating alone, was, however, found to have a low activity compared with the fuel additives. The poor contact between a catalytic coating and the accumulated particulates must be held responsible for this low activity, as has also been described elsewhere (9,15).

4.6 DISPERSION OF THE ADDITIVE IN THE SOOT PRODUCED

The soot oxidation rate is not a function of the amount of copper present in the segments, but rather a function of the dispersion of the copper and the contact between copper and soot. The TEM/EDX experiments show that the copper is dispersed on a (sub)nanometre scale in the soot particulates. Electron diffraction as well as XRD would have shown a diffraction pattern if significant amounts of crystalline copper particles had been present. It can, therefore, be concluded that the copper is present in an amorphous state, or that copper particles are not larger than about 1 nm. These findings are corroborated by results published by Li (20), who also used EDX to show that CuO was present in diesel particulates, and TEM to show that the CuO was very well dispersed (invisible in TEM).

These results show that the copper or copper oxide particles that are formed from the copper fuel additives after combustion of the diesel fuel are smaller compared with the metal or metal oxide particles from combustion of diesel fuel containing iron, lead or platinum fuel additives. Sizes of such particles have been reported to be in the range of a few to 25 nm (7,8), the only exception being Li (20) who reported iron particles to be also very small (undetectable by TEM). These findings explain why copper fuel additives (and the iron fuel additives of Li (20)) have been found to be more active than lead fuel additives (21,22), in spite of the higher activity of lead compared with copper in both tight and loose contact model studies (9).

4.7 MEMORY EFFECT UPON CHANGING FUEL ADDITIVE CONCENTRATION

TEM showed that in some soot particulates many more, and larger, copper particles are present compared with the bulk of the soot particulates. This suggests long residence times of the soot, prior to collection. An explanation for this observation is the collection of soot on the walls of the exhaust pipe line. The high temperatures of the diesel exhaust gases (about 600-650 K) in combination with the insulated and heated exhaust gas pipe lines cause this soot to oxidize slowly. If eventually such particulates are re-entrained by the exhaust gases they contain more copper which apparently has sintered to a large degree. This picture also explains the very long times needed to clean the exhaust system from the copper additive.

4.8 OCCURRENCE OF A MAXIMUM IN THE PRESSURE DROP CURVE

An interesting point is the occurrence of the maximum in the pressure drop curves (Figures 7, 8, and 9). The first part of these curves can be easily understood because the pressure drop levels off due to a rate of soot combustion which comes closer and closer to the rate of particulate collection. The subsequent decrease in pressure drop, followed by a new and durable equilibrium pressure drop, is less straightforward to comprehend.

A possible explanation for this observation is given elsewhere (12). This explanation is based on mobility of the copper catalyst, and as this mobility is thought to be a key-point in the development of catalysts for the oxidation of diesel soot (9,15), this maximum in pressure drop will be studied in more detail in the future.

4.9 FUTURE STUDY

No attention has been paid in this study to the role of other diesel exhaust gas components (H_2O, NO_x, SO_x) on the oxidation activity of diesel soot. Preliminary experiments in our laboratory reveal that H_2O and NO_x can significantly influence the oxidation rate of copper catalyzed soot oxidation. A possible explanation for the role of NO_x might be the oxidation of NO to NO_2 by the copper catalyst, followed by the oxidation of soot by NO_2. This mechanism has been earlier proposed to explain the high activity of platinum catalysts which are not in direct contact with soot (23,24). This reaction is an example of a reaction other than the oxidation of carbon with oxygen. As such reactions might influence the soot oxidation rate to a large extent, they will be investigated in a further study.

5 CONCLUSIONS

The use of a copper coating was found to increase the activity of soot, formed from the combustion of diesel fuel containing a copper fuel additive. The activity of such a coating was, however, small compared with the activity of the copper fuel additives.

The activation energies of catalytic soot oxidation were found to be the same for powdered copper oxide catalysts and for copper from the copper fuel additive catalyst.

A temperature of 365°C was found to be sufficient to establish an equilibrium between the rate of soot collection in a wall-flow monolith and the rate of soot oxidation. It was considered to be difficult, however, to extrapolate this result to other practical conditions (a full-size wall-flow monolith, an engine with lower particulate emissions, lower copper fuel additive concentrations). On the basis of this study, copper fuel additives are considered to have a high feasibility for application in practice. In service experience (3,4) confirms this.

Based on the samples supplied, the use of silicon carbide wall-flow monoliths is advantageous over the use of cordierite wall-flow monoliths in terms of (*i*) lower pressure drops over the monoliths as a result of a better filter texture design, and (*ii*) maximum temperatures reached during batch-wise regeneration of accumulated soot are lower due to better thermal conductivity properties of silicon carbide compared with cordierite.

6 REFERENCES

(1) J.B. Howard and W.J. Kausch Jr., *Soot control by fuel additives*, Prog. Energy. Combust. Sci. 6, 263-76 (1980).

(2) N. Miyamoto, Z. Hou, A. Harada, H. Ogawa and T. Murayama, *Characteristics of diesel soot suppression with soluble fuel additives*, SAE Paper 871612 (1987).

(3) D.T. Daly, D.L. McKinnon, J.R. Martin and D.A. Pavlich, *A diesel particulate regeneration system using a copper fuel additive*, SAE Paper 930131 (1993).

(4) J.A. Saile, G.J. Monin and D.T. Daly, *On the road experience of passive diesel particulate filters using copper additive*, In: *Worldwide engine emission standards and how to meet them*, IMechE, London, 171-81 (1993).

(5) B. Krutzsch and G. Wenninger, *Effect of sodium- and lithium-based fuel additives on the regeneration efficiency of diesel particulate filters*, SAE Paper 922188 (1992).

(6) J.P.A. Neeft, M. Makkee and J.A. Moulijn, *Diesel particulate emission control*, Fuel Process. Technol., Accepted for publication (1995).

(7) C. Wong, *Characterization of metal-soot systems by transmission electron microscopy*, Carbon 26, 723-34 (1988).

(8) R.C. Peterson, *The oxidation rate of diesel particulate which contains lead*, SAE Paper 870628 (1987).
(9) J.P.A. Neeft, M. Makkee and J.A. Moulijn, *Catalysts for the oxidation of soot from diesel exhaust gases. I. An exploratory study*, Appl. Catal. B: Environ. 8, 57-78 (1996).
(10) U. Hoffmann and J. Ma, *Study on regeneration of diesel particulate filter using a laboratory reactor*, Chem. Eng. Technol. 13, 251-8 (1990).
(11) V.D.N. Rao, H.A. Cikanek and R.W. Horrocks, *Diesel particulate control system for Ford 1.8L Sierra turbo-diesel to meet 1997-2003 particulate standards*, SAE Paper 940458 (1994).
(12) J.P.A. Neeft, *Catalytic oxidation of soot. Potential for the reduction of diesel particulate emissions*. Ph.D. Thesis, Delft University of Technology, Delft (1995).
(13) H.E. Kissinger, *Reaction kinetics in differential thermal analysis*, Anal. Chem. 29, 1702-6 (1957).
(14) D.M. Heaton, R.ter Rele, P. Tap, J.de Rijke and P.van Gompel, *Gaseous and particulate exhaust emissions measurements on a Yanmar L90E diesel generator set*, IW-TNO, Delft, 15 pages (1991).
(15) J.P.A. Neeft, O.P.van Pruissen, M. Makkee and J.A. Moulijn, *Catalysts for the oxidation of soot from diesel exhaust gases. II. Contact between soot and catalyst under practical conditions*, Submitted for publication in Appl. Catal. B: Environ (1996).
(16) S.C. Sorenson, J.W. Høy and P. Stobbe, *Flow characteristics of SiC diesel particulate filter materials*, SAE Paper 940236 (1994).
(17) A. Itoh, K. Shimato, T. Komori, H. Okazoe, T. Yamada, K. Niimura and Y. Watanabe, *Study of SiC application to diesel particulate filter (part 1): Material development*, SAE Paper 930360 (1993).
(18) H. Okazoe, T. Yamada, K. Niimura, Y. Watanabe, A. Itoh, K. Shimato and T. Komori, *Study of SiC application to diesel particulate filter (part 2): Engine test results*, SAE Paper 930361 (1993).
(19) J.W. Høj, S.C. Sorenson and P. Stobbe, *Thermal loading in SiC particle filters*, SAE Paper 950151 (1995).
(20) Q. Li, *Mechanismen und Einflüsse der additivunterstützten Partikelzündung und -verbrennung im Rußfilter von Dieselmotoren*. Ph.D. Thesis, Rheinisch-Westfälischen Technischen Hochschule Aachen, 108 p. (1993).
(21) R.W. McCabe and R.M. Sinkevitch, *A laboratory combustion study of diesel particulates containing metal additives*, SAE Paper 860011 (1986).
(22) H. Ise, K. Saitoh, M. Kawagoe and O. Nakayama, *Combustion modes of light duty diesel particles in ceramic filters with fuel additives*, SAE Paper 860292 (1986).
(23) B.J. Cooper and J.E. Thoss, *Role of NO in diesel particulate emission control*, SAE Paper 890404 (1989).
(24) B.N. Hawker, *Diesel emission control technology. System containing platinum catalyst and filter unit removes particulate from diesel exhaust*, Platinum Met. Rev. 39, 2-8 (1995).

Fuel property effects on polyaromatic hydrocarbon emissions from modern heavy-duty engines

DOEL BSc, PhD
Research Limited, Chester, UK

SYNOPSIS

Recent years have seen an increasing concern over the levels of polyaromatic hydrocarbons (PAH) emitted to the environment. Exhaust emissions from diesel vehicles are one such source of atmospheric PAH and the influence of changes in engine design, engine maintenance practises, exhaust aftertreatment technologies and fuel properties are all of interest.

This work addresses the influence of diesel fuel properties, specifically fuel PAH content, on the levels of the PAH associated with the exhaust particulate matter (PM), for a modern heavy-duty engine running over the European legislated R49 cycle. It shows that the exhaust particulate matter includes significant PAH, both for fuels which themselves contain virtually zero PAH, as well as for fuels which meet the current European EN 590 specification. It is shown that, in the above tests, the majority of the PAH associated with the PM do not originate from the fuel PAH, but may be attributed to material pyro-synthesised in the combustion process. The lubricating oil was shown to make a negligible nett contribution to exhaust PAH over the duration of the regulated cycle.

These findings are in line with increasing evidence in the recent literature that, for modern heavy-duty diesel engines, running on fuel meeting current specifications, surviving fuel PAH molecules are playing a diminishing role as a source of exhaust PAH. This evidence is in contrast to previous beliefs and will contribute to an informed assessment of exhaust PAH emissions from diesel engines allowing the contribution of fuels to be put in context with other factors, such as engine design and aftertreatment technologies.

1. INTRODUCTION

The yield, and distribution by species, of the polyaromatic hydrocarbons (PAH) present in diesel vehicle exhausts are potential targets for legislation driven by health and safety concerns. In this context, the PAH associated with particulate matter (PM) in ambient air is of interest, although there is currently no evidence of a direct link with any health effects.

In order to ensure appropriate fuel specifications the origin of these molecules must be known: they may be original fuel components that survive combustion, or they may be formed during combustion (i.e. pyro-synthesised), or they may arise from the lubricating oil. This work reports experimental results concerning the relationship between fuel PAH content and PM PAH levels for a modern heavy-duty engine running on commercial type fuels over the European R49 cycle. The results permit an evaluation of the contribution of the various sources of PAH (surviving fuel components, pyro-synthesised molecules and lubricating oil components) to the exhaust. This, in turn, allows an assessment of the likely impact on compression ignition engine exhaust PAH emissions of any variation in fuel composition and a comparison with the impact of changes in engine design and aftertreatment technologies.

Below, we briefly review the literature before describing our experimental programme (Section 3) and analysing the results (Section 4). Our conclusions are presented in Section 5.

2. LITERATURE REVIEW

Despite a lack of clear evidence there is a widespread assumption that a reduction in automotive gasoil (AGO) PAH content will lead to a *proportionate* reduction in exhaust PAH emissions. The literature supports a range of interpretations, although more recent work increasingly finds that a significant amount of PAH pyro-synthesis occurs during combustion.

Researchers at Leeds University has published extensively in this field (1a - 1e). From experiments on single cylinder engines of now dated design (Petter AV1, Perkins 4.236) they concluded that exhaust PAH were overwhelmingly comprised of surviving fuel components (1e), although PAH were detected, albeit at reduced levels, in the exhausts of near-zero PAH fuels derived from vegetable oils. The interpretation of these results, in terms purely of the effects of fuel PAH levels, is made difficult as the fuels used had widely differing physical properties (e.g. cetane number, density, calorific value and back-end distillation temperatures) and chemical characteristics (e.g. C/H ratio, presence/absence of oxygenates).

In order to minimise the differences in properties (other than PAH levels) of test fuels, a fuel sample may be 'doped' with a quantity of PAH and its emissions compared to those of an untreated sample of the same fuel. Henderson (2) used this technique (again employing a single-cylinder engine of old design). However, severe problems still remain. Firstly, at the doping levels used ($\approx$ 1%wt. [i.e. 10 000 ppm] of a single species), it is questionable whether the dominant processes governing the fate of the PAH molecules remain those that operate at the levels typical of PAH in current market fuels. (Individual three and four ring PAH species are generally found at 1 - 100 ppm levels, larger ones at 0.1 - 10 ppm.) Secondly, such experiments may suffer severely from 'carry-over' of the dopant (e.g. in the exhaust/sampling system) to such an extent as to seriously corrupt subsequent results.

A group at Plymouth University have developed an experimental technique that avoids these problems (3a - 3g). Firstly, their exhaust sampling apparatus avoids both carry-over and sample ageing (as may occur when PM is collected on a filter held in the exhaust stream). Secondly, their use of ^{14}C radio-labelled PAH permits doping in sufficiently low concentrations to minimise any changes in other fuel properties. Additionally, ^{14}C labelled PAH allow definitive fuel component survival rates to be calculated. Steady-state tests on a light-duty DI engine (Perkins Prima) showed, for four (non ^{14}C labelled) PAH, that yields in the exhaust were equivalent to 0.15 - 0.3% of the amount in the fuel (3h). As the exhaust yields of the different PAH were similar, it was suggested that surviving fuel components were the most likely source. (This is an initially persuasive proposition. However, if the *relative* rates of pyro-synthesis of the different PAH within the cylinder are similar to those that determine the fuel PAH profile, then the argument is flawed.) This work was extended by doping the fuel with various ^{14}C labelled PAH (3i, 3j). The survival rates of these PAH covered a much wider range of values: 0.04% for benzo(a)pyrene to 0.87% for fluorene. Interestingly no evidence was found for modification of one ^{14}C PAH in to another (e.g. benzo(a)pyrene to pyrene), although small quantities of ^{14}C labelled polar species were detected. Data yet to be published also indicated a significant proportion of the exhaust PAH were pyro-synthesised (3k).

As a consequence of the limitations and difficulties associated with the doping methods discussed above, an experimental design relying on a statistical method of interpretation has become a favoured technique by which the dependency of PAH emission levels on fuel properties may be addressed.

An early investigation (4) in to the dependency of the exhaust PAH levels on the fuel PAH content illustrated the necessity of using fuels in which, as far as possible, only the PAH content is varied. In this programme, cetane number (ranging from 28 to 65) was strongly correlated with PAH content (0.9 - 3.7%). For the fuels of low cetane number it was recognised that the resulting late ignition had led to incomplete combustion and consequent high emission levels of unburnt fuel. The amount of PAH in the exhaust was thus dominated by factors other than the fuel PAH content per se. It was not possible to conclude from this work to what extent exhaust PAH were comprised of pyro-synthesised molecules, or surviving fuel components.

Another early programme (5), in which six US commercial grade diesel fuels were run in a then-current (mid '80's) production model heavy-duty engine, paid great attention to correlations between fuel properties and consequently produced more applicable results. Five PAH species were quantified, both in the fuels and in the exhaust particulate matter, for three operating conditions (idle, cruise and high power). The results showed "no clear relationship between fuel PAH concentrations and the PAH emission rates". In particular, the fuel with the lowest PAH concentrations did *not* produce significantly lower PAH emission rates. Rather, it was found that: "the concentration of PAH compounds in diesel particulate samples was highly dependent upon engine operating conditions". A further indication of the relatively minor influence of fuel PAH levels on PAH emissions was the observation that the repeatability of tests on a single fuel exceeded the fuel-to-fuel variability.

The question of the influence of engine design and running conditions on PAH emissions has been studied by Zierock et al. (6) for three very different light-duty engine designs (direct injection, pre-chamber and swirl-chamber) at 28 different combinations of speed and load. The PAH emission rates varied in a complex, but different, manner for each engine, according to the operating conditions. The twin conclusions were that the PAH levels "of the three engines were very different [when run on the same fuel]" and were "greatly influenced by the operating conditions". This work clearly demonstrated the impact of engine design, and the importance of optimising the design to match the intended operating range.

A recent SAE paper (7) investigated the effects of both fuel PAH and total aromatics levels on exhaust PAH levels in one heavy-duty and one medium-duty engine (both 1994 designs) running on the US heavy-duty transient cycle. The four fuels tested were blended from both synthetic and conventional crudes and a cetane improver was used where necessary to meet the US specifications. Also, the medium-duty engine could (optionally) be fitted with a catalyst (two designs were available). The results indicated "no relationship between total fuel aromatics and total measured PAH emissions" although "there was a correlation found between the level of polyaromatics in the fuel and the PAH emissions". Within this correlation it was apparent that "a fairly high proportion of the PAH emission is independent of the fuel quality [i.e. fuel PAH content]". The results were fitted to a linear regression indicating that, for these modern engines running on low sulphur, commercial quality fuels, over the US heavy-duty cycle, PAH emissions could be described by the following equations:

<u>Heavy-duty case:</u> $PAH_{exhaust} = 50 + 10\ PAH_{fuel}$ \hfill (1)
<u>Medium-duty case:</u> $PAH_{exhaust} = 44 + 33\ PAH_{fuel}$ \hfill (2)

where: $PAH_{exhaust}$ is the exhaust tri+ PAH content in µg/bhp-hr;
       PAH_{fuel} is the fuel tri+ PAH content in %wt.

These equations indicate several highly important phenomena. Firstly, both engines emit significant (and similar) levels of PAH which cannot be attributed to surviving fuel PAH. Secondly, the fraction of the fuel PAH that enters the exhaust varies strongly with engine design. Thirdly, the split between the exhaust PAH that may be attributed to fuel PAH and those that may be attributed to pyro-synthesis was such that, for typical (US) commercial fuels (of say 3%wt. tri+ PAH), between 30 and 60% (depending upon engine design) of the exhaust PAH *cannot* be attributed to surviving fuel PAH.

Finally, we turn to the work of Westerholm et al. This Swedish group's work on the relationship between fuel properties and exhaust PAH levels (in both CI and SI engines) has recently been summarised (8a), but we focus here on work (8b) in which eight fuels (with tri+ aromatic levels ranging from 0.05 - 1.1%vol.) were tested in two vehicles using a transient cycle representative of public transport vehicles in city traffic. A synthetic lubricating oil was used to minimise the contribution to PAH from that source. Again a quantitative interpretation was made using a linear regression fit:

<u>Vehicle 1 (bus):</u> $PAH_{emitted} = 81\ (\pm 23) + 7.7 \times 10^{-4}\ (\pm 1.3 \times 10^{-4})\ PAH_{consumed}$ \hfill (3)
<u>Vehicle 2 (lorry):</u> $PAH_{emitted} = 90\ (\pm 27) + 5.9 \times 10^{-4}\ (\pm 1.5 \times 10^{-4})\ PAH_{consumed}$ \hfill (4)

where: $PAH_{emitted}$ is the PAH emitted over this cycle in µg/km;
       $PAH_{consumed}$ is the PAH consumed as part of the fuel, again in µg/km.

It was concluded that, over this low load, low speed "bus cycle", PAH in the fuel were responsible for 60 to 80% of the exhaust PAH when running on fuels containing 0.6 to 1.1% PAH. In comparing these results with those of others (e.g. 5, 6, 7) we note that test cycles involving a lot of transients and time spent at low load/idle will likely lead to a higher *proportion* of the emitted PAH being attributable to surviving fuel components.

3. EXPERIMENTAL PROGRAMME

3.1. Experimental Concept

The majority of PAH in the exhaust are taken to arise from one of two mechanisms: they may either be surviving fuel components, or they may be pyro-synthesised in the flame front. (A large amount of experimental and theoretical work [e.g. see (9) and (10) respectively] indicates that PAH are readily formed from small hydrocarbon radicals created when any hydrocarbon is pyrolysed.) We assume that there is relatively little nett modification of one PAH in to another (as evidenced by the ^{14}C work at Plymouth (3i, 3j)) and that the lubricating oil neither contributes, nor removes, a significant amount of PAH (see sub-section 3.4 below).

An estimate of the exhaust PAH yields due to surviving fuel components and due to pyro-synthesis may then be obtained by running test cycles with fuels specially blended to have a range of polyaromatic levels, but little variation in other significant parameters (particularly cetane number, density and distillation properties). If, for each of these fuels the PAH content is known, and for each test the fuel consumption and exhaust PM PAH levels are measured, then a plot showing the variation of exhaust PM PAH levels as a function of fuel PAH content may be produced. It must be appreciated that this experimental methodology requires the PM (and associated PAH) sampled to be representative of that produced by the fuel under test: we assume that, in tests with current quality fuels that have not been 'doped', no significant interchange of exhaust PM or PAH with material on the walls of the exhaust and sampling system occurs when following the legislated procedures.

There is no universally accepted definition of "PAH". The IP391 method quantifies species according to the number of 6-membered aromatic rings, using *o*-xylene, 1 methyl-naphthalene and phenanthrene as standards for the mono-, di- and tri+ aromatic classes. Under this classification, the US EPA "priority pollutant" list of PAH contains 4 di- and 12 tri+ species (see fig. 1). In discussing our own data we refer to those species for which yields were individually measured as "targeted" species, and the sum of the tri+ targeted species as "total targeted PAH": we take these tri+ species to be representative of PAH in general.

3.2. The Fuels

The inspection properties of the fuels, and their PAH levels for the 16 targeted species are shown in table 1. The fuels (all of which were unadditivated) may be described as:

A: An RF73 look-a-like reference fuel.
B: An extreme low aromatics fuel. This fuel has a much higher cetane number and a slightly lower (about 5%) density than the others.
C: A "base case" fuel blended from refinery streams and near the 0.05% S specification.

D: Blended from refinery streams to have high di-aromatics, but low tri+ aromatics.
E: Blended from refinery streams to be high in both di- and tri+ aromatics.

Note that fuel B, included to assess the PAH emissions from a near-zero PAH fuel, does not meet the requirements discussed above of having a similar cetane number and density to the other fuels. However, as its cetane number is higher, and its density lower its combustion should be improved and its PAH emissions tend to be *lower* than would otherwise be the case.

3.3. The Engine

The engine tested was a modern heavy-duty unit built to meet the 1996 "Euro2" emissions standards, see table 2 for key engine details. A heavy-duty engine was tested (rather than a light-duty vehicle) as the former dominate the market, consuming 70% of AGO in Europe.

3.4. The Lubricating Oil

A fresh, high performance oil, Shell Myrina 15W40, with very low PAH levels was used. Tests show that, over the duration of several test cycles, a negligible nett change in the PAH content of the oil occurs (see table 3).

3.5. The Test Procedure

Prior to every test, the engine was pre-conditioned under the R49 mode 8 conditions, using the fuel to be tested. During the R49 cycle the regulated emissions, BSFC and power were measured. The repeatability of these measurements was very good, and well within ACEA's recommended limits (table 4). PM was collected on two membranes, placed in series. Only material from the primary membrane (which generally accounted for 90 - 95% of the PM) was analysed for PAH content. In calculating the PM borne PAH emissions (table 5) the relative abundance of PM PAH on a secondary membrane was assumed to be as that on the primary.

4. RESULTS AND DISCUSSION

The PM borne PAH levels (table 5) show a remarkable lack of variation with fuel PAH levels. The test-to-test variability exceeds the fuel-to-fuel variability (as reported in ref. 5). This remains the case even when the targeted PAH are grouped, in an effort to reduce measurement errors. In fig. 2 we plot PAH emission against consumption for the sum of the targeted tri+ PAH (i.e. excluding naphthalene, acenaphthalene, acenaphthylene and fluorene which are all measured as di-aromatics by IP391). This figure shows that the reduction in exhaust PM PAH obtained by moving from a fuel with 3.2%wt. tri+ aromatics, to one with < 0.1%wt. tri+ aromatics, is negligible (about 15%, but this is comparable with the experimental scatter). A similar plot is obtained for total targeted tetra+ PAH, see fig. 3. Note especially that the exhaust from the near-zero PAH fuel (B) contains significant PM associated PAH. If exhaust PAH were primarily or solely surviving fuel PAH, these data would necessarily form a line through the origin. For those tri+ and tetra+ PAH targeted in this work, the exhaust PM associated PAH may be described by the following regression fits:

$$\text{Tri+ PAH}_{exhaust} = 0.10\ (\pm 0.01) + 7\times10^{-5}\ (\pm 9\times10^{-5})\ \text{Tri+ PAH}_{fuel} \quad (5)$$
$$\text{Tetra+ PAH}_{exhaust} = 0.022\ (\pm 0.02) + 2\times10^{-4}\ (\pm 2\times10^{-4})\ \text{Tetra+ PAH}_{fuel} \quad (6)$$

where: $\text{PAH}_{exhaust}$ is the exhaust PM total targeted tri+ (or tetra+) PAH content in mg/kWh;
PAH_{fuel} is the total targeted fuel tri+ (or tetra+) PAH consumed in mg/kWh.

As we have noted in the literature, tests on engines of increasingly modern design show surviving fuel components to be playing a diminishing role in exhaust emissions whilst PAH created by pyro-synthesis are more important than generally thought (1, 3, 5, 7). Within this context we rationalise our results as follows. The pyro-synthesised PAH (described by the constant in eqs. 5 and 6) are created in the flame front. Our data show that their creation is independent of the aliphatic/(poly-)aromatic nature of the original fuel molecules: a factor which must not exert much influence on the controlling chemical kinetic parameters of pressure, temperature and chemical species (probably small $[C_2 - C_4]$ hydrocarbon fragments (9)) found in the flame front. Surviving fuel component PAH found in the exhaust may arise from crevices or gas pockets either too rich, or too lean, for complete combustion. As engine design advances, significant reductions may be made in that fraction of the fuel surviving combustion: however alterations to engine design cannot greatly effect the physical/chemical conditions found in the flame front where PAH molecules may be pyro-synthesised. Present heavy-duty technology has reached the point where, for current commercial fuels, driven over regulated cycles, a very significant majority of the PM associated PAH are created through pyro-synthesis, rather than being surviving fuel components.

5. CONCLUSIONS

We conclude that fuel PAH levels have minimal influence on exhaust PM PAH levels for modern designs of heavy-duty compression ignition engines running on current quality fuel over the ECE legislated R49 cycle.

An experimental programme has been carried out to determine the relationship between fuel polyaromatic hydrocarbon (PAH) content and exhaust particulate matter associated PAH levels. The results indicate that, for a modern design of heavy-duty engine, running over the legislated R49 cycle on fuels representative of current quality low sulphur commercial gasoils, the PM PAH levels are virtually independent of the fuel PAH content. Even an extreme near-zero tri+ aromatics fuel shows a similar level of PM borne PAH to that from a fuel containing 3.2%wt. tri+ aromatics. Our data suggest that the majority of exhaust PAH are pyro-synthesised molecules, originating in equal measure from all fuel components. It was also shown that the repeatability of PM PAH level measurements from tests on a single fuel exceeded the differences between fuels (a result that has also been noted by others (5)).

A review of the literature indicates that the PAH emissions from modern engine designs are increasingly less dependent on the PAH levels in typical market fuels and that the PAH emission rates vary from engine to engine to a greater extent than they do between fuels.

Acknowledgements
The author would like to acknowledge the work of M. Brown and N. Bailey in carrying out the engine tests, of Ricardo Ltd. in performing the PAH analyses and the assistance of R. Stradling and A. Hudson in planning the work and interpreting the results.

6. REFERENCES

1a. Andrews, G.E. et al. Proc. I. Mech. E., 1983, C73/88.
1b. Williams, P.T. et al. J. H. Res. Chrom., 1986, **9**, 39.
1c. Abass, M.K. et al. SAE 872085, SAE 892079, SAE 901563, SAE 910487.
1d. Williams, P.T. et al. SAE 870554.
1e. Williams, P.T. et al. Comb. Flame, 1989, **75**, 1.
2. Henderson, T.R. et al. Env. Sci. Tech., 1984, **18**, 428.
3a. Petch, G.S. et al. Proc. I. Mech. E., 1987 C340/87.
3b. Trier, C.J. et al. Proc. I. Mech. E., 1988, C64/88.
3c. Petch, G.S. et al. Proc. I. Mech. E., 1988, C65/88.
3d. Rhead, M.M. et al. Sci. Tot. Env., 1990, **93**, 207.
3e. Trier, C.J. et al. Proc. I. Mech. E., 1990, C394/003.
3f. Rhead, M.M. et al. I. Mech. E. Seminar: Fuels for Automotive & Industrial Diesel Engines, 1991, 27.
3g. Trier, C.J. et al. Proc. I. Mech. E., 1991, C433/010.
3h. Collier, A.R. et al. Fuel, 1995, **74**, 362.
3i. Tancell, P.J. et al. Sci. Tot. Env., 1995, **162**, 179.
3j. Tancell, P.J. et al. Env. Sci. Tech., 1995, **29**, 2871.
3k. Tancell, P.J., private communication.
4. Mills, G.A. and Howarth, S.J. J. Inst. Energy, 1984, **March**, 273.
5. Wall, J.C. and Hoekman, S.K. SAE841364.
6. Zierock, K.-H. et al. SAE830458.
7. Mitchell, K. et al. SAE942053.
8a. Westerholm, R.N. et al. Env. Health Pers., 1994, **102(4)**, 13.
8b. Westerholm, R.N. and Li, H. Env. Sci. Tech., 1994, **28**, 965.
9. Frenklach, M. and Wang, H. 23rd. Symp. (Int.) Comb., 1990, 1559.
10. Stein, S.E. and Fahr, A. J. Phys. Chem., 1989, **89**, 3714.

© Shell Internationale Research Maatschappij BV

	Fuel Description	A RF73 look-a-like	B Extreme low arom.	C Base case	D High di-, low tri+	E High di- high tri+
Property	Unit					
IBP	°C	168.0	211.0	188.0	170.5	173.0
10 % Vol.	°C	192.0	240.0	222.5	215.0	198.0
20 % Vol.	°C	207.0	257.0	239.5	236.5	223.5
30 % Vol.	°C	227.0	272.5	254.0	253.0	255.5
40 % Vol.	°C	248.0	287.5	268.0	267.5	274.0
50 % Vol.	°C	265.0	301.5	281.5	280.0	289.0
60 % Vol.	°C	279.0	314.0	295.0	291.5	302.0
70 % Vol.	°C	290.0	325.5	309.0	303.5	316.5
80 % Vol.	°C	303.0	336.5	323.5	319.0	331.5
90 % Vol.	°C	323.0	350.0	343.5	340.5	348.5
95 % Vol.	°C	339.0	361.0	NA	357.5	NA
FBP	°C	356.0	367.0	380.5	381.5	382.0
Density @ 15°C	kg/l	0.8330	0.8016	0.8496	0.8484	0.8488
Cetane Number		49.9	71.5	49.2	48.1	49.2
Viscosity. @ 40°C	cSt	2.256	3.901	3.525	3.283	3.044
Sulphur Content	ppm	513	8	582	612	1145
Aromatics (IP391)						
Mono-Aromatics	%wt.	24.3	0.3	20.5	20.2	21.3
Di-Aromatics	%wt.	2.8	<0.1	2.0	5.3	6.1
Tri+ Aromatics	%wt.	0.4	<0.1	0.4	0.3	3.2
Total Aromatics	%wt.	27.5	0.3	22.9	25.6	30.6
PAH species	Abbreviation					
Naphthalene	Naph µg/g	158.74	23.13	108.28	197.25	110.05
Acenaphthalene plus acenaphthylene	Aces µg/g	20.96	3.01	128.62	0.00	880.92
Fluorene	Fl µg/g	64.64	6.36	40.61	0.00	29.91
Phenanthrene	Phen µg/g	179.35	19.55	119.46	559.36	655.62
Anthracene	Anth µg/g	2.95	0.27	1.27	3.35	2.03
Fluoranthene	Fluo µg/g	14.53	1.86	16.11	27.52	79.43
Pyrene	Py µg/g	9.10	1.22	30.00	18.93	161.55
Benz(a)Anthracene	BaA µg/g	1.77	0.24	6.53	1.37	49.83
Chrysene	Chry µg/g	10.68	1.12	21.42	2.68	122.34
Benzo(b)Fluoranthene	BbF µg/g	0.55	0.09	0.97	0.39	7.64
Benzo(k)Fluoranthene	BkF µg/g	0.26	0.04	0.25	0.20	1.22
Benz(a)Pyrene	BaP µg/g	0.15	0.04	0.24	0.18	0.85
Dibenz(a,h)Anthracene	DahA µg/g	0.37	0.08	0.12	0.00	1.58
Benzo(ghi)Perylene	BghiP µg/g	0.74	0.05	0.40	0.77	2.07
Indeno(123,cd)Pyrene	IcdPy µg/g	0.00	0.15	0.00	0.00	0.00
Totals						
4 Di aromatics	Naph to Fl µg/g	244.34	32.50	277.51	197.25	1020.88
12 Tri+ aromatics	Phen to IcdPy µg/g	220.45	24.71	196.77	614.75	1084.16
9 Tetra+ aromatics	Py to IcdPy µg/g	23.62	3.04	59.93	24.52	347.07
All 16 species	µg/g	464.79	57.21	474.28	812	2105.04

Table 1. Inspection properties and targeted PAH concentrations of the test fuels. (A combined abundance is given for acenaphthalene and acenaphthylene.)

Engine Specification	Heavy-duty, built to meet "Euro2" emissions standards
Design Details	DI, TC, Aftercooled, no exhaust after-treatment
Swept Volume	6 litres
Cylinders	6, in line
Max. Power	180 kW at 2600 rpm
Max. Torque	850 Nm at 1400 rpm
Oil Capacity	20 litres

Table 2. Details of the engine under test.

	Sample	Fresh Oil	Sample 1	Sample 2	Sample 3	Sample 4
	No. of engine hours	0	27	33	40	51
	No. of tests	0	8	11	14	19
PAH species	**Abbreviation**	µg/g	µg/g	µg/g	µg/g	µg/g
Naphthalene	Naph	6.35	4.73	5.62	4.85	5.63
Acenaphthalene and acenaphthylene combined	Aces	0.94	0.81	0.71	0.89	0.63
Fluorene	Fl	0.62	0.56	0.72	0.96	0.76
Phenanthrene	Phen	8.29	5.06	6.02	6.69	4.66
Anthracene	Anth	0.20	0.13	0.18	0.16	0.15
Fluoranthene	Fluo	0.85	0.63	0.79	0.96	0.72
Pyrene	Py	4.44	1.10	2.49	1.09	1.14
Benz(a)Anthracene	BaA	0.22	0.21	0.22	0.29	0.22
Chrysene	Chry	0.94	0.80	0.79	0.94	0.71
Benzo(b)Fluoranthene	BbF	0.18	0.10	0.15	0.12	0.10
Benzo(k)Fluoranthene	BkF	0.07	0.04	0.06	0.05	0.04
Benzo(a)Pyrene	BaP	0.12	0.04	0.07	0.08	0.07
Dibenz(a,h)Anthracene	DahA	0.06	0.06	0.06	0.05	0.05
Benzo(ghi)Perylene	BghiP	0.13	0.14	0.07	0.16	0.14
Indeno(123,cd)Pyrene	IcdPy	0.14	0.00	0.06	0.00	0.04
Totals						
4 Di aromatics	Naph to Fl	7.91	6.10	7.05	6.70	7.02
12 Tri+ aromatics	Phen to IcdPy	15.64	8.31	10.96	10.59	8.04
9 Tetra+ aromatics	Py to IcdPy	6.30	2.49	3.97	2.78	2.51
All 16 species		23.55	14.41	18.01	17.29	15.06

Table 3. The PAH content of the fresh lubricating oil, and at various stages through a test programme. (A combined abundance is given for acenaphthalene and acenaphthylene.)

Measurement	Test programme covariance (%)	ACEA recommended limits covariance (%)
CO	4.0	9
NOx	1.6	2
HC	3.4	10
PM	6.2	7
BSFC	0.4	0.5
Power	0.1	0.6

Table 4. The covariance of the test measurements, and the ACEA recommended limits, for regulated emissions, BSFC and power.

Fuel/Test PAH	A/1 µg/kWh	A/2 µg/kWh	A/3 µg/kWh	A/4 µg/kWh	A/5 µg/kWh	B/1 µg/kWh	B/2 µg/kWh	B/3 µg/kWh	B/4 µg/kWh	C/1 µg/kWh	C/2 µg/kWh	C/3 µg/kWh	D/1 µg/kWh	D/2 µg/kWh	D/3 µg/kWh	E/1 µg/kWh	E/2 µg/kWh	E/3 µg/kWh
Naph	185.71	220.09	177.09	76.32	493.79	221.19	210.69	124.21	60.51	42.65	132.02	116.03	255.90	89.45	72.02	67.99	74.11	115.04
Aces	18.81	9.31	4.48	5.20	46.70	14.65	4.19	5.78	2.90	4.99	4.35	5.03	3.88	4.98	4.51	8.66	6.83	7.81
Fl	17.87	24.46	8.80	9.68	12.77	24.26	18.63	19.12	8.61	38.12	9.59	14.63	10.04	11.56	12.13	8.80	13.49	15.62
Phen	62.62	83.79	34.24	45.44	57.48	80.79	50.93	52.27	39.96	45.38	35.05	74.13	39.18	66.41	73.20	87.37	66.24	60.22
Anth	4.57	3.34	0.94	0.96	4.23	2.75	1.40	1.93	0.64	1.29	0.96	1.71	1.04	1.26	1.32	1.48	1.04	1.35
Fluo	16.53	17.65	9.03	9.36	13.44	19.45	16.46	11.00	7.77	20.49	9.37	13.10	9.97	8.50	11.79	14.13	12.85	13.52
Py	14.92	14.45	8.32	9.20	59.47	17.97	15.06	14.31	7.13	65.49	17.56	19.74	14.88	12.75	11.37	18.50	16.14	17.50
BaA	1.68	2.29	1.30	1.44	1.74	2.23	1.71	1.58	0.60	2.57	2.25	2.47	1.73	1.69	1.32	4.81	3.61	4.21
Chry	4.97	5.35	2.32	3.44	4.98	4.81	3.49	4.06	1.52	2.65	3.06	3.90	2.84	2.29	2.88	10.88	8.99	10.44
BbF	2.62	2.50	1.14	0.64	0.62	2.40	1.94	2.06	0.42	1.55	1.66	0.72	1.42	0.76	0.76	2.15	0.96	0.98
BkF	0.67	0.70	0.47	0.28	0.25	0.80	0.62	0.83	0.14	0.64	0.63	0.27	0.49	0.27	0.28	0.74	0.28	0.26
BaP	0.40	0.63	0.43	0.24	0.95	0.63	0.62	0.55	0.18	1.40	0.52	0.31	0.45	0.43	0.38	0.93	0.24	0.30
DahA	0.13	0.70	0.24	0.00	1.41	0.34	0.00	0.28	0.00	0.00	0.30	0.27	0.21	0.13	0.00	0.22	0.24	0.23
BghiP	1.48	1.39	0.71	0.40	10.78	1.14	2.02	1.24	0.28	7.41	0.96	0.27	1.38	0.53	0.49	0.81	0.40	0.45
IcdPy	1.21	0.83	0.71	0.00	0.50	1.37	1.55	1.65	0.49	1.51	1.11	0.00	0.62	0.40	0.83	1.63	0.40	0.45
Totals																		
4 Di Ar.	222.39	253.86	190.36	91.20	553.26	260.10	233.52	149.11	72.02	85.76	145.97	135.69	269.81	105.99	88.66	85.45	94.42	138.47
12 Tri+ Ar.	111.80	133.60	59.84	71.40	155.86	134.68	95.80	91.75	59.13	150.38	73.43	116.89	74.20	95.43	104.61	143.64	111.41	109.90
9 Tetra+ Ar.	28.08	28.83	15.63	15.64	80.71	31.70	27.02	26.55	10.77	83.22	28.04	27.95	24.02	19.26	18.30	40.65	31.27	34.80
All 16 PAH	334.19	387.45	250.20	162.60	709.12	394.78	329.31	240.86	131.15	236.14	219.40	252.57	344.02	201.42	193.27	229.09	205.83	248.37

Table 5. The yields of the targeted PAH on the R49 cycle particulate matter. (Acenaphthalene and acenaphthylene could not be distinguished: their combined abundance is given.)

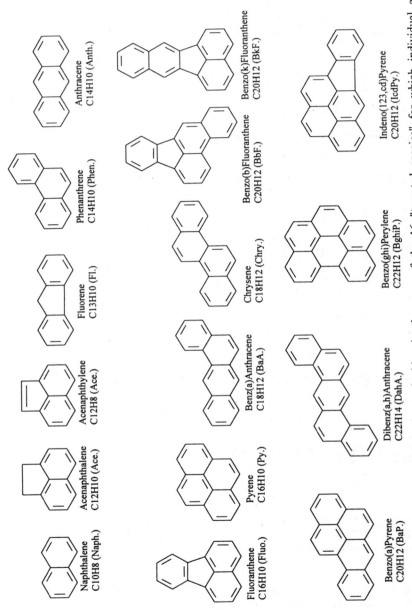

Figure 1. The structures, formulae and abbreviated names of the 16 "targeted species" for which individual abundances (except acenaphthylene and acenaphthylene for which a combined value was obtained) in the fuels and exhaust emissions were measured. We regard the first 4 types as di-aromatic species and the last 16 as representative of all tri+ PAH.

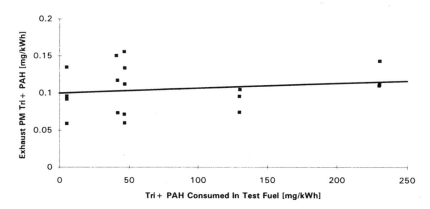

Figure 2. Exhaust particulate matter associated PAH as a function of fuel PAH, for the totalled targeted tri+ aromatic PAH species. The best fit linear regression (see eq. 5) to the data is also shown. If surviving fuel PAH were the sole source of exhaust PM PAH these data would form a line through the origin.

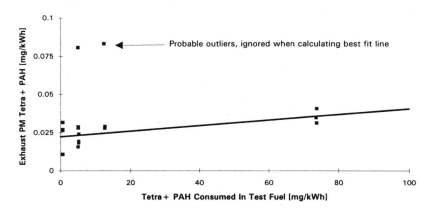

Figure 3. Exhaust particulate matter associated PAH as a function of fuel PAH, for the totalled targeted tetra+ aromatic PAH species. The best fit linear regression (see eq. 6) to the data (ignoring two probable outliers) is also shown. If surviving fuel PAH were the sole source of exhaust PM PAH these data would form a line through the origin.

An experimental characterization of the formation of pollutants in DI diesel engines burning oxygenated synthetic fuels

C BEATRICE, C BERTOLI, N DEL GIACOMO, and M LAZZARO
Istituto Motori CNR, Napoli, Italy

SYNOPSIS.

In order to meet the regulations scheduled for the end of the century, further development of the combustion system of direct injection diesel engine is required. It appears strictly linked to the understanding of the main mechanism of soot formation and oxidation process during the combustion phase.

This study aims at to investigate on the role of some oxygenated synthetic fuels in a conventional technology DI diesel engine, as regard the pollutants formation.

In particular the following fuels were tested:
- Butyl Ether
- Pentyl Ether
- Ethylene Glycol Dimethyl Ether
- Diethylene Glycol Dimethyl Ether
- Diethylene Glycol Diethyl Ether

The experimental technique adopted to investigate the pollutants formation is the two colours pyrometry. The use of this technique permitted a comparison between evolution of soot volumic fraction burning the different fuels. The experimental tests have demonstrated that the oxygen content of the fuel has a strong influence on soot formation process.

Simultaneously, due to the particular molecular structure, these fuels present very high cetane numbers. This leads to a reduction of the NOx, HC and CO emissions, and notwithstanding the high oxygen content also the Aldehydes emissions are relatively low.

To confirm the main trends observed in the research engine, a comparison regulated emissions between Pentyl Ether, Ethylene Glycol Dimethyl Ether, and two reformulated Diesel fuels of a modern four cylinders DI diesel engine for passenger car application, is presented.

1 INTRODUCTION.

The further reduction of diesel engines emissions limits planned for the end of the century represent the major research challenge for engine designers. Proposals for future legislation lay large emphasis on the control of nitrogen oxide emissions. The drastic reduction of in cylinder emissions formation is normally hindered by the well known trade off between NOx and particulate. So the development of diesel engines led to the manufacturing of complex combustion systems featuring high pressure fuel injection system electronically controlled, cooled exhaust gas recirculation (E.G.R), turbocharging and complex after treatment devices [1,2].

On the other hand the constraints imposed on NOx emission level led to the proper use of E.G.R, if the DeNOx catalysts technology will not reach the required efficiency (up to 50%) and reliability in next few years.

Because of the reduction of fuel sulphur content, planned in next years, and of the actually very low hydrocarbons emissions, carbonaceous compounds (soot) will represent the major contribution (up to 80 %) to the diesel engine particulate emissions. In order to largely reduce or suppress the in cylinder soot formation, the effect of flame temperature on this process as well as the way to obtain an efficient use of the oxygen available must be deepen.Using E.G.R in turns the role of oxygen "starvation" typical of the ethereogenous combustion is strongly enhanced.

An other way to improve diesel combustion, especially with the use of high E.G.R rates, is to use some pure oxygenated fuels, so reducing the tendency to generate soot from the core of fuel jets either vapour or liquid. Alternative fuels as alcohols were considered in the literature for this purpose as they exhibit good emission and power potential. However these fuels require, for diesel operation, some ignition aids which have limited durability.

Others diesel fuel substitutes are today under examination, like dimethylether (DME). In particular [3,4] experimental engine tests demonstrated the potential of DME to achieve smokeless combustion keeping direct injection diesel thermal efficiencies and ultra low emission (ULEV) level. To obtain the liquid injection of DME the injection apparatus (that can work at a maximum injection pressure quite low) must be modified, as well as the vehicle fuel storage tanks, although such technology, typical of l.p.g. applications, is well assessed. So, from a conceptual point of view, if the advantages of the DME combustion can be achieved with a synthetic fuel liquid at the ambient pressure and temperature, this hypothetical fuel will represent the best solution to obtain "clean" combustion, not only in "dedicated engines" but also in existing diesel powered vehicles for close fleet operation (retrofit solution).

Some oxygenated like ethers or methyl carbonates have been tested as additives to improve petroleum derived diesel fuel performances [5,6,7]. However studies on emission potential and combustion characteristics of pure synthetic oxygenated compounds for diesel engine application are very limited in literature.

Starting from previous considerations in the present paper a characterisation of the sooting tendency, and exhaust emissions, of some oxygenated fuels will be presented in comparison with Tetradecane combustion. Experiments were carried out using the two colour pyrometry method on a single cylinder direct injection diesel engine of conventional design, equipped with a cooled E.G.R. laboratory system.In addition some of these fuels were also tested on a modern direct injection light duty engine equipped with an electronic E.G.R. system and variable geometry turbocharger. Tests were carried out in steady state conditions (2 bar b.m.e.p. and 2000 rpm) and oxygenated fuels emission potential was compared with that obtained by burning two reformulated diesel fuels of different quality.

2 EXPERIMENTAL APPARATUS.

Tests were carried out with two different diesel engines. The first is a single cylinder Direct Injection (DI) Diesel engine, 95mm stroke and 100 mm bore, described in the Tab.1. The combustion system is swirl supported type with a toroidal combustion chamber having an equivalent diameter/height ratio of about 4.6. The required swirl level is generated by a shrouded intake valve, whose angular position can be changed during the measurements. In the present test series it was positioned to produce an intake swirl ratio (measured on a steady state device) of about 1.8.

The injection apparatus comprises a Bosch P type in line pump and a four hole injector (0.28 mm diameter), with a cone angle between the sprays of 160°. The injection pump was fitted on a in-house machined device in order to vary the injection timing during the engine running. With a proper choice of plunger diameter and of injection pipe bore to length ratio, a very stable operation at low test speed was obtained, thus allowing to control the cyclic variation of the injection law below 1%.

Table 1 Single cylinder engine characteristics.

Engine	Ruggerini RP 170
Bore (mm)	100
Stroke (mm)	95
Compression ratio	18
Injector	4/0.28/160
Swirl ratio	1.8
Combustion chamber d/h ratio	4.6

The engine was equipped with a combustion pressure transducer, a needle lift transducer and an injection pressure transducer. The optical access for the sensing of radiance signal has been suitably located on the engine cylinder head and directed into the chamber (15 degree downstream of one of the sprays), where the soot concentration in the combustion chamber is expected to be relatively high.

The theoretical a priori uncertainty of the measurements, optical characteristics and calibration method are fully described in [8, 9].

The engine was equipped with a cooled Exhaust Gas Recirculation laboratory system. Such system is an improvement of an elder recirculation device [10]. In order to minimize drag losses, grouping many elements of the plant is suitable. To this aim a fluid bed was adopted to obtain simultaneously the filtering of particulate, the dumping of engine pulsed flow and its cooling. The derivation line of recirculating gas was built by an isocinetic probe. In such operation a maximum EGR rate of 65% was reached with a PMI reduction of less than 5% (with respect to the no EGR), and an intake mean temperature of 35° C.

The second engine is a four cylinder 1929 cm^3 (DI) diesel engine equipped with an variable geometry turbocharger and an E.G.R. system electronically controlled. Its main characteristics are described in Tab.2.

The exhaust analysis bench, for both engines, comprises a FID HC analyser, a chemiluminescence NOx analyser and two infrared analysers for CO_2 and CO. On the exhaust manifold a sampling probe is positioned for an high resolution smoke meter. Moreover, only for the single cylinder engine, aldehyde emissions were analyzed with HPLC chromatographer.

Particulate emission of four cylinder engine were measured by gravimetric standard method with particulate sampling by a diluition tunnel.

Table 2. Four cylinder engine characteristics.

Engine	FIAT
Bore (mm)	82.6
Stroke (mm)	90
Compression ratio	19.8:1
Displacement (cm^3)	1929
Rated power (kW)	68 at 4200 rpm
Rated torque (Nm)	196 Nm at 2000 rpm
Injection system	Distributor-type pump Bosch VE-R 493
Turbocharger	Garret VNT 25 with variable geometry turbine
E.G.R.	Electronically controlled

3 TESTED FUELS.

Previous preliminary experiments evidenced that addition of Pentyl-Ether (12% in volume) to straight run diesel fuel leads to an improvement of the cetane number and to a diminution of soot emission. Thus we considered to test Pentyl-Ether and Butyl-Ether as pure fuels in a diesel engine suitably equipped. Moreover three others oxygenated fuels (Ethylene-Glycol-Dimethylether, Diethylene-Glycol-Dimethylether and Diethylene-Glycol-Diethylether) was chosen, taking into account their high oxygen content and auto-ignition quality. Tetradecane was tested in the two color pyrometry measurements as "reference" fuel.

Table 3. Fuels characteristics. RF= Reference Fuels- DF= Diesel Fuels.

Fuel	Formula	Density [kg/m^3 @ 15°C]	Boiling Point [°C]	Heat content [kJ/kg]	Cetane Number RF-DF
Butyl-Ether	$(C_4H_9)_2O$	767	141		91-100
Pentyl Ether	$(C_5H_{11})_2O$	785	184	43600	111-130
Ethylene-Glycol-Dimethylether	$CH_3O(CH_2)_2OCH_3$	870	85	31560	90-98
Diethylene-Glycol-Dimethylether	$(CH_3OCH_2CH_2)_2O$	943	162	30340	112-130
Diethylene-Glycol-Diethylether	$(C_2H_3OCH_2CH_2)_2O$	910	189	33200	113-133
Tetradecane	$C_{14}H_{30}$	762	252	47300	93

Pentyl-Ether and Diethylene-Glycol-Dimethylether (Di-Glyme) was tested also in four-cylinder engine, in comparison with a highly hydrotreated fuel -HDT (at cetane level of 66 and sulphur and polyaromatic free), and Fisher Tropsch fuel (blend of pure paraffines at cetane level of 83.3). In the syngle-cylinder experiments, only for aldehyde emissions, also HDT fuel was tetsted. Preliminary engine tests showed that oxygenated fuels exhibit a very high tendency to self ignition and the problem to determine their cetane number arose. To this aim a correlation between cetane number and ignition delay time measurements was carried out (see Figure 1), both for diesel fuels having a known cetane number in the range 35-93, and for blends of the secondary reference fuels for ASTM D639 procedure. By measuring ignition delay time of oxygenated fuels their cetane number was obtained by extrapolation.

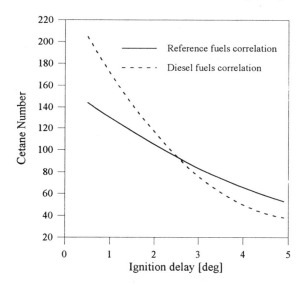

Fig 1 Correlation between the measured ignition delay and cetane number for single cylinder diesel engine. r.p.m.=1250, i.m.e.p.=4.5 bar, start of combustion=TDC.

However this method could produce dubious results about ignition characteristics for high cetane number synthetic fuels [11]. Therefore the extrapolated cetane number values for this ethers and glycol-ethers must be taken only as a self ignition index on our test engine, with respect to aforesaid reference fuels. Table 3 shows the main fuels characteristics.

4 RESULTS AND DISCUSSION.

4.1 Single cylinder engine tests

Single cylinder tests were performed at constant engine speed (1250 r.p.m.) and indicated mean effective pressure (4.5 bar i.m.e.p.). The injection timing was set to obtain, for each fuel tested, the same crank angle of combustion start, in particular at the top dead center (TDC). This setting allows to investigate soot formation process in the about same starting conditions as to the in-cylinder pressure and charge temperature. The crank angle of combustion start was evaluated on the heat release patterns as the first point in witch the rate of heat release goes to zero after the decrease due to injected fuel evaporation. Such set of injection timing was also retained when engine ran at different E.G.R. rates. In Figure 2 the heat release patterns obtained burning all oxygenated fuels are shown in comparison with Tetradecane combustion. In the diagram heat release curves of Butyl-Ether and Ethylene-Glycol-Dimethylether (Mono-Glime) combustion as well as these of Diethylene-Glycol-Diethylether (Diethyl-Diglycol) and Di-Glyme are grouped together because no significant differences in these cases were found. Due to the very high value of the cetane number, the combustion evolution is mainly controlled by the diffusive phase for all fuels tested.

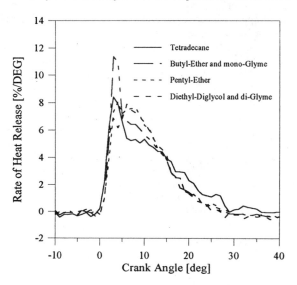

Fig 2. Single cylinder diesel engine. Rate of heat release of Tetradecane and oxygenated fuels combustion. r.p.m.=1250, i.m.e.p.=4.5 bar, start of combustion=TDC.

Figure 3 shows the evolution of soot volume fraction for Tetradecane, Butyl-Ether, Penthyl-Ether, Diethyl Diglycol, Di-Glyme combustion without E.G.R. When the fuel oxygen content rises the whole fv curve is lowered. In particular the Pentyl Ether and Butyl-Ether fv peak is lower by 50% with respect to the Tetradecane one, while the Diethyl-Diglycol fv peak is one magnitude order lower. In this condition Di-Glyme shows radiance levels so low that soot volumic fraction measurements are only indicatives, while Mono-Glyme radiance level is under the response threshold of detector system at both measurement wave lengths (600 and 1000 nm).

It must be noticed that all fuels present big differences in terms of density, viscosity and boiling point that affect the spray characteristics (atomisation, tip penetration, angle, liquid and vapour fractions). In any case a very high cetane number of the fuel reduces the influence of fuel properties on spray characteristics. Thus the oxygen effect on soot formation and oxidation process is enhanced. The peak values of flame temperature related to fv curve are labelled in the same figure while in the bars chart of the figure 4 smoke, HC, NOx, CO exhaust emissions are reported. A decrease of soot loading when the soot peak temperature rises is clearly detected.

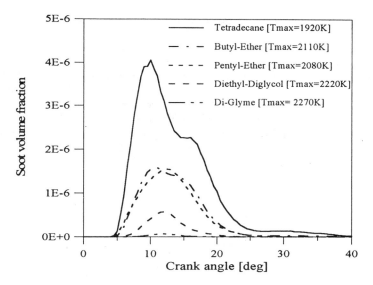

Fig 3. Single cylinder diesel engine. Soot volume fraction evolution for Tetradecane and oxygenated fuels combustion. *Mono-Glyme radiance level is under response threshold of detector system.

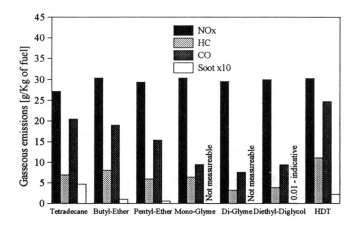

Figure 4. Single cylinder diesel engine. NOx, HC and CO emissions for Tetradecane, HDT and oxygenated fuels. r.p.m.=1250, i.m.e.p.=4.5 bar, start of combustion=TDC.

The pyrometry results are validated by exhaust smoke measurements. Moreover the HC and CO emissions decrease with cetane number, as well known. Vice versa the NOx emission doesn't scale with cetane number, probably because of the over control of fuel oxygen content and flame temperature on NOx formation process.

One of the main known disadvantage of the oxygenated fuel is related to their tendency to increasing aldehydes emission. In the bar chart of the figure 5 the exhaust aldehydes emission of some of the oxygenated fuels tested with respect to tetradecane and HDT fuel are compared.

The sampling and the analytical procedures for the aldehydes measurements are detailed in [12]. The Acetaldehyde and the Formaldehyde exhaust emissions of the oxygenated fuels combustion are similar to the tetradecane combustion and anyway lower with respect to HDT combustion. This fact confirms that the Aldehydes emissions follows about the same trends of unburned HC emissions. In fact the short ignition delay times leads to relatively longer combustion times, then to lower unburned products emission. The effect of the very high cetane number of this oxygenated fuels under testing probably over-controls the aldehydes emissions also, overcoming the tendency of these fuels to produce carbonilic compounds. Tests aimed at investigate the fuels' behaviour with E.G.R. were also performed.

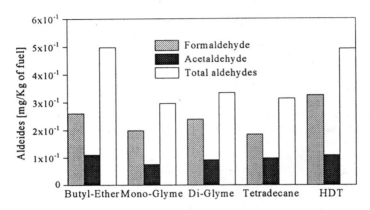

Figure 5. Single cylinder diesel engine. Aldehydes emissions for Tetradecane, HDT and oxygenated fuels. r.p.m.=1250, i.m.e.p.=4.5 bar, start of combustion=TDC.

In Figure 6 the evolution of soot volume fraction for Tetradecane combustion, at three different E.G.R. rates is shown. In the same figure NOx exhaust concentration, smoke emission and maximum flame temperature during combustion cycle are also labelled. The increase of the peak of soot volume fraction when the E.G.R. rate increases up to 45% is quite evident. At higher E.G.R. rate (60%) a strong decrease of both fv peak and the whole fv curve values are clearly detected. The formation process of soot, in these conditions, is shifted later in the expansion stroke with respect to the operation without E.G.R. Therefore the exhaust smoke emission is similar to those obtained without E.G.R., even if the fv curve peak is quite low; in any case lower with respect to the case of intermediate E.G.R. rates. This last effect is probably due to the oxygen lack, at highest E.G.R. rate, that bring down the efficiency of oxidation process.

In turns, at highest E.G.R. rate, the exhaust NOx emission is practically avoided. Looking at the flame temperature peak values, the well-known bell shaped behaviour of the soot volume fraction versus flame temperature can be clearly detected for Tetradecane combustion. This fact appears as a useful key to overcame the constraints of lowering both in cylinder soot and NOx emissions during diesel combustion. As matter of fact, to achieve soot and NOx free combustion appears in principle possible, if stable engine operating conditions can be reached at very high E.G.R. rates, by a suitable control of the maximum flame temperature value.

The synthetic fuels previously described were tested at the same i.m.e.p. value as Tetradecane into the experimental engine, to assess the capabilities offered by oxygenated fuels in order to achieve soot free combustion process. As previously stated, during Tetradecane combustion the E.G.R. effect on soot formation is similar to that of diesel fuels, showing a rise of soot loading at intermediate E.G.R. rates followed by a decrease when the limit of engine stable operation is approached (E.G.R.≥65%). Unlike Tetradecane, the oxygenated fuels present, during high E.G.R. operation, a flame radiance too low for the response threshold of detector system. Exhaust smoke measurements also validated such results, as the instrument was not able to detect any smoke. On the other hand the very low level of NOx emissions always exhibited under high E.G.R operation with all fuels hinder the possibility to discriminate any difference in flame temperature level by indirect way.

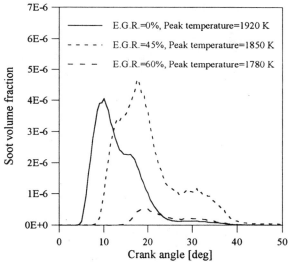

Fig 6 Single cylinder diesel engine. Soot volume fraction evolution for Tetradecane combustion at three different E.G.R. rate. The E.G.R. rate, Soot and NOx exhaust emissions, in g/Kg of fuel, are reported: E.G.R.=0% - Soot=0.47 [1.5 BSU], NOx=27.2. E.G.R.=45% - Soot=0.60 [2.3 BSU], NOx=6.9. E.G.R.=60% - Soot=0.21 [1.6 BSU], NOx=0.46. r.p.m.=1250, i.m.e.p.=4.5 bar, start of combustiuon (without EGR)=TDC.

4.2 Four cylinder engine tests

To confirm the results obtained by the single cylinder engine tests, Pentyl-Ether and di-Glyme were tested also in a modern four cylinder d.i. diesel for light duty applications. Their behaviour was compared with that obtained burning the HDT and FT high quality diesel fuels previously described. Tests were carried out in steady state conditions setting the load at 2 bar B.M.E.P. and 2000 rpm. As well known this test point is very representative of the engine emission global performance over the transient EEC 15 schedule. In this range of load and speed, the electronic E.G.R on board control system provides the full opening of the E.G.R. proportional valve. So the steady state tests were performed opening fully the valve too. In this condition the E.G.R. rate (measured by carbon dioxide method) is about 25%. As in the single cylinder engine tests, the dynamic timing injection was settled, for each fuel, to obtain the same crank angle of the combustion start, actually 3 C.A. before the top dead center. The gaseous exhaust emissions for all fuels tested are reported in the figure 7, while in the figure 8 the particulate emissions splitted into the insoluble (IOF) and the soluble (SOF) fractions by dicloromethane extraction are shown.

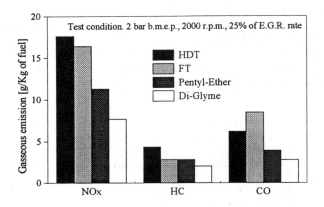

Fig 7 Four cylinder diesel engine. Gaseous emissions for HDT, FT, Pentyl-Ether and DI-Glyme fuel.

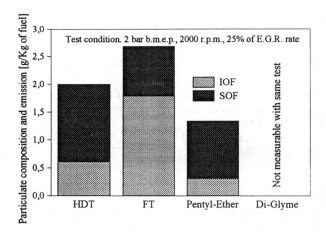

Fig 8 Four cylinder diesel engine. Particulate composition and emission for HDT, FT and Pentyl-Ether fuel.

The higher cetane number of the oxygenated fuels produces a reduction of all gaseous emissions at a very low level. Looking at the particulate emissions, it is easy to observe that the four cylinder engine tests fully confirms the results obtained with the single cylinder engine. As matter of fact the Pentyl-Ether particulate emission is about 50 % lower with respect to the FT fuel. This last fuel is practically a blend of pure paraffins and its main properties are very close to the tetradecane. Because all fuel tested are practically sulphur free, the IOF fraction of the particulate is mainly composed by carbonaceous compounds (soot). The effect of a favourable Carbon/Oxygen ratio of Pentyl-Ether and Di-Glyme on soot emissions is quite evident. As matter of fact, despite of high cetane number of these fuels, the soot fraction of the particulate is very low or not detectable. Unlike this, burning paraffinic diesel fuels, the soot emission strongly rise when the cetane number increase. Anyway it is interesting to notice that the Di-Glyme combustion is practically particulate free, if the procedure for particulate measurements is the same as in the case of diesel fuels.

5 CONCLUSIONS.

The emission potential of synthetic oxygenated fuels was studied both in a single cylinder d.i. diesel engine and in a four cylinder d.i. light duty engine. As our test conditions shown, soot tendency of synthetic fuels appears strictly correlated with oxygen content of the parent fuel molecule. In particular the soot loading peak decreases of 50% for Pentyl-Ether and Butyl-Ether with respect to Tetradecane combustion and of an order of magnitude for the Diethyl-Diglycol combustion, whereas for Di-Glyme and Mono-Glyme it is negligible.

The E.G.R. rate effect on sooting tendency of all tested fuels is quite similar: increasing the E.G.R. rate an increase of fv was firstly observed followed by a remarkable decrease. This has been interpreted by considering the bell shaped behaviour of the fv curve vs. the flame temperature. The E.G.R. increase, as it diminish the temperature of the flame, moves the working point from the right side of the bell to the left one passing for the fv maximum.

Part of the oxygen content in the fuel molecules is probably available to oxidise pyrolysis products and this effects is probably responsible for soot lacking when tested in diesel engine with E.G.R.

The Ethers and Glycol-Ethers fuels tested present a high cetane number that corresponds to a greater time available for oxidation. So the Aldeydes emissions, despite of the strong tendency of oxygenated fuels to produce they, are of the same order of magnitude or lower with the respect to diesel fuels.

This oxygenated fuels present a very low soot emission and a very high cetane number, with the corresponding favourable effects on HC, CO and NOx emission. Moreover, some physical-chemical characteristics (like, viscosity, density, heat content, etc.) are quite close to those of commercial automotive fuels; so they appear as a practical alternative (both like pure fuels and in high percentage blend) if a commercial way to produce they will be available.

REFERENCES.

(1) HERZOG, P.L. BURGLER, L. WINKLHOFER, E. ZELENKA, P.and CARTELLIERI, W.P. NOx reduction strategies for DI diesel engines. SAE Paper 920470, 1992.

(2) HVENITH, C. NEEDHAM, J.R. NICOL A.J. SUCH, C.H. Low emission heavy-duty diesel engine for europe, SAE Paper 932959, 1993.

(3) SORESON, S.C. MIKKELSEN, S.E. Performance and emissions of a 0.273 liter direct injection diesel engine fuelled with neat dimethyl ether, SAE Paper 950064, 1995.

(4) FLEISCH, T. MCCARTHY, C. BASU, A. UDOVICH, C. CHARBONNEAU, P. SLODOWSKE, W. MIKKELSEN, S. MCCANDLESS, J. A new clean diesel

techonolgy: demostration of ULEV emissions on a navistar diesel engine fueled with dimethyl ether, SAE Paper 950061 1995

(5) MURAYAMA, T. ZHENG, M. CHIKAHISA, T. OH, Y.T. FUJIWARA Y. TOSAKA, S. YAMASHITA, M. YOSHITAKE, H. Simultaneous reductions of smoke and NOx from a DI diesel engine with EGR and dimethyl-carbonate, SAE Paper 952518, 1995.

(6) SPREEN, K.B. ULLMAN, T.L. AND MASON, R.L. Effect of cetane number, aromatics, and oxygenates on emissions from a 1994 heavy-duty diesel engine with exhaust catalyst SAE Paper 950250, 1995.

(7) LIOTTA, F.J. MONTALVO, D.M. The Effect of Oxygenated Fuels on Emissions from a modern Heavy-Duty Diesel Engine, SAE Paper 932734, 1993

(8) DI STASIO, S. MASSOLI, P. Influence of the soot property uncertainties in temperature and volume -fraction measurements by two-color pyrometry, Meas, Sc. Technol. 5, 1994, pp.1453-1465.

(9) BERTOLI, C. BEATRICE, C. DI STASIO, S. DEL GIACOMO, N. In-Cylinder Soot and NO_x Concentration Measurements in D.I. Diesel Engine fed by Fuels of Varying Quality, SAE Paper 960852, 1996.

(10) BEATRICE, C. BERTOLI, C. CIRILLO, N.C. DEL GIACOMO, N. INNOCENTE, R. The influence of high EGR rate on emissions of a DI diesel engine, ASME Fall Technical Conference Lafayette, 1994, Vol. Heavy duty engines: A look at the future pp 193-202.

(11) HARDENBERG, H.O. EHNERT, E.R. Ignition quality determination problems with alternative fuels for compression ignition engines, SAE Paper 811212, 1981.

(12) BERTOLI, C. DEL GIACOMO, N. IORIO, B. PRATI, M.V. The influence of fuel composition on particulate emissions of DI diesel engines, SAE Paper 932733, 1993.

New Generation Fuels

fuel injection system concept for dimethyl ether

OFNER, D W GILL, and **T KAMMERDIENER**
AVL LIST GmbH, Graz, Austria

SYNOPSIS

DiMethylEther (DME) has been shown to be an interesting alternative fuel for diesel cycle engines. Due to its low boiling point under normal ambient conditions (-25°C), DME has to be handled as a liquified gas and special measures are necessary in order to inject it directly into the cylinder of the engine. This presents challenges to the fuel system designer as leakage and the properties of the fuel mean conventional systems are unsuitable. The paper presents the work done to design and layout a suitable system for engine test work which is safe and allows the combustion development to be carried out for both passenger car and truck size engines.

The fuel system concept is presented, together with the reasons for the choice of the system, background information covering the properties of DME which are of importance to the fuel injection system designer and initial simulation results.

1. INTRODUCTION

DiMethyl Ether (DME) has recently been shown to be an interesting alternative fuel for compression ignition engines running on the diesel cycle /1, 2, 3, 4/. Recent developments in catalyst technology /5, 6/ have also shown that it is possible to produce DME in large quantities at an acceptable cost. The engine tests have proved that Ultra Low Emission Vehicle (ULEV) exhaust emissions levels are attainable for both truck and passenger cars without significant additional exhaust gas aftertreatment and with low (gasoline engine) combustion noise levels. Sofar, these results have only been demonstrated on the engine test bed under steady state conditions due to the difficulties of injecting DME under transient conditions with the conventional injection systems used for these tests.

Therefore, work has been continuing at AVL to develop a concept for a fuel injection system which is specifically designed for injecting DME into diesel cycle engines. These activities also include the layout of the complete fuel system and emphasis was put on overcoming the difficulties encountered in the initial phases, namely:

- Preventing fuel leakage into the environment (engine, surroundings etc...).

- Preventing uncontrolled fuel injection into the combustion chamber of the engine during engine operation and during standstill.

- Introduction of control features for the complete fuel system that include basic engine operations such as engine start and stop.

- Improving aspects for the handling of DME e.g. filling and emptying the fuel system.

This paper describes the work done to define the fuel system and its principle of operation with DME. For designing the system, studies were also undertaken to determine the thermodynamic properties of DME. Furthermore, a test system has also been designed and assembled in a laboratory and tests are being carried out to assess the new system before engine tests begin.

2. EXPERIENCES WITH VARIOUS TYPES OF FUEL INJECTION SYSTEM

There are currently a wide variety of fuel injection systems available for diesel engines which can be divided into cam driven systems and hydraulically actuated systems. Cam driven systems are such types as pump-line-nozzle systems and unit injectors. The hydraulically actuated systems consist basically of high pressure accumulator type and hydraulically intensified systems.

The first tests at AVL were carried out using a hydraulically intensified common rail system on the Navistar 7.3l V8 engine /1/. During these tests various difficulties were experienced such as:

- Leakage past plunger barrel seals.
- High compressibility of DME.
- High vapour pressure at elevated temperatures.

As a result of these difficulties, significant modifications to the fuel supply and injection system were necessary in order to obtain meaningful results.

The next series of tests /2/ were carried out using a unit injector on a single cylinder FM538 engine having a bore and stroke of 124 x 165 mm. Due to the experience gained with the previous tests which showed DME's tendency to leak past any clearance seals, the unit injector was modified with an oil block arrangement which prevented the DME leaking past the plunger. This arrangement worked well while the engine was running, but the DME leaked out of the system as soon as the oil pressure was removed.

For the tests on a high speed direct injection (HSDI) diesel /4/, an in-line pump was used. This system suffered from a large amount of leakage past the plungers and barrels and due to the excessive compressibility of the DME when warm, the maximum injection quantity for full load running was hard to obtain.

3. PREREQUISITES FOR A DME FUEL SYSTEM

Based on the experience described above a study was carried out to define a fuel injection system for use with DME /7/. Because of the leakage problem, it was clear that all arrangements which relied on fine clearances such as those between plunger and barrel, rotor and barrel, etc. should be avoided . The logical alternative was to consider contact seals, but with these, tightness cannot be guaranteed with both the high pressures used within fuel injection systems and with the high plunger velocities of standard fuel injection pumps. Therefore conventional pumping systems with slip fit clearances were considered unattractive for use with DME. In addition the variation of compressibility with temperature exhibited by DME makes "jerk pump" systems unsuitable due to the large differences in pumping stroke necessary for "relatively" small temperature fluctuations.

Within common rail systems the high pressure can be produced by a continuous pumping and a variety of pumping concepts can be applied. However, this is only one advantage among the following items which also support a common rail concept:

- Several high pressure pump concepts as e.g. used in chemical industries can be considered. Designs which are considered unconventional within automotive applications can be state of the art in other fields (e.g. special contact seals or diaphragms).

- Beside the pump, other readily available components (valves, fittings, pressure regulators etc.) as used in other fields of the hydraulic and chemical industries can also be applied.

- Common rail systems best overcome the characteristic properties of DME such as high vapour pressure, low viscosity, high compressibility and considerable change of properties versus temperature.

- Their application to different engine sizes is relatively easy.

- They are electronically controlled systems which will be required anyway for achieving ultra low emission levels.

Beside the design points of view, a common rail system also looks attractive with respect to the engine requirements . The work in /1,2,4/ demonstrated that even low injection pressures (200 to 300 bar) lead to excellent engine performance with DME and it has also been shown that combustion reacts less sensitively than diesel fuel to long injection durations and slow injection pressure decays at the end of injection. Any control of the initially injected fuel rate (pilot injection or rate shaping) improved both the NOx emissions and combustion noise.

Finally these engine requirements make clear that a common rail DME injection system looks different to those which are currently under development for diesel fuel. The high pressure part of the system must withstand pressures no higher than 300 bar (compared to 1500 bar in diesel fuel systems!). However, it must be considered that the advantages, which originate from the physical and chemical propertys of DME, are partly compensated by the undesirable property that DME immediately evaporates into the environment or combustion chamber due to any uncontrolled leakage. Thus it forms an explosive mixture with air at concentrations higher than 3.4 % (Vol.). To avoid that, precautionary measures must be considered during the concept phase.

4. DME FUEL SYSTEM FOR VEHICLE APPLICATION

4.1 Basic Concept

Even if clearances at all moving parts within the system can be replaced by safe and reliable designs, the poor sealing of the needle seat in conventional fuel injection nozzles is still a concern. It must be considered that DME is stored in a pressurized tank in order to keep it a liquid. Similar to systems for Liquified Petroleum Gas (LPG), DME is held in the saturated state in the storage tank. This means that the lowest pressure level available in the DME system is the saturation pressure, which at 20°C ambient temperature, lies at 5 bar (see fig. 1). If this pressure would remain active in a conventional diesel injection nozzle during engine shutdown, DME would certainly leak into the combustion chamber as the needle seal is a metal to metal contact. Compared to LPG this leakage is more of a problem with DME as it would selfignite uncontrollably during the next engine start.

Therefore the demand was made to depressurize the injector during normal engine shutdown. As soon as the injector's trapped DME has evaporated, the majority of the mass is removed and the remaining gas does not represent a risk any more. The problem is that for emptying the injector, a pressure level below saturation pressure is required.

This shutdown problem is not the only reason why a low pressure reservoir is desirable. When considering any contact seal as shown in figure 2, the design considerably improves in terms of safety if two seals are specified in series and the chamber between them is ventilated. For the prototype systems used in vehicles, the design principle as shown in figure 2 is used as the standard for all contact seals even if no relative movement occurs between the parts sealed. Standards like this are considered a "must" as long as no experience, durability tests or specially designed components for DME systems within automotive applications are available.

The basic concept for the DME fuel system is outlined in figure 3. According to all considerations stated above, the complete system can be split into three individual parts - the fuel injection system (figure 3 - I), the fuel storage and supply system (II) and a low pressure part which in the following is called the "purge" system (III).

Basically, these individual parts can be characterized by their operational pressures which are the injection pressure (200 to 300 bar) in the injection system (I), saturation pressure in the supply system (II) and approximately ambient pressure in the purge system (III). In both the purge system and the decompressed parts of the injection system, the DME is in a gaseous state. A compressor is required to recompress the gas into the storage tank. This compressor is also used to control the purge tank pressure and to keep it below a certain level. It is not driven by the engine but by an external drive which is also able to operate during engine still stand.

4.2 Features of the Common Rail Fuel Injection System

With respect to the items listed in chapter 3 the most appropriate fuel injection system is considered to be a common rail injection system as schematically outlined in figure 3. Within this system a fuel rail contains liquid Dimethyl Ether at a pressure between 200 and 300 bar. From the rail, injection pipes lead to the individual cylinders. Solenoid valves are directly fitted in these pipes and switch the main fuel stream from the high pressure rail to the injectors. For the fuel injectors, conventional nozzle holders and nozzles are applied as commonly used e.g. in pump-line-nozzle injection systems.

This concept entirely differs from the diesel fuel high pressure common rail systems e.g. /8/. The latter actuate the injector needle by means of a command piston. Consequently extremly small, fast and accurate solenoid valves can be used as only the pressure of a small actuation chamber is controlled. For DME, a concept as described above can be applied because of the relatively low rail pressures and the lower demands for the injection rate shape.

The main advantage of this concept is considered to be that the injectors are pressurized only for the injection events. During the remaining time between the injections the residual pressures are low and preferably correspond to the storage tank pressure. This can be most easily realized if three way solenoid valves are used as outlined in figure 3. With respect to safety reasons, these valves keep the injection pipes closed from the valve springs and open them only for "power on". Even if one of the high pressure valve seats leaks, the injector is not pressurized because of a return line into the storage tank. The injectors' needle opening pressure is considerably higher than the storage tank pressure. If using two way solenoid valves this concept still holds as the injectors' fuel pressure is relieved via suitable orifices.

The following items are considered further advantages of the common rail system as introduced by figure 3:

- The fuel injectors are standard type nozzle holders and injection nozzles. Consequently minimum changes in cylinder head designs are required.

- Several types of initial rate shaping devices can be applied to the system, e.g. Stanadyne RSNs (Rate Shaping Nozzles) /9/, two spring injectors, pintle nozzles (with a convenient characteristic of flow rate versus needle lift), retraction type split injection devices etc.

- In the case of engine shutdown, the fuel injectors can be purged independantly from the remaining system. As in this case only small volumes need to be purged, the injectors fuel pressures can be relieved very quickly which is important within emergency shutdowns.

The high pressure pump preferably contains a diaphragm sealing concept. Such pumps are state of the art in the chemical industry and diaphragms which are sufficiently flexible and resistant to DME are available. However, not all pumps are suitable for automotive use and development work is continuing with a pump manufacturer to develop a suitable pump.

4.3 Fuel Storage System

The low pressure system is based upon LPG technology which has been modified to be compatible with DME. These systems are laid out to provide the fuel as a subcooled liquid to the injection system. Any evaporation at the intake to the high pressure pump must be avoided as otherwise fuelling problems occur and the cavitation causes erosion or damage (e.g. if diaphragm pumps are used cavitation could destroy the diaphragms).

Within the storage tank the fuel is usually kept in the saturated state, thus the tank pressure is automatically controlled by the vapour-liquid equilibrium behaviour of the fuel. A subcooled liquid state is achieved by increasing the pressure of the saturated liquid fuel using a pump which preferably is located inside the tank (because of the leakage).

The tank pressure can be estimated from the vapour pressure behaviour given in figure 1. Depending on the ambient temperature it usually lies between 3 and 8 bar. The pressure increase from the tank mounted pump should amount to about 5 bar.

For dimensioning the DME tank it must be considered that a maximum of 80% of the storage tank volume may be filled with liquid. The remaining gas volume is necessary to stay within reasonable pressures if the ambient temperature changes.

4.4 Purge System

Purge systems are provided for prototype systems used in vehicles. The question as to whether they can be avoided and a sufficient saftey standard still can be reached without them cannot be answered at this time as both the know how and experience are not yet available with DME fuel systems. The purge tank absorbs all the leakages as well as the DME released from purging the fuel injectors. It is also required for basic control features (engine stop, emergency stop) and makes the filling and emptying of the DME fuel system easier.

The purge tank pressure will be kept close to atmospheric pressure. If liquid DME expands into the purge tank it cools down to -25°C and partly evaporates. It is considered that after this expansion the heat from the environment evaporates the remaining liquid and as a consequence the pressure in the purge tank increases. After a certain amount of time the temperature of the completely evaporated DME will reach the environmental temperature.

For dimensioning the tank volume both this thermodynamic change of state must be considered and the delivery rate of the compressor which recompresses the gas into the storage tank. The recompressed gas is liquefied again. If it is directly returned into the storage tank, the heat of condensation is automatically transferred to the tank and the environment. If the quantity of returned gas is higher than that which condenses inside the tank, the tank pressure increases. In order to avoid that, an external condenser could be necessary which uses the environmental air to condense the DME, however layout calculations done so far do not indicate that a condenser is required.

The design targets of both the purge tank and the compressor are to keep the purge pressure below two bar even if both the injectors, high pressure rail, pipes and supply lines are purged. The recompression of the released DME should require no more than one to two minutes.

5. THERMODYNAMIC PROPERTIES OF DME

For dimensioning the fuel system as described, thermodynamic data of DME is required. In contrast to diesel fuel knowledge of the phase behaviour and the vapour state of the fuel are also required as changes of states like expansion and evaporation and the during purging, recompression and condensation of DME need to be investigated. On top of this the data for subcooled liquid DME must be determined for layout of the high pressure part of the flow and for computer simulation of the system dynamics.

Some liquid properties of DME have already been published in /1/. These were identified from dynamic measurements using numerical identification methods developed at AVL /10/. This data was sufficient for the layout of th fuel injection equipment used for the engine tests as published and also some flow phenomena were analysed in more details. However, it was clear that a more comprehensive investigation of thermodynamic data was necessary and studies were carried out to determine the most essential data /11/.

As both the properties of liquid DME and the vapour pressure must be considered for temperatures up to more than 100°C, difficulties come up with respect to approaching the critical point (t_{crit} = 126.9°C, p_{crit} = 52.36 bar). At this point both the isothermal compressibility and isobaric heat capacity approach infinity, consequently the prediction of properties in its vicinity is extremely sensitive.

Thermal properties and velocity of sound can be calculated from an equation of state and functions for saturated vapour pressure and heat capacity using thermodynamic laws. The model introduced provides physically meaningful data for the liquid phase except for the temperature range close to the critical temperatures. The state behaviour of the saturated vapour is described using the ideal gas equation. Therefore these values should be treated with care as they are only estimations even down to temperatures well below the critical point.

The figure 4 shows the density and modulus of elasticity of liquid DME as functions of pressure and temperature. These are the most important properties with respect to simulating the flow characteristics in the injection system. The modulus of elasticity shows a large variation dependent upon both the pressure and temperature. This extreme variation in the modulus of elasticity is one of the major factors influencing the choice of a common rail injection system for DME.

For a rough layout of the fuel storage and purge systems some saturated vapour phase properties were also calculated as shown in figure 1. This data is subject to the assumptions stated above and should be treated with care.

6. INJECTION SYSTEM LAYOUT BY MEANS OF COMPUTER SIMULATIONS

Figure 5 shows the simulated characteristics of an injection system as outlined in figure 3. These results are based on a system laid out for a truck engine of about 1.5 liter/cylinder with a 2000 rpm rated speed.

For this example, a rail pressure of 250 bar is assumed and the injection is controlled by a three way solenoid valve. The lower diagram in figure 5 shows the simulated switching of the cross sectional flow areas at the valve seats. The effective flow areas are considered to be 3 mm^2 and they open and close within 1 ms.

The centre diagrams show the pressures in both the injection line and the injector and the rate of injection. It can be seen that the injection pressure very rapidly increases as soon as the solenoid valve opens. At end of injection the pressure decay is relatively slowly.

With respect to a diesel fuel injection the shape of the injection rate as given in figure 5 looks problematic because of the high initial rate and the moderate decay at the end of injection. While the latter is assumed to be acceptable with DME, the high initial rate needs to be controlled by a rate shaping device in order to achieve a smooth start of combustion.

7. TEST EQUIPMENT IN LABORATORY

In order to assess and test a DME fuel system as described, a system has been assembled in an explosion proof laboratory at AVL. For this system, the goal was to keep it as flexible as possible. Thus, e.g. components can be interchanged without any system modification, the number of injectors can be varied, the common rail fuel volume can quickly be changed etc. This philosophy lead to the design of several "external" rails which are characterized by short collecting tubes which contain all the fittings and measuring adaptors.

The "external" rails are also schematically outlined in figure 1. E.g. within the fuel storage system the tank is provided with only one supply line and one return line. Both the lines to the injection system and all return lines lead into small boxes. These symbolize the "external" rails.

Similar strategies are pursued for the common rail and the purge system. It should be mentioned that another reason for providing the "external" rails was the use of LPG tanks as commonly available on the market. These are certified tanks for use within vehicles and consequently should not be provided with additional threaded holes for check valves etc.

The photograph in figure 6 clearly shows the "external" rails. Starting from the bottom the picture shows the rail for the fuel flow out of the storage tank, the rail for all return lines, the common rail and the external purge tank rail. It can be seen that for most of the connections hoses are used and that all fittings and valves can be easily accessed. If no more space is available for additional adaptors or components on an "external" rail, it can be lenghtend.

As it is expected that the prototype systems frequently will have to be filled and emptied again, special equipment was also provided for making these jobs easier. Additionally, parts of the system can be seperated and purged individually.

The system as demonstrated on the photograph in figure 6 still contains some provisional arrangements. E.g. as no LPG fuel delviery pump was available because of material compatibility, the storage tank currently is pressurized by Nitrogen in order to prevent cavitation on the intake to the high pressure pump. The DME compressor used to recompress the purge tank vapour into the storage tank is a pneumatically actuated piston compressor as used by the chemical industry within explosion proof areas.

8. SUMMARY

The design of a fuel injection system suitable or use with DME must primarily consider safety and handle any possible leakage from the system. Therefore, a purge system has been introduced and tests will be carried out to determine the extent to which it will be needed in practice.

As the traditional technology used within the automotive industry is not entirely suitable for handling DME, techniques from such areas as the chemical industry have been used and must be adapted if production is to be seriously considered later.

Tests are currently being carried out with the system, initially with diesel fuel, to ensure that the system is leak free, before tests with DME are started.

References

/1/   McCarthy, C. et al.: A New Clean Diesel Technology: Demonstration of ULEV Emissions on a Navistar Diesel Engine Using a New Alternative Fuel, SAE-Paper 950061

/2/   Kapus, P.; Ofner, H.: Development of Fuel Injection Equipment and Combustion System for DI Diesels Opeated on Dimethyl Ether; SAE-Paper 950062

/3/ Sorensen, S.C.; Mikkelsen, S.E.: Performance and Emissions of a 0.273 Liter Direct Injection Diesel Engine with a New Alternative Fuel, SAE-Paper 950064

/4/ Kapus, P.; Cartellieri, W.: ULEV Potential of a DI/TCI Diesel Passenger Car Engine Operated on Dimethyl Ether, SAE-Paper 952754

/5/ Hansen, J.B. et al.: Large Scale Manufacture of a New Alternative Diesel Fuel from Natural Gas, SAE-Pape 950063

/6/ Fleisch, T.; Meurer, P.C.: DME The Diesel Fuel of the 21st Century? Presented at AVL Conference "Engine and Environment 1995" Graz, Austria

/7/ Concept Study on DME Fuel System, AVL internal R&D work, 1995

/8/ Krill, W.: New BOSCH solenoid valve-controlled fuel injection systems IMechE Seminar Publication 1995-3

/9/ Johnson, P.: Fuel nozzle designed to cut emissions, DIESEL PROGRESS Engines & Drives, August 1995

/10/ Ofner, H.; Egartner, W.: Identification of flow phenomena in fuel injection systems for diesel engines from dynamic measurement, C499/053 IMechE 1996

/11/ State and phase behaviour and thermodynamic properties of Dimethyl Ether; a preliminary study; AVL internal R&D work, 1995

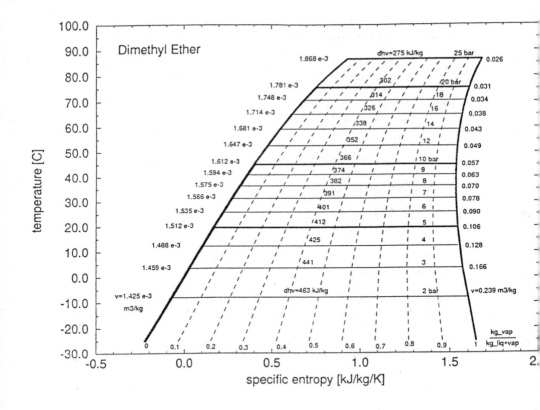

Figure 1   Thermodynamical properties of Dimethyl Ether at the saturated state

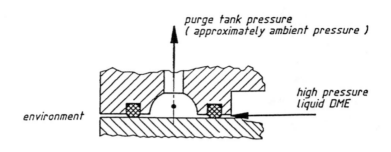

Figure 2   Concept of a contact seal

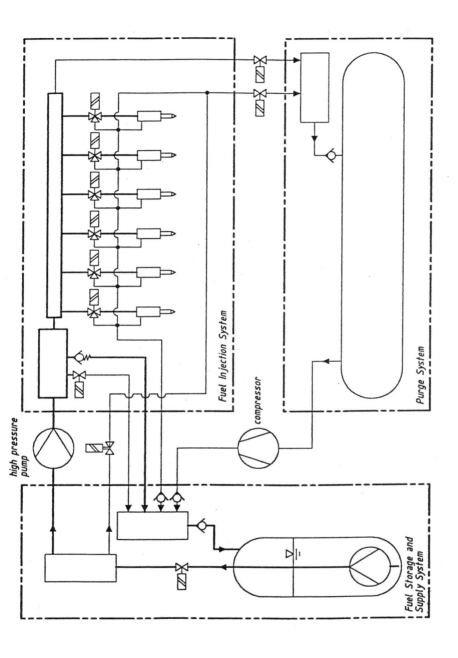

Figure 3 Scheme of Dimethyl Ether fuel system

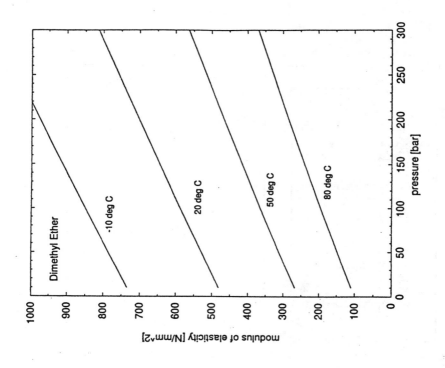
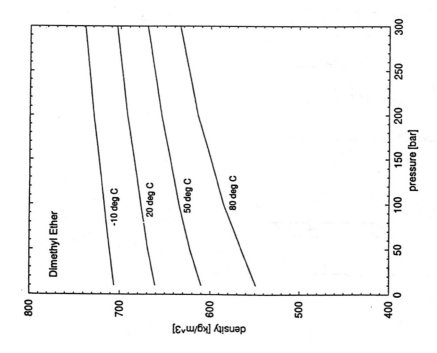

Figure 4 Propeties of liquid Dimethyl Ether as function of pressure and temperature

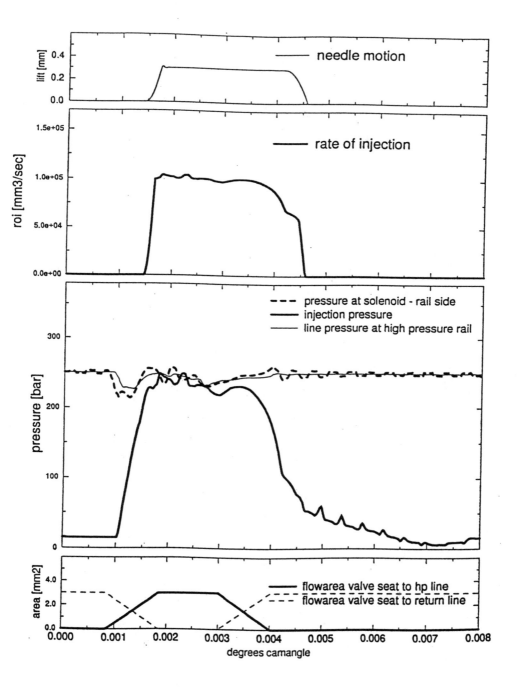

Figure 5 Computer simulation of common rail injection system as outlined in figure 3

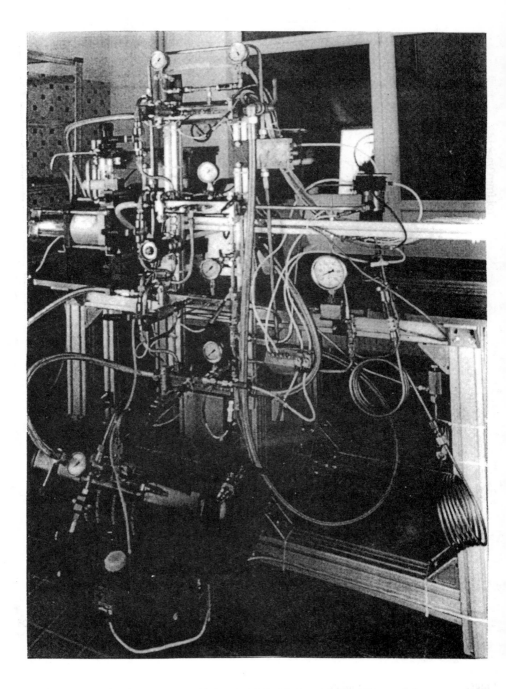

Figure 6   Test equipment of Dimethyl Ether fuel system in AVL test laboratory

Dimethyl ether as an alternate fuel for diesel engines

E MIKKELSEN and **J B HANSEN** MSc
Haldor Topsøe A/S, Denmark
S C SORENSON BS, PhD, MSAE
Technical University of Denmark

Synopsis

Experimental studies have demonstrated that neat dimethyl ether (DME) exhibits very favourable characteristics as an alternative to conventional diesel fuel. These benefits include low NO_x emissions, no smoke/low particulate emissions, lower noise and lower exhaust gas reactivity, without a loss in engine thermal efficiency, using essentially conventional diesel combustion technologies.

Detailed analysis of the organic components in the exhaust will be presented and compared with that of normal diesel exhaust. Insignificant amounts of photo-reactive components are emitted compared with diesel operation, thus leading to a low potential for ozone formation.

Although DME is not a commodity, it can be manufactured from nearly all carbon-bearing substances at a price not unduly above that of standard diesel. A brief discussion of the production processes from the most attractive feedstock, natural gas, will be outlined.

1. Introduction

Within a short period of time, dimethyl ether (DME) has emerged as an attractive alternate fuel for diesel engines. Initial studies on a small, single cylinder diesel engine (1) demonstrated the benefits of DME: low NO_x emission, no soot formation, favourable response to EGR and low engine noise. The development of a new process for DME production has shown that it can be produced in an economically competitive manner from natural gas or other carbonaceous materials (2). Furthermore, DME is an environmentally friendly, non-toxic fuel (3,4).

These results prompted subsequent studies that have quantified the performance of DME in larger engines for vehicular applications (5,6). These studies have shown that DME has the potential for achieving ULEV emission levels in different engine sizes. These achievements have been made without extensive development programmes, which suggests that even lower emission levels should be attainable with more intensive development programmes.

Since the use of DME as a diesel fuel is still in its infancy, this paper will provide a brief introduction to the properties of DME, its manufacture and achievable emission levels. In addition, detailed analyses of the composition of the organic material in the exhaust of a DME powered diesel engine will be presented.

2. DME Properties

DME is the most simple ether, as methanol is the simplest alcohol and natural gas (methane) is the simplest alkane.

From a physical point of view, DME very much resembles liquefied petroleum gas (LPG). This can be seen in Table 1. It appears that due to its low boiling point, DME is a gas at ambient conditions. Contained in low pressure vessels DME is both in gas and liquid phase, like LPG. Although it has a low viscosity, like LPG, studies have shown that long-term operation on DME does not present a significant, technical obstacle (1,7).

Table 1. The physical and chemical properties of DME as compared to those of propane, methane, and typical diesel fuel

Physical and Chemical Properties of DME, Propane, Methane and Diesel Fuel					
		DME	*Propane*	*Methane*	*Diesel*
Chemical Formula		CH_3-O-CH_3	C_3H_8	CH_4	$C_nH_{1.8n}$
Lower Heating Value	MJ/kg	28.8	46.3	50.0	42.5
Liquid Viscosity	kg/m-s	0.15	0.15	NA	2 - 4
Liquid Density	kg/m^3	668	501	720 @-162°C	840
Cetane Number		>> 55	< 10	< 10	38 - 53
Boiling Point	°C	-25	-43	-162	180-360
Stoichiometric A/F Ratio	kg/kg	9.0	15.6	17.2	14.6
Autoignition Temperature	°C	235	470	650	250
Explosion Limits,% gas in air	vol %	3.4 - 18	1.9 - 9.5	5 - 15	0.6 - 7.5
Carbon Content	wt %	52.2	82	75	87
Hydrogen Content	wt %	13.0	18	25	13
Oxygen Content	wt %	34.8	0	0	0

The overall chemical formula, C_2H_6O, is equal to that of ethanol, but DME has an 8.4% higher Lower Heating Value (LHV). The liquid density at ambient temperature is between that of propane and normal diesel fuel, and is roughly the same as natural gas (LNG) at -162°C. It has a somewhat wider flammability range than the hydrocarbon fuels.

The most significant property of DME, in terms of operation in a diesel engine, is its low self-ignition temperature, which is close to that of normal diesel fuel. Since it is injected at a pressure above its critical pressure (53.7 bar) and has a high vapour pressure, DME readily vaporises upon injection, which combined with its low ignition temperature gives it a very high cetane number, in excess of 55 (5).

The high oxygen content and lack of carbon-carbon bonds result in smokeless operation in diesel engines of a wide range of sizes. The only source for small amounts of PM comes from the lube oil (1,5,6). Figure 1 illustrates the molecular structure and chemical composition of various fuels.

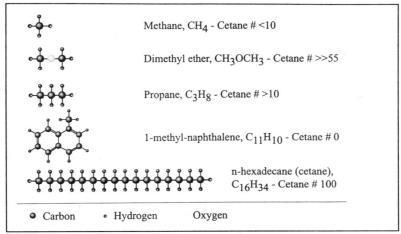

Fig.1 Chemical composition and cetane rating

3. Manufacture of DME

DME does not occur naturally in any significant amount. To date, the relatively small amount of DME used in aerosol cans in Europe (100 000-150 000 TPY) is produced by dehydration of methanol. This process is relatively expensive, and the company of authors 1 and 3 have developed a new production process for DME (2). Although DME can be made from any carbonaceous material such as coal, crude oil or biomass, the currently most attractive option is production from natural gas from remote sources or locations, where pipelines are not economically feasible thus making the gas available at a low price.

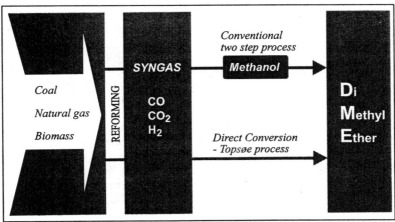

Fig. 2 Schematic diagram of the production process for DME from natural gas

The production process, shown schematically in Figure 2, involves the reforming of natural gas to syngas consisting of mainly CO, CO_2 and H_2. Initially, the process was developed to produce DME as an intermediate in the production of synthetic petrol (8). The process has been successfully demonstrated in pilot scale. Syngas has been used to make methanol, from which DME can be produced through dehydration. However, the new process involves the direct production of DME in a single conversion step. Introduction of this DME process gives more favourable thermodynamic conditions, which results in a lower synthesis pressure with lower energy consumption than for the production of methanol. An important step in the total process is the production of syngas, which is accomplished through the use of a specially designed autothermal reformer. This new technology permits the large scale production of DME at reduced cost. Details of the production process are given in (2).

4. Emission Potential

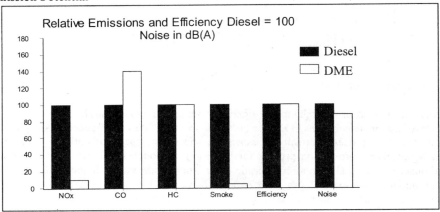

Fig. 3. Relative emissions for a small DI diesel engine for diesel and DME operation with EGR and no after-treatment

Figure 3 shows the relative emissions and noise (combined mechanical and combustion) from a small, direct injection diesel single-cylinder engine operating on diesel fuel as compared to those for DME with 30% EGR without exhaust gas after treatment. It is significant to note that the NO_x and smoke emissions have been drastically reduced, without a reduction in the thermal efficiency of the engine.

It should also be noted that operation with an oxidation catalyst on the above engine resulted in additional, significant reductions in the HC and CO emissions. The injector opening pressure was reduced from 200 to 80 bar for the DME engine. The original injector with four 200μ holes was maintained in the DME engine. This indicates that these low emissions can be achieved at much lower injection pressures than currently used in diesel engines; a result which has also been confirmed in larger engines (5). The overall CO and VOC (Volatile Organic Compounds) emission levels are similar in magnitude to those of diesel-fuelled engines. The term VOC is used here, since DME is an oxygenate and not a hydrocarbon, but it still counts at the traditional Flame Ionisation Detector. The measurements of DME-exhaust gas reported here as HC were made using this FID type detector.

5. Organic Compounds in the Exhaust Gasses

In previous studies it was reported that the composition of the organic material in the exhaust of engines operating on DME consists primarily of DME, with a smaller amount of methane being the only other large organic contributor to the exhaust gasses (1). Additional studies have been undertaken to further investigate the detailed composition of exhaust gas organic compounds in a diesel engine operated on DME. These studies have been conducted on a new Yanmar single cylinder, direct injection diesel engine, the specifications of which are given in Table 2.

Table 2.

Test Engine Type	Single Cylinder 4-stroke DI Diesel Engine
Rated Power	4.5 kW
Cooling System	Forced Air
Bore	68 mm
Stroke	75 mm
Displacement	0.273 l
Compression Ratio	21:1
Maximum Speed	3600 rpm
Injector Opening Pressure on diesel fuel	200 bar

Of importance with respect to the local environment is the wide range of hydrocarbons and derivatives emitted from engines operated on standard diesel fuel. It is well known that diesel exhaust contains a very large number of organic compounds. In recent work, the European Programme on Emissions, Fuels and Engine Technologies, EPEFE (9), has reported the typical levels of a large number of organic substances in exhaust from modern diesel engines and fuels. A large portion of hydrocarbon emissions from diesel fuel consists of heavy ($C > C_{13+}$) hydrocarbon, predominantly from unburned fuel. This is comparable to the unburned DME in the exhaust of DME powered diesel engines. The second large group of organic substances in diesel exhaust according to EPEFE is monoalkenes, $C_2 - C_5$. These substances are known to have a high degree of photochemical reactivity. Light aldehydes C_{1-3}, compose the third large group of organic compounds in diesel exhaust. Significant amounts of 1,3- butadiene, benzene, and other aromatic compounds are found in diesel exhaust, in spite of the absence of these species in the fuel.

In order to compare the composition of the organic materials in the exhaust gases, engine tests were made using DME and a commercial "light diesel" fuel. The diesel fuel had a sulphur concentration of 412 ppm, while the 99.99+% pure DME was sulphur free. Exhaust gas samples were collected in PTFE analysis collection bags, and hydrocarbon analyses were conducted by gas chromatography. The engine was tested at 3 loading conditions; idle, 1 kW and 2 kW. Preliminary investigations had shown that hydrocarbons above C_4 could not be found in DME exhaust, so the gas analysis was conducted for the light hydrocarbons only. This was also confirmed by the present analysis. Figures 4 and 5 show the amount of light hydrocarbons produced from the diesel and DME fuels as a function of load. Particularly at light load, the diesel engine produced relatively large amounts of the light hydrocarbons. Ethene is produced in the largest relative amounts for diesel fuel. Ethene is also the predominant hydrocarbon formed during DME operation but it is formed in the amount of a magnitude lower. Besides ethene, ethane is seen but in small quantities, especially at increased load. Higher hydrocarbons were not seen when operating on DME.

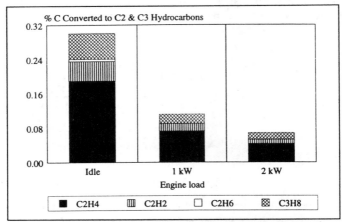

Fig. 4 Hydrocarbon species in exhaust gas from diesel operation

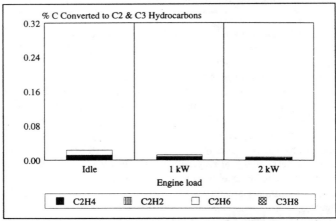

Fig. 5 Hydrocarbon species in exhaust gas from DME operation

Figures 6 and 7 show a comparison of the emissions of methane from operation on the diesel and DME fuels.

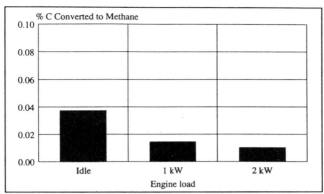

Fig. 6 Methane formation from diesel operation

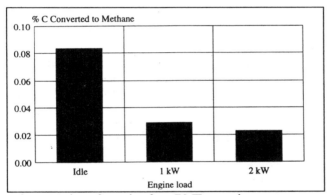

Fig. 7 Methane formation from DME operation

It appears that DME exhaust contains more methane than the diesel exhaust, however the quantity in both cases is very low. Common for both fuels is that the methane amount is lowest at high engine load (highest temperature).

As shown in Figure 3, the total VOC emission levels of the diesel and DME are the same. The remainder of the VOC from the DME engine consists solely of DME, while in the case of the diesel engine, it is composed of the wide variety of organic compounds described above. In excess of methane, the only other hydrocarbons produced from DME are very small amounts of ethene and ethane. The amounts of light hydrocarbons in DME exhaust are about an order of magnitude lower than that of the diesel fuel.

Table 3. Relative emissions of light hydrocarbons from DME and diesel fuel as a function of load.

	Ratio of DME Emission to Diesel Emission		
Component	Idle	1 kW	2 kW
Methane	2.2	1.75	2
Ethane	2.4	*)	*)
Ethene	0.06	0.1	0.12
Acetylene	0	0	0
Propane	0	0	0
Propene	0	0	0
Butenes	0	0	0
Total carbon < C_4	0.061	0.22	0.305

*) The amount of ethane was under the detection limit with diesel fuel at 1 kW and 2 kW.

Figure 8 shows that for operation on neat DME at a speed of 3200 rpm the relative amount of unconverted DME appearing in the exhaust gasses decreases from 1.9% at idle to 0.2% at a load of 2 kW. This relative amount is consistent with other studies (10).

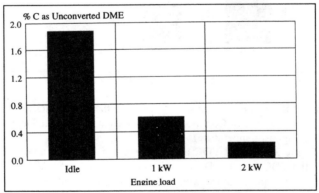

Fig. 8 DME in exhaust gas from DME operation

There are no other references for quantitative amounts of DME in the HC fraction from diesel engines operated on DME. One could expect the conversion/combustion to be increased in a larger engine due to the higher volume to surface ratio, but this has to be confirmed. It is expected that DME and the small C_2 hydrocarbon fraction can be converted to CO_2 and water from the exhaust gas by means of a simple oxidation catalyst, at a high conversion rate, but this has to be studied. It is known that methane is difficult to remove using oxidation catalysts (11). In Table 3 the results have been summarised and presented as the ratio of DME emission to diesel emission.

Another area of concern is that of aldehyde formation. It has long been acknowledged that there is a significant amount of aldehydes present in diesel engine exhaust gasses. Their presence increases the odour problems with diesel fuel. In the current investigation measurements have indicated that the formaldehyde emissions from DME operation are in the order of 3 - 5 ppmv. A 10-HDD transient cycle emission simulation with DME concluded that DME was able to meet the 1998 California ULEV standard for medium-duty trucks. Formaldehyde formation was found to be 0.022 g/bhp-hr (5).

6. Final Remarks and Conclusion

A large number of studies have shown that the environmental characteristics of DME are quite benign. These studies have been done in order to justify the use of DME as a replacement for CFC aerosol propellant for the cosmetic industry. The main conclusions regarding environmental aspects are that DME is non-toxic; it does not have a high photochemical reactivity; it has a short lifetime in the atmosphere; and it does not produce formaldehyde during atmospheric oxidation (3). Since methane is also quite benign in the local environment, the appearance of these two substances as the most prevalent organic substances in DME exhaust, shows it to be a very clean fuel in terms of hydrocarbon reactivity, as well as the more commonly reported areas of NO_x and particulate emissions.

The limited number of DME studies presented to date indicate that with only a minimal development time, engines can be changed to operate on DME at or below ULEV emissions levels for heavy duty engines and for passenger cars (5,12).

References

(1) Sorenson, S. C., Mikkelsen, S.-E, "Performance and Emissions of a 0.273 Liter Direct Injection Diesel Engine Fuelled with Neat Dimethyl Ether", SAE Paper 950964, 1995.

(2) Hansen, J. B., Voss, B., Joensen, F., Sigurðardóttir, I.D., "Large Scale Manufacture of Dimethyl Ether - a New Alternative Diesel Fuel from Natural Gas", SAE Paper 950063, 1995.

(3) Japar, S.M., Wallington, T.J., Richert, J.F.O. and Ball, J.C., "The atmospheric chemistry of oxygenated fuel additives: t-Butyl alcohol, Dimethyl Ether, and Methyl t-butyl Ether. International Journal of Chemical Kinetics, Vol. 22, 1257-1269, 1990.

(4) Bohnenn, L.M. "Safety Aspects of Dimethylether Pure". Presented at the symposium "Science in the service of safety", Bath, England, Nov. 1978.

(5) Fleisch, T., McCarthy, C. Basu, A., Udovich, Charbonneau, P., Slodowske, W., Mikkelsen, S.-E, McCandless, J., "A New Clean Diesel Technology: Demonstration of ULEV Emissions on a Navistar Diesel Engine Fueled with Dimethyl Ether", SAE Paper 950061, 1995.

(6) Fleisch, T.H., Meurer, P.C., "DME - The Diesel Fuel for the 21st Century ?", Engine and Environment 1995, Graz Austria, 1995.

(7) Kapus, P., Ofner, H., "Development of Fuel Injection Equipment and Combustion System for DI Diesels Operated on Dimethyl Ether", SAE Paper 950062, 1995.

(8) Bibby, C.D., Chang, R.F., Howe, and S. Yurchac (Editors), "Methane conversion". Elsevier Science Publishers B.V., Amsterdam No. 36.

(9) European Programme on Emissions, Fuels and Engine Technologies (EPEFE).

(10) Schulz, H., Bandeira, G., Ziwey, U., Geng, J., Mattes, P., Neisecke, P. "VOC-Emissions from Diesel Engines". Presented at the International Symposium on Automotive Technology and Automation, Aachen, September 13-14, 1993.
(11) Poulsen, J.H., Wallance, J.S. "Operating parameter effects on the speciated hydrocarbon emissions from a natural gas fueled engine". SAE Paper 942007, 1994.
(12) Kapus, P., Cartellieri, W., "ULEV Potential of a DI/TCI Diesel passenger Car Engine Operated on Dimethyl Ether", SAE Paper 952754, 1995.

A new generation of LPG and CNG engines for heavy-duty applications

C G DE KOK
DAF Trucks NV, The Netherlands

Synopsis:

Urban areas show clearly the pollution problems we are facing today. Traffic is a significant player in creating and more importantly, solving this issue. Legislators are supporting technical solutions, meeting extremely severe emission limits compared to the near future emission, and noise legislation on heavy duty trucks. Within the technical solutions to meet these demands, a heavy duty LPG or CNG SI engine is a practical and realistic way to go, for the next 15 years. Although creating technical challenges on thermal load and some compromising on fuel consumption, the stoichiometric concept plus 3-way catalytic converter was adopted. Extremely low emission levels can be reached. To avoid the shortcomings of the conventional LPG / CNG systems on reliability, serviceability and driveability, new multipoint injection systems have been developed. Development, hardware and test results of the engine introduced in 1994 and the engine to be introduced at the end of 1996 will be discussed.

1 INTRODUCTION

In Europe, there is increased concern about air quality in large urban areas. Use of alternative fuels for heavy duty applications, for the improvement of this air quality, has been limited over the last few decades to demonstration projects, with one exception : the city of Vienna. The main causes for the reticence were the additional vehicle costs, the (un)reliability and the maintenance requirements of the systems available, and the availability and price of fuels In the field of emissions, over the last few years, diesel technology has made huge steps forwards. However, it will be at least ten more years before the diesel engine can compete on emissions with an SI lambda = 1 engine + 3-way catalytic converter. Using this SI technology, it is possible to achieve NO_x emissions, of far less than 1 g/kWh 13 mode. However the heavy duty SI technology was not sufficiently developed to make economically feasible operation possible. With systems originating from developments in the passenger car industry, the technicians now have succeeded in developing designs which have eradicated the shortcomings in respect of reliability and service life. It is now up to government to create political and financial parameters, which will make the wide-scale application of alternative fuels possible. Examples of this are the cancellation of the extra road tax for heavy duty LPG and CNG in the Netherlands, and the project support provided by the Dutch Ministry for Transport to DAF, for the development of LPG and CNG engines.

2 CHOISE OF ALTERNATIVE FUEL.

Over the last few years, concepts have been developed for a wide range of fuels, and tested in practice. Many of these fuels are subject to massive operational problems and disadvantages, which far outweigh the profit which could be gained in emission. . Of all the alterative fuels looked into effectively only LPG and CNG are suitable for practical application in the short to medium term. The choice between LPG and CNG is often determined by the infrastructure available; the minimum required range of operation, and the willingness to make investments in filling stations.

3 AREA OF APPLICATION / ENGINE RANGE

Due to the disadvantages in respect of diesel, from the point of view of range of operation, fuel infrastructure and power density, the application of LPG and CNG engines will be limited to those sectors where these disadvantages play little or no role, and where the advantages in emission weigh far more heavily, namely in urban traffic. It is therefore no surprise that DAF started LPG development on the 11.6 litre horizontal engine (LT160LPG), specially developed for bus applications. The follow-up to this generation of bus engines is the 9 litre horizontal diesel engine in production since '95. End '96 an turbo intercooled LPG version will be introduced, the GG170LPG. This 9-litre engine contains all the experience accrued with the LT160LPG engine. Even more than its predecessor, this 170 kW 9-litre engine will therefore meet the requirements imposed on heavy duty applications. *(photograph 1 : engine)*

4 DEVELOPMENT TARGETS FOR ENGINE AND VEHICLE

(table 1) Emissions subject to legal standards are NO_x, CO, HC, Particulates and acceleration smoke. In an LPG / CNG engine (lambda=1 plus 3-way catalytic converter), NO_x and Particulates are several times lower than the future EURO III emission level, whilst acceleration smoke is totally absent. In respect of a number of other performance aspects not regulated by law, such as noise and smell, the LPG / CNG engine also scores significantly better than the EURO III diesel engine. However, in the field of efficiency, the LPG / CNG clearly does not measure up to the diesel engine.

4.1 Objectives for the engine

The objectives for this engine development programme are: extremely low emission values, far in advance of the expected EURO III requirements; low noise level; and for SI (lambda=1) standards, a relatively high best point performance of 36%.
(table 2) The low emission level target was the determining factor in selecting the stoichiometric principle (lambda=1) as opposed to the Lean Burn principle.
Although Lean Burn has clear advantages on efficiency and thermal load of the engine and engine compartment, with the technology available now and in the near future, it is not possible to achieve an NO_x level of less than 3.5 g/kWh without negatively influencing the driveability

and the risk of backfire and misfire of the engine. These were considered sufficient reasons by DAF for selecting the stoichiometric combustion principle. As a consequence of this selection, in the development programme, considerable attention was paid to the thermal loadability of the engine.

4.2 Objectives for the vehicle

For the vehicle as a whole, additional requirements are imposed including high service life, high reliability of the LPG and CNG installation, and of course low operating costs.

5 CHOICE OF TECHNOLOGY

The older LPG and CNG concepts are all based on venturi systems plus, in the case of LPG, a vaporizor. Venturi systems can only be controlled with difficulty, within the target lambda=1 window. Also, the transient behaviour and driveability in combination with low emission targets using this system cannot be brought to a level acceptable for this day and age. For this reason, DAF opted to use a form of technology which has fully proven itself in the passenger car SI engine, namely timed multipoint injection plus 3-way catalytic converter.

5.1 LPG injection system

(fig 1) Following the extensive survey of various system concepts, the eventual choice went to the "timed multipoint liquid LPG injection system" as developed by GENTEC / VIALLE. In the engine, the injectors are placed in the individual ports of the inlet manifold. In this way, the volume in the inlet system, where the mixture is located, is kept to a minimum. As a result, backfiring is eradicated. The other advantages of timed multipoint liquid LPG injection are clear:
1 Thanks to the evaporation of the liquid-injected LPG, there is a significant fall in the mixture temperature, which acts positively on volumetric - , and thermodynamic efficiency.
2 A cooling effect is also achieved, on the inlet valve.
3 Extremely rapid transient responses without misfiring

The GG170LPG has a cross-flow cylinder head, facilitating the use of the straight forward configuration of one injector per cylinder.
A feature of multipoint injection which must not be ignored is the possibility of employing Fuel Cut Off during deceleration, without immediately causing problems in drive-ability, misfiring and/or backfiring. FCO has a positive effect on fuel consumption, and HC emissions in applications with a highly transient character, such as city buses.

5.2 CNG injection system

For the CNG engine, the same design guidelines are maintained, as for the LPG engine. This engine, too, is developed using an MPI system. Unlike the LPG, however, liquid injection is not possible with CNG, making it impossible to use the already fully developed LPG injectors. Specially for this application, a CNG injector must be developed. In the case of gaseous CNG injection, the volume injected at every injection is many times greater than with liquid LPG. This immediately becomes clear in the dimensions of the injector. In a limited number of repetitive development stages, a CNG injector concept was determined, and the hardware produced. The service life target of this injector is one billion injections. The location of the injector in the inlet manifold is identical to that of the LPG engine. The results of the durability test on the injector test bed and the first functional results for an engine equipped with these CNG injectors would appear to be highly promising.

5.3 Electronic unit and sensors

Multipoint injection cannot be considered, without an electronic engine management system. In addition to engine management, the 'EMS3' system also monitors the fuel tanks, communication with the gearbox, and contains a range of diagnostic features on OBD2 level.

6 FUEL TANKS AND VEHICLE INFRASTRUCTURE

(table 3) The storage of alternative fuels in the vehicle is more complex than the storage of diesel. LPG and CNG are no exceptions. The difference in specific weight calls for a much larger tank capacity, in order to guarantee a comparable range of operation.

6.1 LPG storage in the vehicle

The storage of LPG, combined with liquid LPG injection, places special demands on the tank configuration. To ensure that the LPG is supplied to the injectors, in the liquid phase, it is vital to fit a fuel pump in every LPG tank. This pump guarantees a supply pressure of 5 bar above vapour pressure. In the LT160LPG design, we were forced to fit 4 tanks, each supplying 2 injectors. The result of this design choice was a limited option for total tank content, and a large number of supply and return lines. In the subsequent development, using the GG170LPG, the layout of the fuel system has been further improved and simplified. The LPG supply and return lines have been integrated in 2 fuel rails (1 for every 3 cylinders), thus considerably simplifying installation in the vehicle. Instead of a fixed number of tanks, various options are possible. If there is a possibility for mounting the tanks on the roof, the preferred choice is a simple configuration of 2 * 300 litre tanks. Each of these tanks will then feed a single fuel rail, generating a minimum requirement for shut-off valves. If roof mounting is not possible, a more complex solution is a larger number of smaller tanks in the chassis. The fuel control strategy ensures that the tanks are emptied in sequence, whereby it is ensured that no demand for fuel is ever placed on an empty tank. This logic is integrated in the engine electronics.

.2 Storage of CNG in the vehicle

The CNG is stored at a pressure of over 200 bar. This places special demands on tank strength. Although there are developments relating to light-weight tanks, using composite materials, in the DAF project a more conventional solution was chosen in the form of steel tanks. The high total tank volume for an acceptable range of operation means that a large number of tanks are necessary. However, the system is less complex than with LPG tanks. Legislation permits these tanks to be interlinked, making a minimum number of shut-off valves necessary. Return lines, essential in the case of liquid LPG injection, are not required here.

After passing through the tank coupling point, the pressure is reduced to the constant supply pressure of 8.3 bar.

THE BASE ENGINE

The SI lambda=1 combustion principle places completely different requirements on design and materials, than those for a comparable diesel engine. The mechanical loading is lower, but thermal loads are significantly higher. The main components have therefore been thoroughly altered, on the one hand in order to ensure the same levels of reliability and service life for the SI engine, as for a standard diesel engine, and on the other hand, to optimise combustion, in order to achieve high efficiency.

7.1 Cylinder head

The inner shape of the cylinder head casting has been adjusted, to house the spark plug and ignition coil. The shape has been designed to keep the thermal stress as low as possible. To prevent inlet valve stem oil consumption, valve seals have been used.

7.2 Valves / valve seats / valve lift

The maximum exhaust gas temperatures under full load are approx. 300°C higher than with diesel engines. This places special demands on the materials of the valves and valve seats. To prevent valve wear, the use of high temperature resistant materials is not sufficient. The opening and closing speeds of the valves must be drastically reduced, but in such a way that the engine performance is not influenced. To reduce sensitivity to backfiring, the valve overlap has been considerably reduced. This also results in very smooth running of the engine at "low idle" speed.

7.3 Turbo / throttle valve

Simply installing the diesel turbo, will lead to a mismatch in the turbo setting. In the LT160LPG, considerable work was carried out on adjusting the turbine and compressor wheels. To achieve the desired torque curve, transient behaviour and driveability, use was made of a double throttle valve unit. In the development of the GG170LPG use has been made of a single throttle valve, plus waste gated turbo, in order to further optimize efficiency and to reduce complexity of the system.

7.4 Piston / piston rings

(fig 2) The diesel piston cannot be used. The combustion study indicated that for the LPG version, a compression ratio of 1:9 generates the best possible compromise between liability to knock, efficiency and fuel quality. The 'HEMI' type combustion chamber has been shown to have a better knock reserve, when lower quality LPG is used than provided by the 'HERON' type chamber. The piston profile and top-land clearance had to be re-optimised, to suit the higher thermal load. The modern piston ring designs for diesel use are not suitable for LPG and CNG engines, as a result of the high negative pressures in the cylinder at part load. In order to avoid oil consumption, the profile of the piston rings was adjusted.

7.5 Exhaust manifold / thermal insulation

The full load exhaust gas temperatures of a Lambda=1 SI engine are high. Special temperature-resistant materials were used, and alterations were made to the tolerances and design. Amongst the characteristics of the manifold material is a very high nickel content. This material adjustment, has no effect on radiation and convection to the engine compartment. In the LT160LPG, a separate cooling tunnel and fan were mounted around the exhaust manifold, to transfer the heat from the exhaust manifold to the outside air. In the GG170LPG engine, an insulated manifold has been developed, with the inner section consisting of a number of sliding sections of high quality material, coated with a ceramic insulation material, cast as a single unit, into the surrounding manifold. Using this technology, it has proved possible to reduce the maximum surface temperature of the exhaust manifold to 250°C.

By using extremely high quality turbine materials, and a water-cooled bearing housing, it has also been possible to mount an insulation cover over the turbo. In this way, all engine hotspots have been eliminated.

7.6 Inlet manifold

The inlet manifold has been adjusted in such a way it can house the LPG and CNG injectors. For the injection of liquid LPG, special attention must be paid to the design of the inlet ports, since otherwise ice deposits would occur.

8 ENGINE PERFORMANCE

In passing, a number of engine performances have been mentioned, which have resulted in alterations to the engine design. Naturally, all performance details of the LT160LPG are known. The GG170LPG engine is still in the development phase, as a result of which the final engine performance figures are not yet available. The first functional tests are currently being carried out on the GG170CNG version.

8.1 Emissions

(table 1) As may be expected with a lambda=1 design plus 3-way catalytic converter, the emissions from both LPG engines are extremely low. For NO_x, values have been achieved

which are more than a factor of 5 lower than the future EURO III 13 mode standards. The particulate emission is negligible. Generally speaking, the odour of the exhaust gases is considered less irritant than the smell of diesel. In addition, multipoint timed injection makes it possible to optimise the transient emissions whereby using the fuel cut off feature. The HC emission during deceleration is totally eliminated. In the development of the GG170 LPG/CNG extra attention will be paid to further reducing the emissions under transient conditions, in order to comply with and exceed even future legislation.

3.2 Noise

The noise level (1 metre distance) measured on the engine test bed, is more than 6 dB(A) lower than that of the comparable diesel version. Because the engine noise encapsulation in modern city bus designs is of such a high level, the effects in the standard drive-by noise test were minimal. However, these levels do mean that large sections of the engine noise encapsulation can be left out, without having any noticeable effect on the vehicle noise level. The LPG/CNG engine scores extremely high when it comes to low interior noise, and bodywork vibrations. The engine as the dominant source of noise has been eradicated.

3.3 Fuel consumption / efficiency of the LT160LPG

graph.1) By optimising the shape of the combustion chamber, the spark plugs and the spark plug position, the combustion process has improved to such an extent that early pre-ignition is possible, without the occurrence of knock. As a result, even with poorer quality fuels (50% propane - 50% butane), good economy values are achieved. In urban conditions the vehicle fuel consumption in **liter/km** is aproxx. 80 % higher compared to the same bus equipped with a diesel engine.

3.4 Cooling

table 4) An LPG/CNG lambda=1 engine has high combustion temperatures compared to the diesel engine. As a result, at the same power level, the volume of heat emitted to the cooling water increases considerably. Naturally, the cooling installation layout of the vehicle must be altered, to counter this problem.

4 ENDURANCE TEST PROGRAMME

One of the project targets was an equivalent 'diesel' service life. To guarantee this result, an extensive number of durability tests, thermal load cycle tests and temperature measurements were carried out, both on the engine test bed and in the vehicle itself. Customer field tests were also started up with the HERMES public transport company in Eindhoven, using a two of DAF Bus International SB220 city buses, equipped with LT160LPG engines. Customer field tests are also now being carried out on the GG170LPG engine, whilst vehicle tests and customer field tests with the CNG engine are being prepared. The experiences with the LT160LPG, has provided us with the assurance that the project targets can be achieved.

10 VEHICLE TEST PROGRAMME

Although positive performances on the engine test bed are a primary requirement for successful operation in the vehicle, extensive vehicle tests remain a vital necessity. The programme consists of : optimisation of the vehicle cooling system; optimisation of the drive-ability; matching the vehicle electronics with the engine electronics; and the confirmation of reliability in a field trial. An additional purpose of field testing is to gain an idea of the operating costs of a bus, equipped with an LPG/CNG engine.

10.1 Cooling system

On the SB220 bus, fundamental changes were carried out on the cooling system. In addition to the use of an extra cooling tunnel over the manifolds, the complete heat management and temperature protection system in the engine compartment were tackled.
With the advent of the GG engine, the additional cooling tunnel can be dropped.

10.2 Driveability

Acceleration smoke and gearshift smoke are limiting factors for driveability, in the diesel engine. In gas engines, smoke plays no role whatsoever. The multipoint injection system, and the related electronics are eminently suitable for optimisation of vehicle driveability. The result is driveability in the bus equipped with the LPG engine, which far exceeds that of the diesel version. This is also confirmed as such, by the drivers.

10.3 Matching vehicle electronics

The technology employed was new for engine manufacturer, bus builder and user. The majority of problems which arose in the beginning of the vehicle tests were caused by the interface between engine electronics, and vehicle electronics. Measures have been taken to improve the vehicle cable harness, for example to eliminate EMC liability. The interfacing between the engine electronics and the electronics of the automatic gearbox have been optimised, and following these actions, the field tests were concluded, with no further major problems.

10.4 Maintenance costs / serviceability

In general maintenance costs accounted for approx. 25% of total operating costs. The target is to bring maintenance costs down to a level equal to those of a bus with a diesel engine. At this moment, the feedback information from the field test is not sufficiently extensive to guarantee this target. Indications currently are that the target is achievable, on condition that the uncertainty still surrounding the service life of the catalytic converters and spark plugs can be cleared up. All experiences in respect of serviceability noted in the LT160LPG-engined bus have been processed in the GG170 LPG/CNG engine. The most noticeable changes relate to the positioning and assembly of the injectors, the use of the two fuel rails, improved accessibility to the spark plugs, integration of the ignition coil in the spark plug shaft thus eradicating the need for high voltage cables, and the simplification of the throttle valve mechanism. The sulated exhaust manifold eradicates the need for the complex cooling tunnel. All these evelopments will make it possible to further reduce maintenance costs.

1 FUTURE DEVELOPMENTS

n the basis of the concluded LT160LPG project and the current development projects for the .7 LPG and CNG engines, it will be clear for all that DAF does not consider the development f gas engines a one-day wonder; rather, this is the start of a new engine philosophy for public ansport applications.

he possible application of gas engines in urban distribution trucks and municipality vehicles is lso being considered. For this purpose, a vertical version of the GG170LPG engine is under evelopment.

N SUMMARISING, WE MAY STATE THAT DAF LPG AND FUTURE CNG ENGINES VILL BE THE URBAN TRANSPORT ALTERNATIVE, FOR THE FUTURE

dK

photo 1 : GG170LPG engine

Emissions R49 (13m)	Target LPG/CNG	DAF GG170LPG+ 3 W.Kat.	EURO 3@
NO_x (g/kWh R49)	< 1.0	< 0.3	5.5
CO (g/kWh R49)	< 1.0	< 0.3	2.0
CH (g/kWh R49)	< 0.6	< 0.03	0.6
Part. (g/kWh R49)	< 0.05	< 0.01	0.12

@ Limits as well as the cycle for EURO 3 is not yet determined

table 1 : Emission targets and test results LPG / CNG engines

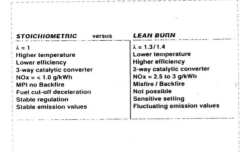

table 2 : Lean burn vs stoichiometric

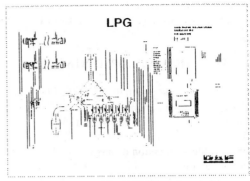

fig 1 : Multipoint Liquid LPG injection system

		DIESEL	LPG	CNG
Combustion heat output	MJ/kg	42.7	66.1	47.7
Storage phase		liquid	liquid	gas
Storage pressure	bar	1	9+	200
Density	kg/m³	840	540	140
Capacity for 500 km range of operation	L	200	600 (480)	1300
Extra weight	kg	---	240	1000

fuel storage

table 3 :   Fuel tanks and vehicle consequences.

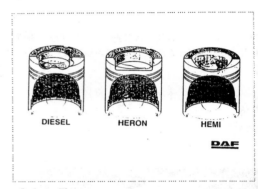

fig 2 :   Pistons

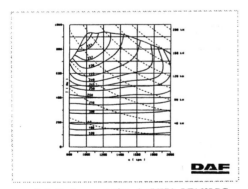

graph 1: Fuel consumption (gr/kWh) LT160LPG

Percentage of invested energy

	LPG		Diesel	
	1200 RPM	2000 RPM	1200 RPM	2000 RPM
water	25.4	27.3	21.0	21.0
exhaust	32.0	36.6	33.9	39.8
radiation	8.4	5.9	5.1	4.0

LPG with 95% propane; Amb. 25 °C

table 4 : cooling performance $\lambda = 1$ LPG engine

EGR for Spark Ignition

Improving fuel consumption of stoichiometric spark ignition engines by the use of high EGR rates and high compression ratio

DE PETRIS, S DIANA, V GIGLIO, and G POLICE
Istituto Motori CNR, Napoli, Italy

SYNOPSIS

This paper describes an experimental study aiming to assess the potential for fuel economy of stoichiometric high compression ratio engines in presence of Exhaust Gas Recirculation (EGR).

Previous experiences have shown that EGR can limit the knock tendency of stoichiometric mixtures even with high compression ratio engines running at full load. The experiments, carried out on a commercial four cylinder stoichiometric spark ignition engine have confirmed the trends obtained on single cylinder engines.

At medium and high loads, efficiency improvements higher than 10% have been obtained with a compression ratio equal to 11.5. Similar improvements have been obtained for CO emissions while (HC + NOx) reductions are ranged between 20% and 30%.

At low loads, cycle by cycle variation limits the effectiveness of EGR. Probably, for these conditions, high swirl or high tumble are required to take full advantage by the present strategy of combustion control of stoichiometric spark ignition engines.

1 INTRODUCTION

Fuel economy is nowadays object of increasing attention because it is the most important way to reduce CO_2 emissions in the field of on road transport. The fuel consumption reduction constitutes a further constraint for engines optimisation. In fact, some technical choices improving fuel economy do not allow to comply with emission limits that are becoming more and more stringent.

For example, lean burn engines with high compression ratio provide good results from the point of view of fuel economy but, in absence of a specific exhaust catalyst, their emissions are not generally acceptable.

On the contrary, catalysts for stoichiometric engines have been very well developed. So nowadays stoichiometric spark ignition engines are probably the best solution to limit emissions. Unfortunately their efficiency is much lower than the typical one of lean burn engines.

Exhaust gas recycle (EGR) may be considered an attractive strategy to obtain some benefits of charge dilution in stoichiometric spark ignition engines (1), (2), (3). In fact, for some aspects EGR determines the same effects due to an excess of air in air/fuel mixture. It can reduce very efficiently NOx emissions because of the lowering of combustion temperature. Moreover, EGR can improve fuel economy by the reduction of dissociation, of thermal losses and of pumping work (4).

According to the experience of the authors of the present paper, further efficiency gains can be achieved if EGR is used, also at full load, as a mean to avoid knock. Experiments carried out on single cylinder spark ignition engines (CFR, AVL 528) have shown that EGR can control knock even at wide open throttle (WOT), with a compression ratio of about 13.5. With higher compression ratios, cycle by cycle variation can be reduced and the EGR tolerance increases substantially (5).

Anyway, the " lean burn - high compression ratio engine" obtained by a stoichiometric engine with the addition of EGR allows the use of three way catalysts with obvious advantages for emission control.

This work aims to assess the effects of EGR and of high compression ratios on fuel economy and pollutant emissions of a real multicylinder engine under steady running conditions.

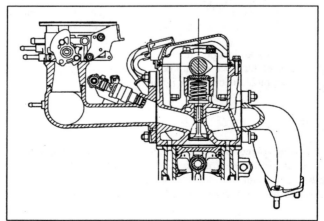

Fig 1 Section of the engine head.

2 EXPERIMENTS

The experiments have been carried out using a commercial four cylinder four stroke spark ignition engine with a bathtub combustion chamber (Fig.1). Engine main characteristics are listed in Table 1.

Engine type	4 cylinder 4 stroke
Bore	70.8 mm
Stroke	78.86 mm
Swept volume	1242 cm^3
Compression ratio	10
Rated torque	106 mN @ 3500 r/min
Rated power	54 kW @ 6000 r/min

Table 1 Test engine specifications.

The original electronic management system has been replaced by a test bench device, developed by Intelligent Control Corporation, that allows a manual or automatic control of air/fuel ratio, ignition timing and EGR. Actually air/fuel ratio has been controlled in closed loop with an Exhaust Gas Oxygen (EGO) sensor, maintaining its value close to stoichiometric (14.7). Basic injection pulsewidths are provided on the basis of engine speed and air mass flow measured by a hot wire flow meter.

Exhaust gas has been introduced before throttle valve to simplify EGR measurement. The recycled gas has been cooled to obtain an inlet temperature not higher than 200 °C. Combustion pressure has been measured by an AVL QC32C quartz transducer. Regulated pollutants and CO_2 have been measured before the catalytic converter.

The experimental investigation has been carried out at 8 different running conditions (two engine speeds and four loads) for both the compression ratios of 10 and 11.5. The highest compression ratio has been obtained modifying the piston without variation of crevices volume.

Ignition timing has been selected by manual control to obtain the maximum brake mean effective pressure (BMEP) with two limiting conditions, that is: absence of knock and NOx level less than 2500 ppm.

EGR has been varied ranging between 0 - 12 % for the compression ratio 10 and between 0 - 18 for the compression ratio 11.5.

3 EFFECTS OF EGR ON PERFORMANCES AT THE STANDARD COMPRESSION RATIO

At the compression ratio 10, tests have been carried out at air/fuel ratio 14.7 and at fixed throttle openings which provide BMEP equal to about 2.5, 5, 7.5, and 9 bar at 0 % of EGR.

BMEP decreases on EGR increasing because the mass of mixture introduced into the cylinder is smaller. That can be observed in Fig. 2 where BMEP is plotted versus EGR at 3500 r/min for the compression ratio 10.

Brake Specific Fuel Consumption (BSFC) decreases slightly with EGR. For example at WOT and at 12% of EGR the decrease of specific fuel consumption is about of 6% (Fig. 3).

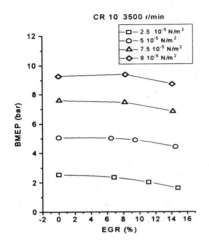

Fig 2 Brake Mean Effective Pressure vs EGR. Compression ratio 10. Air/fuel ratio 14.7.

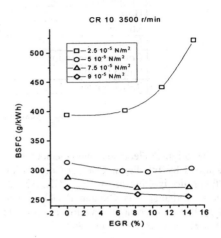

Fig 3 Brake Specific Fuel Consumption vs EGR. Compression ratio 10. Air/fuel ratio 14.7.

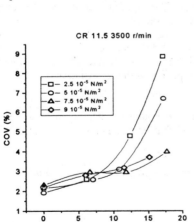

Fig 4 Cyclic variation vs EGR. Compression ratio 10. Air/fuel ratio 14.7.

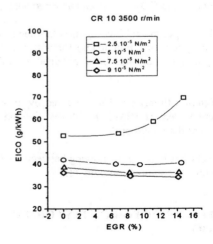

Fig 5 CO emission index vs EGR. Compression ratio 10. Air/fuel ratio 14.7.

At partial load the EGR tolerance becomes smaller. At 5 bar of BMEP the optimum value of EGR for fuel consumption is about 8 % while for very low loads BSFC increases continuously with EGR. This fact is probably due to cyclic variation that at low loads sharply increases with EGR (Fig. 4). Note that the cyclic variation has been evaluated as the standard deviation of Indicated Mean Effective Pressure (IMEP) divided by the mean IMEP.

The CO concentration in the exhaust depends mainly on the air/fuel ratio and not on EGR. Therefore CO index curves (EICO) versus EGR are very similar to the ones of specific fuel consumption (Fig. 5).

The unburned hydrocarbons index (EIHC) rises with EGR (Fig. 6). The strongest increasing of EIHC can be observed at the lowest load. The rise of EIHC with EGR can be ascribed partly to the lower flame speed, partly to the increased cyclic variation and lower exhaust temperature.

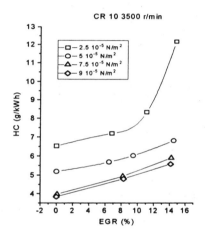

Fig 6 HC emission index vs EGR. Compression ratio 10. Air/fuel ratio 14.7.

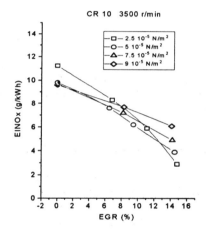

Fig 7 NOx emission index vs EGR. Compression ratio 10. Air/fuel ratio 14.7.

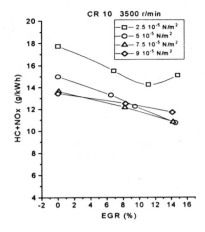
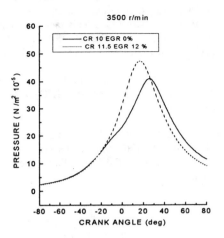

Fig 8 HC + NOx emission index vs EGR. Compression ratio 10. Air/fuel ratio 14.7.

Fig 9 Combustion pressure vs crank angle for compression ratios 10 and 11.5. Air/fuel ratio 14.7.

On the contrary NOx emission index (EINOx) decreases with EGR (Fig 7) However it is interesting to observe that the sum of EIHC and EINOx generally decreases with EGR (Fig. 8).

4 EFFECTS OF COMPRESSION RATIO ON PERFORMANCES

At the compression ratio 11.5 throttle openings have been selected to provide, with 0 % of EGR, the same air mass flow rate measured with the compression ratio 10 and 0 % of EGR at loads equal to 2.5, 5, 7.5, and 9 bar. Air/fuel ratio has been still maintained close to stoichiometric (14.7).

The increasing of compression ratio has some beneficial effects on spark ignition combustion in presence of EGR. In particular it is easier to select spark timing for the best torque, in absence of knock with the constraint of 2500 ppm of NOx.

In Fig. 9 it is possible to observe that, at WOT, the peak of pressure obtained with the higher compression ratio in presence of EGR increases and it is advanced with respect to the one obtained with the lower compression ratio in absence of EGR.

Exhaust gas temperature is lower during the power stroke and therefore efficiency increases (Fig. 10). The rising of compression ratio increases combustion flame. That can balance the effect of EGR that normally causes combustion angle to became longer (Fig. 11). The efficiency gains obtained with higher compression ratios provide the possibility to maintain at full load the original power in presence of EGR.

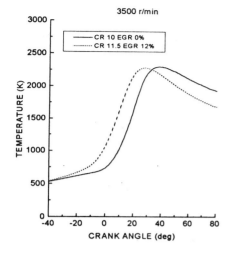

Fig 10 Combustion temperature vs crank angle for compression ratios 10 and 11.5.

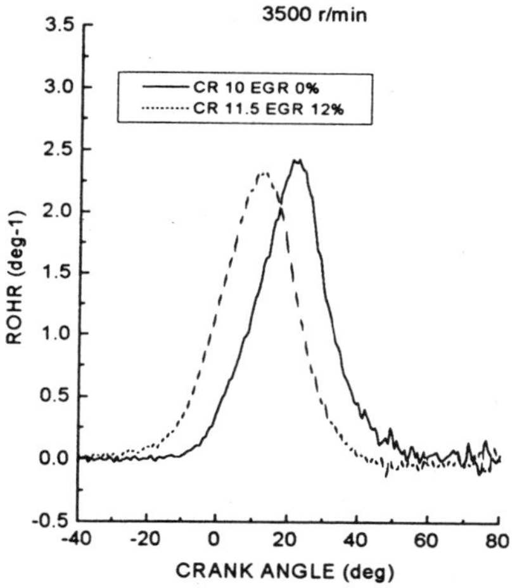

Fig 11 Rate Of Heat Release vs crank angle for compression ratios 10 and 11.5.

It is interesting to compare emissions and efficiency that can be obtained at the same IMEP with the two adopted compression ratios. In particular the attention is focused on specific fuel consumption and the sum of HC and NOx emission index. Fig. 12 and 13 show BSFC versus BMEP for the compression ratios 10 and 11.5 respectively. In both diagrams the baseline curve represents the condition relative to compression ratio 10 and 0 % of EGR.

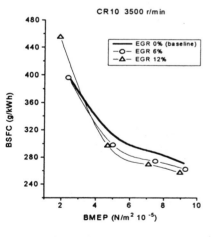

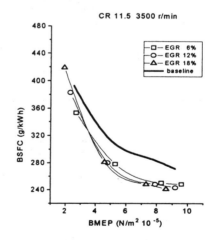

Fig 12 BSFC vs BMEP for compression ratio 10. Air/fuel ratio 14.7 .

Fig 13 BSFC vs BMEP for compression ratio 11.5. Air/fuel ratio 14.7 .

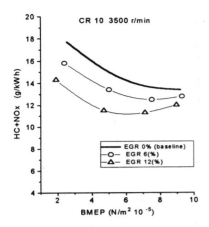

Fig 14 HC + NOx emission index vs BMEP. Compression ratio 10. Air/fuel ratio 14.7.

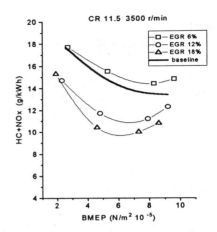

Fig 15 HC + NOx emission index vs BMEP. Compression ratio 11.5. Air/fuel ratio 14.7.

It is possible to observe that, for BMEP values higher than 4 bar, specific fuel consumption decreases up to 5 ÷ 8 % with respect to the 'baseline' condition. With the compression ratio 11.5 the reduction of specific fuel consumption ranges between 10 - 13%. Fig. 14 and 15 show the sum of HC and NOx, expressed as emission index versus BMEP. For both compression ratios, the pollutants reduction is comprised between 10% and 20%.

	Compression ratio 10 EGR 12 % - 3500 r/min		Compression ratio 11.5 EGR 12 % - 3500 r/min	
BMEP (N/m^2 10^{-5})	HC+NOx Δ (%)	BSFC Δ (%)	HC+NOx Δ (%)	BSFC Δ (%)
9	-12.5	-5.9	-9.7	-11
6	-13.4	-1.3	-23.4	-13.8
3	-23.5	0	-20	-7.5

Table 2 Summary of the percent differences of BSFC and HC + NOx emission with respect to the reference conditions of compression ratio 10 and 0 % of EGR at 3500 r/min. Air/fuel ratio 14.7.

	Compression ratio 10 EGR 12 % - 4500 r/min		Compression ratio 11.5 EGR 12 % - 4500 r/min	
BMEP (N/m^2 10^{-5})	HC+NOx Δ (%)	BSFC Δ (%)	HC+NOx Δ (%)	BSFC Δ (%)
9	-23.6	-10.9	-4.2	-10.9
6	-26.8	-3	-24.4	-9.6
3	-29.3	0	-29.8	0

Table 3 Summary of the percent differences of BSFC and HC + NOx emission with respect to the reference conditions of compression ratio 10 and 0 % of EGR at 4500 r/min. Air/fuel ratio 14.7.

Tables 2 and 3 summarise the percent differences of BSFC and HC + NOx respectively for 3500 and 4500 r/min. The percent differences are calculated for three load conditions, namely low, medium and high load, with reference to the conditions of compression ratio 10 and 0 % of EGR. The performances obtained at 4500 r/min are similar to the ones shown at 3500 r/min.

5 CONCLUSIONS

Tests have been carried out on a commercial SI four cylinder stoichiometric engine, varying its compression ratio and EGR levels. The experiments have confirmed some trends obtained in previous experiences with single cylinder research engines. The results can be briefly summarised as follow:

- EGR is an effective mean to control knock occurrence even at WOT.

- A proper selection of spark timing allows to preserve the maximum power also in presence of EGR.

- With the compression ratio 11.5 and an EGR of 12% it is possible to obtain gains in efficiency higher than 10 % at medium and high loads.

- CO reduction is roughly proportional to the efficiency gains. HC + NOx reduction is comprised in the range 10 - 20%

- At low load, because of the absence of a proper air motion, cycle by cycle variation causes a worsening of fuel consumption on EGR increasing. Probably, the High Compression Ratio - High EGR engine concept requires a proper swirl /tumble motion to be effective also at low loads

REFERENCES

(1) KURODA, H. et al 'The Fast Burn with Heavy EGR, New Approach for Low Nox and Improved Fuel Economy', SAE Paper 780006.

(2) NEAME, G.R. et al. 'Improving the Fuel Economy of Stoichiometrically Fuelled S.I. Engines by Means of EGR and Enhanced Ignition. A Comparison of Gasoline, Methanol and Natural Gas', SAE Paper 952376.

(3) MICHIHIKO, T. 'Improving NOx and Fuel Economy for Mixture Injected SI Engine with EGR', SAE Paper 950684.

(4) HEYWOOD, J.B. Internal Combustion Engine Fundamentals, 1988, (McGraw Hill).

(5) DE PETRIS, C. et al. 'High Efficiency Stoichiometric Spark Ignition Engines', SAE Paper 941933.

Enhanced exhaust gas recirculation using exhaust throttling for NOx reduction at WOT

OSSES, P CLARKE, and **G ANDREWS** MIMechE
University of Leeds, UK
UNZAIN and **G ROBERTSON**
Powertrain Research, Ford Motor Company Limited, UK

SYNOPSIS

An exhaust back pressure valve (EBPV) was used to generate exhaust gas recirculation (EGR) for NOx control at wide open throttle (WOT) low speed medium power conditions. Infinitely variable transmission (IVT) applications may use an operating strategy where low speed WOT is used at medium power outputs (20 KW+). This then generates high engine-out NOx at the low power of the US and ECE cycles and EGR generated by a back pressure valve is a feasible method of NOx control. A Ford Zetec 2.0 litre SI engine was used with a back pressure valve mounted downstream of the 3-way catalyst. Up to 70% NOx reduction was demonstrated at 20 kW power with WOT and lambda 1, but there was a maximum specific fuel consumption increase of 18% for the 8% EGR generated by the EBPV. However, this increased SFC was 8% less than for a conventional transmission system at the same power output. The same or lower engine out NOx was demonstrated compared with the standard engine transmission operation at higher speeds and partially closed throttle for the same power output. Thus for IVT applications an EBPV is a viable method of achieving emission levels the same as or lower than the same engine with a standard transmission, whilst still retaining a fuel economy advantage for the IVT.

1. INTRODUCTION

Figure 1 shows the specific fuel consumption (SFC) and specific NOx emission mapping of the 2.0 litre Zetec engine, as well as one possible IVT control drive line based on constant speed operation until 21.5 kW power output and then higher speeds at wide open throttle for higher powers. This is an engine operating schedule for minimum SFC, as shown in Fig.1.a. However, Fig.1.b shows that in comparison with a power output of say 20 kW in third gear for a conventional transmission the IVT would have of the order of 4 times the engine out NOx. In order to reduce NOx in the WOT region of the IVT operation the use of an exhaust back pressure valve to generate EGR was investigated in this work [1]. The desirable engine out NOx emissions was less than 5 g/kWh throughout the proposed constant speed IVT driveline. This was the minimum engine out NOx in the engine map in Fig.1.b. This would ensure that the IVT transmission would achieve lower NOx emissions downstream of the catalyst over the regulation test cycle than for the conventional transmission as well as retaining the advantage of the improved SFC.

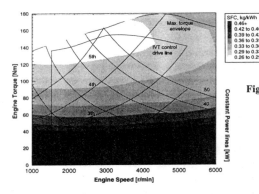

Figure 1.a. SFC mapping from 2.0 litre Zetec engine vs IVT control drive line, conventional gear ratio curves and constant power lines.

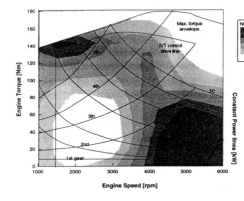

Figure 1.b. NOx mapping from 2.0 litre Zetec engine vs IVT control drive line, conventional gear ratio curves and constant power lines ($\lambda=1$).

The most commonly used non-catalytic method of NOx reduction for internal combustion engines is exhaust gas recirculation [2,3]. The 2.0L Ford Zetec engine used in this work has a homogeneous pre-mixed EGR system designed to operate at part throttle, controlled by the manifold vacuum. The EGR is switched off during high/full load operation so as to maximise the power [4]. In the present work the high NOx emissions of the constant speed IVT driveline in the 15kW+ region were reduced using EGR. However, at WOT there is no manifold vacuum to achieve the EGR flow and normally no EGR is possible at this condition. To achieve EGR at WOT an exhaust back pressure valve was fitted down stream of the catalyst, this was an exhaust throttle and was closed at WOT to increase the back pressure above the standard exhaust pipe back pressure. This was coupled with the use of a wide open EGR valve so that the EGR could be controlled by the back pressure valve. At conditions where there was sufficient manifold vacuum for the EGR to be controlled by the EGR valve the EBPV was set fully open, so as to minimise the fuel economy penalty.

The use of an EBPV is only one method to overcome the potentially high NOx emissions of the constant speed IVT driveline. An alternative technique would be not to operate the IVT at WOT but always keep sufficient manifold vacuum to drive the EGR through the external EGR pipe and flow meter system. This strategy is known as manifold vacuum modulation (MVM) and is based on keeping the EGR valve wide open for maximum EGR flow at the minimum manifold vacuum. Fig.1.a shows that the SFC penalty for this would be relatively small for this engine and a possible driveline is shown in Fig.1.a for speeds higher than 1500 r/min and powers above 20 kW. This method of achieving EGR for reduced NOx will be compared with the EBPV technique in the present work so as to establish whether the EBPV technique has a definite advantage. At powers lower than 20 kW the EGR would be controlled by the EGR valve for both systems and there would be little difference in the NOx.

Alternatively the NOx could be controlled by later spark timing, but this would have an SFC penalty, which could be worse than that due to the use of the EBPV. In the present work the engine autocalibration spark timing was used, this automatically corrected the spark timing for the measured EGR level determined by the venturi in the external EGR flow pipe. However, optimisation of spark timing for further reductions in NOx is possible without significant SFC penalty, this work will be reported separately.

Previous work with an EBPV has been published by Lotus [5] and offered as a solution during engine warm-up leading to a reduced time to catalyst light-off due to the faster rise in the exhaust gas temperature than with the ambient pressure system. This was one of the reasons why the EBPV was placed downstream of the catalyst in the present work. Although the use of the EBPV was completely different in the present work, the valve assembly was similar and if an EBPV was used for NOx control with an IVT system, then it could also be used for enhanced catalyst light off.

At fully warmed-up engine conditions the utilisation of an EBPV downstream of the catalyst enables the catalyst to operate at the higher exhaust temperature created by the back pressure, which will improve the catalyst efficiency at low powers, where the exhaust temperatures are in the 400-450°C region of developing catalyst efficiency in the 90%+ region. In the present work the EBPV typically increased the catalyst inlet gas temperature by 20-50°C, with the higher increase at the lower temperatures of lower power conditions. However, most of this increase was due to the presence of the EGR rather than a direct back pressure effect, as shown by varying the EGR using the MVM technique without the use of the EBPV at low power conditions.

2. EXPERIMENTAL

A Ford Zetec 2.0 litre engine, 84.8 mm bore, 88 mm stroke, 10:1 compression ratio was used [6]. This was coupled to a Froude hydraulic dynamometer. The engine was equipped with an electronic engine control (EEC) unit which was interfaced to a computer for data recording of all the engine monitored parameters and calibration settings. The engine parameters recorded were the speed, torque, power, spark timing, lambda value, EGR mass flow, throttle position, air mass flow rate and coolant temperatures. The present work was carried out at steady state and lambda 1 throughout and the autocalibration lambda, which gave rich mixtures at near to WOT, was bypassed. Extra pressure transducers were fitted to record the inlet manifold vacuum and the exhaust back pressure

The WOT maximum power as a function of speed is shown in Fig.2 for lambda 1 and 0.9 operation. The richer mixture gives the maximum power point for this engine. The engine was new and had 100 hours test time prior to these measurements. The rich mixture maximum powers are about 1 kW below the standard Ford values for this engine across the speed range. However, this engine has an adaptive control system and normally requires a period of operation to optimise the control parameters. Over the period of the present investigations the maximum power at 1500 r/min was checked periodically and gradually increased to equal the standard value for this engine. However, maximum power data at lambda 1 is not published for this engine and the objective of the results in Fig.2 was to demonstrate the magnitude of the power loss for WOT lambda 1 operation, which was 0.8 -1.2 kW (approximately 5%) over the range of engine speeds. For an IVT transmission, which for fuel economy reasons, operates at low speed WOT, lambda 1 operation is essential as the three way catalyst must still operate for emissions control. However, this power loss is simply recovered by increasing the speed, typically by approximately 50 r/min or 3% and with an IVT system this is accommodated automatically as it delivers the required power at WOT, automatically adjusting the speed. Only at the maximum speed maximum power condition would the IVT system need to go to rich operation, as for conventional transmissions, but this forms no part of legislated emissions test cycles.

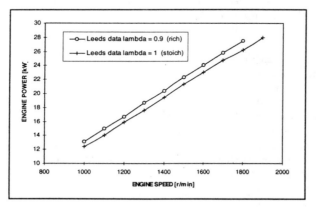

Figure 2. Comparison of maximum power curves from 2.0 litre Zetec engine, at both rich mixture and stoichoimetric operation at wide open throttle.

The homogeneous inlet air manifold pre-mixed EGR system fitted to the engine has a venturi type flow meter in the external EGR flow pipe. The EGR flow rate is controlled by the manifold vacuum applied to lift the EGR valve with the valve position and EGR flow rate controlled by an exhaust vacuum regulator valve (EVR) in the vacuum line. There is an electronic feedback control on the EGR valve position using the measured EGR flow venturi delta pressure compared with the calibrated EGR for that engine condition. It is normally used to provide EGR only at part throttle low power conditions. The EGR valve is a tapered pintle ported diaphragm valve. The pressure difference across the orifice in the EGR pipe is measured by a ceramic capacitive transducer sensor (DPFE). Control of the EGR flow is provided by modulation of the EGR valve which regulates the downstream pressure of the control orifice, and therefore the pressure difference between inlet and exhaust manifolds. [7,8]

In order to provide EGR either at WOT, where there is insufficient inlet manifold vacuum, or higher than calibrated EGR at part throttle conditions, an external vacuum pump was used to open the EGR valve. This external vacuum was used to carry out experiments over a range of EGR in the present work and the EBPV was only used to increase the EGR when the EGR valve was fully open. The EBPV was a conventional high temperature throttle made from a conventional flap valve, which was a modification of a production exhaust diverter for a close coupled light-off catalyst. Most work was carried out at constant power and autocalibration spark timing. This involved adjustments to the air inlet throttle setting at each EGR level to keep the same power output at WOT by increasing the speed as the EGR flow increased. At conditions where the power was less than 20 kW the throttle was partially closed and the speed was kept at 1500 r/min by adjustment on the throttle. Most of the present work was carried out at WOT with the speed varying.

Gas analysis was undertaken at the exhaust pipe sample point of the engine upstream of the catalyst and transferred via a heated sample line maintained at 180±5°C to an on-line gas analysis system. The heated line maintained the higher molecular weight hydrocarbons in the vapour phase at the 180°C sample line temperature and these were analysed as total hydrocarbons using a heated FID analyser. The gases were also passed via a heated line to a chemiluminescence analyser for NOx measurement. NDIR instruments for CO and CO_2 analysis and a Servomex paramagnetic oxygen analysers were also used on a dry gas basis. A computer programme computed by carbon balance the air/fuel on line and a display of the computed specific emissions in the engine control room. There was good agreement between the engine measure lambda and that based on the gas analysis.

All conditions were stabilised before data collection commenced. Data acquisition systems were used for recording the engine parameters and gas analysis, collecting 60 samples per test point at 5 second intervals. The figures used for analysis were simply the average of these samples. The EGR was increased in the tests until NOx levels below 5 g/kWh was achieved at each test condition, this was the minimum NOx on the engine map in Fig.1.b as discussed above. Engine speed was kept between 1000 and 2000 r/min at different constant power output settings, principally at WOT conditions. The powers investigated were representative of those required for urban driving, high speed high power operation was not investigated.

3. RESULTS AND DISCUSSION

The experimental programme was comprised of five parts: a baseline variation of the EBPV back without EGR; a variation in EGR for different EBPV positions and throttle adjustments; a variation in EGR for a fixed exhaust valve position; and a maximum possible variation in EGR using increased exhaust back pressure for a 15 kW optimised condition from above tests.

3.1 Baseline back pressure EGR=0

The influence of the exhaust back pressure alone with the EGR valve closed was investigated to determine any influence of the EBPV on the internal EGR and hence on the emissions. The results showed that there was little influence of increased back pressure on the emissions which implies that there was little influence on the internal EGR. The main effect was that of the additional load put on the engine by the back pressure, which decreased the power output. At constant 1500 r/min-WOT closed loop operation the exhaust back pressure was raised from the standard 4 kPa up to 30 kPa, with the EGR valve closed. This resulted in a 6.1% power reduction, a 4.3% fuel consumption increase with little effect on NOx emissions. Similar results were found at 1000 and 1200 r/min, which are summarised in Table 1.

Table 1. Effect of closing the EBPV at different engine speeds on power output, fuel consumption and nitrogen oxides emissions, WOT, EGR=0 and $\lambda=1$.

Engine Speed [r/min]	Power range [kW]	Initial exhaust back pressure [kPa]	Final exhaust back pressure [kPa]	Power output reduction [%]	Fuel consum. increase [% mass]	NOx reduction [% vol]
1000	12	2	20	5.9	4.5	1
1200	15	3	30	5.0	5.7	0
1500	21	4	30	6.1	4.3	2

This situation is unrealistic in terms of representing the influence of the EBPV as it is never intended to close this valve without the EGR valve being open. The purpose of this part of the programme was to demonstrate that changes in the internal EGR were small as the EBPV was closed, so that the influence of the EBPV induced EGR at WOT was dominated by the external EGR effects. When the EGR valve was opened with the EBPV closed there was a reduction in the back pressure due to the reduced air flow through the engine and hence a reduced power. For the fully opened EGR valve a typical result reported below is for a 15 kW power output at 1500 r/min where the EGR was 8.8%, created by the back pressure valve, the back pressure to achieve this was 13 kPa and the NOx reduction was 63%. The equivalent power condition in Table 1 for the EBPV open was at 1200 r/min and the fuel consumption differences for the same power were negligible. Thus the influence of the EBPV with the EGR valve open was no change in the fuel consumption for the same power output.

3.2 EGR sweep at variable EBPV position and MVM

With the EGR valve fully open and the EBPV partially closed to generate the EGR flow at WOT, it was possible to achieve incremental steps of EGR at WOT for different back pressures and therefore establish the reduction in NOx emissions as a function of the %EGR. The results were compared with those for the same EGR created by manifold vacuum modulation (MVM), where EGR flow was achieved by closing the throttle to induce a manifold vacuum and increasing the speed to recover the power loss.

Equivalent tests were carried out using both the exhaust back pressure valve (EBPV) and manifold vacuum modulation (MVM) systems, at constant 20 and 15 kW power output, increasing engine speed when either throttling the exhaust pipe or closing the intake throttle. Figs.3.a and 4.a show the normalised NOx and SFC as a function of the EGR. The base condition is the same for the EBPV and MVM and is for the EBPV and EGR valves fully open. The two sets of test were carried out on different days with different ambient conditions and the base EGR was slightly different, as shown in Figs. 3 and 4.

The engine computed EGR results in Figs. 3 and 4 do not include any influence of the back pressure and the EGR venturi calibration assumes a standard upstream pressure at the standard back pressure. In the present work increasing the back pressure changes the upstream EGR density. At WOT the pressure differences across the venturi were not large and the flow was subsonic. This gives a square root correction on the upstream EGR density which for the maximum back pressure of 20 kPa in Fig. 3 was a 10% correction on the EGR. This would increase the maximum indicated EGR from 8% to 8.8% and this effect is shown as a correction to the EGR data in Fig.3. At 15 kW the maximum back pressure was 13 kPa and the EGR correction was 6%. In spite of these corrections the greater NOx reduction for the same EGR with the EBPV system needs further investigation

Comparison of the results in Figs. 3 and 4 shows a better NOx reduction for the EBPV operation than the MVM, but a higher SFC penalty. However at 20 kW the MVM system could not achieve the desired low NOx emissions without moving to much higher speeds. In order to obtain a NOx reduction of 50% or more the required speed was higher for MVM operation than for EBPV. At 20 kW 5 g/kWh NOx was measured at 1960 r/min using the EBPV system, whereas the MVM technique gave 9.5 g/kWh NOx at the same speed (see Figure 3.b). At 15 kW 5 g/kWh NOx was achieved at 1430 r/min using the EBPV, but 1710 r/min was required for the same NOx reduction utilising MVM (see Figure 4.b).

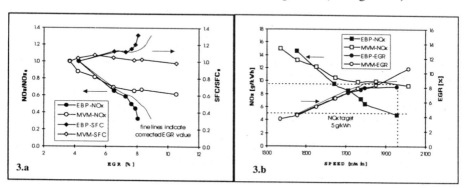

Figure 3. Exhaust back pressure (EBP) throttling and manifold vacuum modulation (MVM) operation at 20 kW, WOT, $\lambda=1$, fully open EGR valve, 2.0 litre Zetec SI engine. **a)** EGR vs normalised NOx and SFC, **b)** engine speed vs NOx and EGR

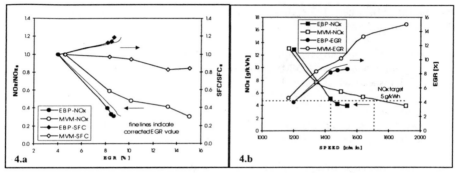

Figure 4. Exhaust back pressure valve(EBPV) throttling and manifold vacuum modulation (MVM) operation at 15 kW, WOT, $\lambda=1$, fully open EGR valve, 2.0 litre Zetec SI engine. **a)** EGR vs normalised NOx and SFC, **b)** engine speed vs NOx and EGR

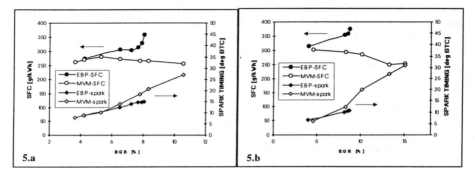

Figure 5. Exhaust back pressure valve(EBPV) throttling and manifold vacuum modulation (MVM) operation at WOT, $\lambda=1$, fully open EGR valve, 2.0 litre Zetec SI engine. **a)** EGR vs actual SFC and spark timing at 20 kW, **b)** EGR vs actual SFC and spark timing at 15 kW.

The reason for the greater NOx reduction for the same EGR for the EBPV system compared with the MVM system is due to the differences in the autocalibration spark timing. With the throttle partially closed for the MVM system the spark timing is quite different than for the WOT EBPV system. This is shown in Figs 5.a and 5.b for the 20 and 15 kW results, which also include the actual SFC data. With the MVM system the spark timing is earlier than for the EBPV system for the same EGR and this increases the NOx and reduces the SFC for the same EGR. There is the possibility that the EBPV system can be further optimised to reduce the SFC penalty by optimising the spark timing strategy rather than using the autocalibration timing. Initial work at constant power has shown that the SFC penalty can be reduced.

3.3. EGR sweep at fixed EBPV position

From the above tests it was possible to determine an optimised EBPV position which produced 20 kPa at 20 kW at WOT and $\lambda=1$, and generated the required EGR level to achieve 5 g/kWh NOx or lower at 2000 r/min engine speed. The following results were obtained at this fixed exhaust back pressure valve position, and the desired EGR value was adjusted by manually varying the intake vacuum to the EGR valve. An alternative system for the control of the EGR valve is electrical actuation which can then be independent of the manifold vacuum. This type of EGR valve is now available, incorporating pintle position feedback for precision flow scheduling, and would be recommended for applications of these enhanced EGR techniques..

This 20 kPa back pressure at 2000 r/min is somewhat arbitrary. However, reference to Fig.1.b shows that at 2000 r/min WOT the specific NOx is at a maximum and decreases at higher speeds. This fixed valve position gave sufficient EGR to achieve 5 g/kWh or lower NOx emissions at all speeds up to 2000 r/min. For higher powers the IVT and conventional transmission systems converge and the NOx disadvantage of the IVT is reduced and hence reduced EGR can be accepted. Although an electrically actuated EBPV would be preferable there is a cost penalty compared with an on-off fixed position valve, which was the objective of the present study.

Two constant power conditions were investigated at WOT, 20 and 15 kW, with the EBPV in the fixed closed position and the EGR varied using the EGR valve position. Another condition at 10 kW was run mainly at part throttle with the EBPV fully open initially and the EGR varied using the EGR valve position as there was sufficient manifold vacuum to generate the EGR. When the EGR was at a maximum for the fully open EGR valve, the EBPV was switched to its fixed partially closed condition and a higher EGR achieved. The results are shown in Fig. 6, which plots the normalised NOx and SFC Vs EGR using the condition with zero EGR (EGR valve closed and EBPV in its fixed position) as the reference. The required EGR value for the NOx to be reduced below 5 g/kWh was between 8 and 9% for all three power conditions, as shown in Fig. 7. The sudden deterioration in the SFC at 10 kW for the highest EGR was due to the closure of the EBPV, the effect of which is included in the baseline for the 15 and 20 kW results as the EBPV was closed to its fixed position at all EGR including zero EGR.

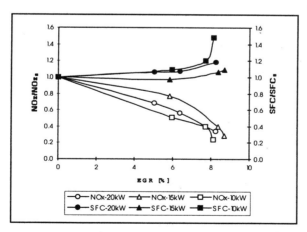

Figure 6. Normalised NOx emissions and SFC at 20, 15 and 10 kW power output when increasing EGR using EBP.

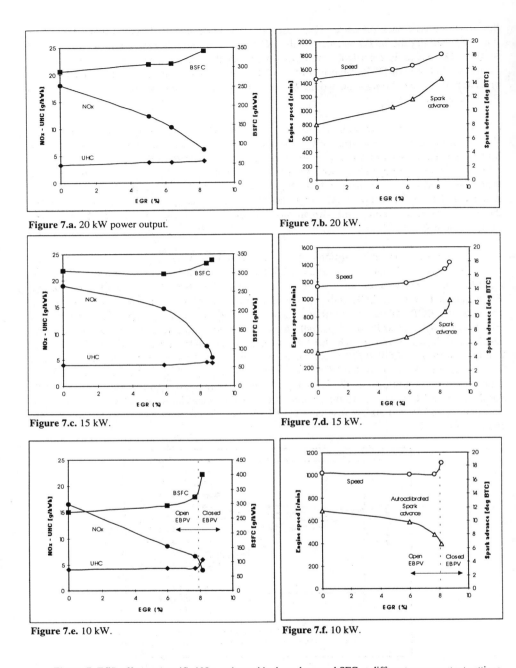

Figure 7.a. 20 kW power output.
Figure 7.b. 20 kW.
Figure 7.c. 15 kW.
Figure 7.d. 15 kW.
Figure 7.e. 10 kW.
Figure 7.f. 10 kW.

Figure 7. EGR effect on specific NOx, unburned hydrocarbons and SFC at different power output settings, showing speed increase in order to keep constant power and the autocalibration spark timing.

The spark was advanced as EGR was added in a consistent manner according to the autocalibration spark timing data for this engine. The spark timing is also affected by the higher speed as well as the EGR as shown in Figures 7.a-7.f. Hydrocarbons emissions were slightly increased, especially when closing the EBPV at 10 kW in the last test point, retaining compatibility with three way catalyst technology. However, the effect of the increased EGR on hydrocarbons was small and the same applied to the CO emissions.

3.4 Enhanced EGR at High Back Pressures

A single condition was tested at high exhaust back pressures. This was achieved by closing the EBPV beyond the optimised EBPV position. The aim was to find out the limitations of the EBPV technique. The chosen engine condition was 15 kW and the exhaust valve was closed until there was an unacceptable deterioration in the engine performance. This was defined as a doubling of the hydrocarbons compared with zero EGR level. This has been found to correspond closely with the condition of unacceptable deterioration in the mean effective pressure covariance. This occurs with the onset of significant misfires which cause the increased hydrocarbons. NOx was further reduced with the enhanced EGR but SFC was increased up to unacceptable limits as shown in Figure 8.

The exhaust back pressure was 12 kPa at the fixed valve position, allowing EGR levels from 0 to 8.7%. After the 5 g/kWh NOx level the exhaust back pressure was increased to 29 kPa, producing 10.9% maximum EGR with NOx emissions of 2 g/kWh. This gives a strategy for extremely low NOx emissions. Some reduction in the SFC penalty could be achieved by optimisation of the spark timing.

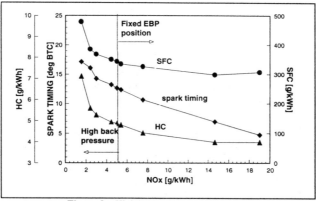

Figure 8. High back pressure effect at 15 kW

3.5 Minimum NOx Emissions Strategy for Urban Driving Conditions

The present results with the exhaust back pressure valve induced high EGR have been used to generate lines of constant NOx levels in Figure 9. This also includes the maximum engine power at λ=0.88 and λ=1. A minimum speed of 1000 r/min was defined for the idle control drive line. The engine would then move smoothly to a speed of 1500 r/min where the IVT transmission would then allow the power to be increased at constant speed. Fig.9 shows that for all urban driving conditions with powers below 15 kW NOx emissions below 5 g/kWh could be achieved, which is an improvement on the conventional transmission results. The results in Fig.9 have been cross plotted onto a torque speed plot with the constant power lines indicated (Fig.10). A minimum NOx IVT operational strategy has been indicated for the EBPV system which offers considerable NOx reductions on the convention results in Fig.1.b. This is not a minimum SFC strategy as this would involve higher NOx emissions, but nevertheless would avoid the greatest SFC penalty region of conventional transmission systems at low power conditions, shown in Fig.1.a.

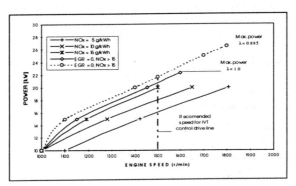

Figure 9. Constant NOx emission lines using exhaust throttling and EGR at WOT.

At WOT and constant 1500 r/min engine speed (IVT control drive line) the maximum power losses were 4%, 14% and 28% for constant NOx of 15, 10 and 5 g/kWh respectively, in comparison with the condition at λ=1, WOT and zero EGR.

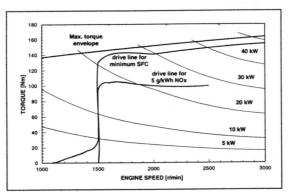

Fig. 10. IVT control drive lines for minimum SFC and 5 g/kWh NOx

4. CONCLUSIONS

The back pressure valve is a suitable means of increasing EGR flow at WOT and when the mass flow required needs a greater pressure difference than that which is attainable with manifold vacuum alone. For the constant power conditions tested, a back pressure of no more than 20 kPa was required to obtain the same or lower engine out NOx levels than the minimum achievable with a conventional transmission at the same power output (5 g/kWh).

Use of the back pressure valve results in a maximum power loss compared with the fully open EBPV and EGR=0. However, the constant 5 g/kWh NOx line is compatible with a reduced fuel consumption schedule for IVT control drive line. There exists a fuel consumption penalty using the EBPV, which could be reduced with spark timing strategy optimisation.

An electronic EGR valve is preferred for the implementation of the exhaust back pressure technique, independent of the manifold vacuum. An electronic EBPV would be preferable for optimisation of the system, but a simple fixed position on-off valve is viable, as demonstrated in the present work.

ACKNOWLEDGEMENTS

We would like to thank Ford Research for a research contract in support of this work. M. Osses would like to thank the Chilean Government for a research scholarship.

REFERENCES

1. SMITH, M.J. Powertrain slimming - courtesy of a viable family of toroidal traction CVT's, Second British/Italian Workshop on Heat Engines, 11 March, 1991.
2. JACKSON, N. Exhaust gas recirculation for reduced NOx and good fuel economy, Spark ignition engine emissions short course, 16-20 January, 1995, Department of Fuel and Energy, Leeds University.
3. NAKAJIMA, Y., SUGIHARA, K., TAKAGI, Y. Lean mixture or EGR - which is better for fuel economy and NOx reduction?, paper C94/79, in Proceedings of Conference on Fuel economy and emissions of lean burn engines, Institution of Mechanical Engineers, London,1979, pp. 81-86
4. BERRY, J., BRUNT, M.F.J. Improved control of EGR during speed/load transients, SAE paper 960068, 1996.
5. WOOD, S.P., BLOOMFIELD, J.H., Clean power - Lotus 2.2L chargecooled engine, SAE paper 900269, 1990.
6. MORRIS, G., LAKE, P. The Ford Zeta engine family, paper C427/243 Institution of Mechanical Engineers, November, 1991.
7. TOELLE, A.D. Microprocessor control of the automobile engine, SAE paper 77008, 1977.
8. REDDY, J.N. Application of automotive sensors to engine control, SAE paper 780210, 1978.

Hydrocarbon Emissions from Spark Ignition Engines

Evaluation of total HCs and individual species for SI engine cold start and comparison with predictions

P EDWARDS BEng and **D TIDMARSH** BSc, PhD, CEng, FIMechE, MIMgt
University of Central England, UK
A WILLCOCK BEng, PhD
TC, AE Goetze Automotive, Bradford, UK

SYNOPSIS The design of engine components to affect reductions in cold start and warm-up engine out emissions is becoming more important in view of future planned regulations. Experimental data characterising cold start HC emissions reveals the importance of component design and how HC species contribute to overall exhaust reactivities during the first minutes of warm-up. The effect of three way catalysts on total and individual HCs is assessed and links proposed with component conditions. The results are also used to assess the performance of a phenomenological model to predict total HC emissions.

1 Introduction

Regulated motor vehicle emissions currently include:
Total hydrocarbons (THC)
Nitrogen oxides (NO_x)
Carbon monoxide (CO)
Particulate matter (compression ignition only)

This paper is concerned with hydrocarbon (HC) emissions only and although emissions of other types are not considered, their adverse affect on the environment is still recognised.

The regulation of HC emissions currently applies to HCs as a single substance and does not discriminate between individual molecules. Experiment has shown that different HC species undergo different reactions in the atmosphere. As there are over 200 individual species of HC in spark ignition (SI) exhaust, the policy of regulating THC looks set to be replaced by the regulation of individual species.

HCs of concern can be divided into two main groups. Firstly, photochemically reactive molecules, including light alkenes, which in the right conditions are a source of ozone (O_3) and contribute to poor air quality, and secondly, carcinogenic molecules such as benzene and 1,3-butadiene which are highly toxic to humans even in moderate concentrations.

Previous research [1,2]* has shown that the concentrations of individual HCs in the exhaust of a SI engine can vary considerably during the warm-up period. Emissions during the first instances of a cold start are typically over twice that for the fully warm condition but can be even higher. These initial emissions are not converted by three way catalysts (TWC) which are by far the most popular form of emissions control device in production. The root of the problem lies with the operational envelop of the catalyst material which is typically above 250°C. Thus TWC do not function at the ambient conditions of a cold start. Advances in catalyst technology have been made resulting with close coupled and electrically heated catalysts. Moving the catalyst closer to the engine together with lower thermal inertia and sometimes fuel addition strategies have succeeded in reducing light off times but not in cleaning up the initial large concentration of HCs effectively. Also these strategies are expensive to implement, a cost which the market may not wish to bear. With the planned reduction of the 40 second period at the commencement of the ECE-EUDC drive cycle during which emissions are not sampled coupled with a possible lowering of the soak temperature, cold start regulations will become harder to meet.

---

* Numbers in parenthesis indicate references at the end of the text.

Another way to address the problems of cold start HCs is at the source. Lean burn technology has progressed recently, but is not yet at the level of refinement that would be required to substantially reduce cold start emissions.

Investigations into the sources of unburnt HCs from homogenous charge SI engines [1,2,3,4] have consistently named combustion chamber crevice volumes and lubricating oil layers as the two major sources. The processes by which HCs are released from these sources are extensively documented in the references. Unburnt fuel represents a major proportion of the emissions for steady state modes of operation. Although the mechanisms are less well understood for cold starting, it would appear that the same release mechanisms as for steady state are also a major factor. Unfortunately, due to the thermal nature of the reciprocating parts in the engine and their lubricational requirements, effecting substantial reductions in initial cold start emissions may be difficult. Crevice volumes are already reducing in an effort to curb emissions, but the reductions are limited by the dimensions and temperature profile of the fully warm piston. The piston rings and skirt operate with modest lubrication in most engines, possibly even boundary lubrication at the limits of each stroke, and further reductions in oil films may compromise the durability of reciprocating parts. Thus, without complimentary progress in materials, optimising combustion chamber crevice volumes and oil films as a strategy to reduce cold start emissions may be applicable, but of diminished value. However, any HC emission reductions achieved during cold starting by simply reaching the limits of the available materials with respect to crevice volumes and oil films must be implemented.

This paper documents experimental engine testing and the data gathered, including the time aligned concentrations of several salient HC species during cold start testing, both before and after the catalyst. This data is used to calculate the reactivity of the exhaust gas using the MIR scale [5] and thus the O_3 forming potential (OFP). The effect of the TWC during this period is assessed. Individual HC concentrations are expressed as a total concentration and correlated to THC levels as measured by a standard bench analyser.

The engine used during these tests was of modern design, incorporating multi point fuel injection and electronic ignition. The control strategy used was to the manufacturers original specification. Therefore, initial enrichment to aid driveability was included. The pistons used for all these tests had a relatively small top land height of 3 mm.

2 Model

The modelling of engine exhaust HC emissions has been attempted by previous researchers [6,7,8], usually describing physical processes in a numerical manner. Numerical representations of physical events can be computationally intensive and it is common to simplify systems to two or three mechanisms which are considered salient. Previous models have considered contributions to HC emissions from ring pack crevice volumes, oil layers and a degree of post flame oxidation (PFO) or a combination of the three. Experimental evidence has also pointed to these mechanisms as possibly the only ones of major importance during fully warm running conditions.

An accurate absolute value for emitted HCs cannot be predicted, due to the incomplete description of events. However, a more useful quantity to deal with is percentage change from a control or baseline situation. This is possible to achieve with only the more salient mechanisms. If a model can correctly predict percentage changes in emissions with varying physical conditions then it is a useful tool in engine development which can indicate in which direction development should proceed without the need for time consuming and expensive testing.

Fewer examples of models predicting transient emissions exist and it is this area that the model described in this paper covers.

3 Absorption / desorption

The basis for modelling the effect of cylinder liner oil films has been developed from previously proposed models [9,10]. Other researchers [8,11] have found this approach suitable. One dimensional diffusion of a dilute solute through a solvent is considered, with Henrys law governing the boundary conditions at the gas / oil interface:

$$C_{eq} = Y_f \left(\frac{P}{H_c} \frac{M_{oil}}{M_{fuel}} \right)$$

where:
C_{eq} equilibrium concentration of fuel in the liquid phase
Y_f mass fraction of fuel in the gaseous phase
P pressure of the gaseous mixture $\left[\frac{N}{m^2} \right]$
H_c Henrys constant for the fuel / oil combination $\left[\frac{N}{m^2} \right]$
M_x is the molecular mass of the suffixed material x $[kg]$

From the equation it can be seen that choosing a valid Henry constant for the solution is critical. Experimental values of the solubility of light HCs (C_1 to iC_4) in larger HC solvents have been published [12] and previous researchers have extrapolated these values to predict the solubilities of larger molecules [7] such as octane (C_8). This approach has been used in this research.

The temperature of one of the cylinder liners during a cold start was monitored using thermocouples at different heights in the thrust and non-thrust faces. An average temperature for the liner was calculated at which the oil film was assumed to be. The Henry constant for the fuel oil combination was then assumed to be:

$$Log(H_c) = 1 \times 10^6 \left[0.013(T_{liner} - 300) - 1.921 \right]$$

where:
H_c Henrys constant $\left[\frac{N}{m^2} \right]$
T_{liner} temperature of the oil $[K]$

A limitation of this method is that at steady state any inaccuracy at this calculated value remains constant (due to constant oil film temperature), however, over a range of temperatures (cold starting), inaccuracies will not be constant but will depend how robust the extrapolation to a larger molecular solute / solvent combination is.

The diffusion of the fuel through the oil film is assumed to be Fickian and is governed by:

$$\frac{\partial c}{\partial x} = D \frac{\partial^2 c}{\partial x^2}$$

where:
c concenration of fuel in the liquid phase
x distance radially from the cylinder liner wall $[m]$
D coefficient of diffusion $\left[\frac{m^2}{s} \right]$

There are both analytical and numerical solutions for this equation. A numerical approach has historically been favoured due to the ease with which boundary conditions can be substituted into the calculations.

Values for the diffusion coefficient have been previously investigated [13] resulting in the proposal of the empirical equation:

$$D = 7.4 \times 10^{-8} \frac{M^{0.5}T}{v^{0.6}\eta}$$

where:
- M moleweight of oil $[\text{gmol}]$
- T temperature of solution $[K]$
- η viscosity of oil $[\text{cp}]$
- v molar volume of solute at bp $[\text{cm}^3]$
- D coefficient of diffusion $\left[\dfrac{\text{cm}^2}{\text{s}}\right]$

Values for the diffusion coefficient for a solute in a solvent can also be derived using the physical parameters of the molecules involved and the mean free path of a solute molecule:

$$D = \frac{RT}{6\pi r N_A \eta}$$

where:
- r molecular radius for octane $\approx 2.25 \text{Å}$ $[m]$
- R universal gas constant $\left[\dfrac{KJ}{KmolK}\right]$
- T liquid temperature $[K]$
- N_A Avagadros constant
- η viscosity $\left[\dfrac{Ns}{m^2}\right]$

The values predicted by these different approaches are compared for a range of temperatures in figure (1). Both these approaches require the correct values for viscosity, which if unknown will result in a variable error over a range of temperatures.

The oil film on the liner is divided into elements axially and radially. The position of an element axially defines the period for which it is exposed to fresh or burnt mixture, or when it is covered by the piston. The number of sections each element is divided into radially controls the accuracy of the diffusion calculation. The finer the mesh, the longer the computational process.

Fuel diffuses into the oil layer during the induction and compression strokes due to the concentration gradient caused by the fresh mixture in the cylinder. This gradient is reversed after combustion, forcing a proportion of the fuel to diffuse back into the cylinder. Previous models have made the assumption that combustion is instantaneous and occurs at top dead centre (TDC). A slightly more realistic approach is adopted in this work proposing that the concentration of fuel in the mixture adjacent to the oil film does not fall to zero until the mass fraction burnt reaches 95%. This is not considered an unreasonable assumption as the spark plug in the engine used is located centrally in the roof of the cylinder head and the flame front would progress spherically into the combustion chamber. The mass burnt fraction is calculated from a heat release model which incorporates data from an in-cylinder piezoelectric pressure transducer. At the point when 95% of the mixture is burnt, the concentration of fuel at the oil film surface is considered to drop instantly to zero.

The mass of the fuel in the oil is calculated from the mass of oil present in the film and the concentration of fuel. Predictions and experimentally determined values for oil film thickness on a cylinder bore have been published [14,15]. The process of mathematically computing cold start oil film thickness is uncertain as several factors are not well understood. Surface finish is known to affect oil layer thickness and emissions [16]. As the lubrication between the piston rings and the cylinder liner at the extents of each stroke can be of the boundary variety, the process of calculation is thus more complicated than the case of simple hydrodynamic lubrication. How this is in turn affected by large changes in oil viscosity and oil supply during cold start conditions is unknown. This author is unaware of published data for cold start oil film thickness.

A simple link between engine speed, lubricant viscosity and cylinder liner surface finish is used as an initial description for oil film thickness in this study. At fully warm condition, the oil film thickness is assumed to be approximately equal to the R_a value of the surface of the liner.

Further improvements to the model will centre around obtaining a more complete description of the oil film behaviour during cold starting.

4 Combustion chamber crevice volumes

Several models describing the influence of combustion chamber crevice volumes and in particular those associated with the piston ring pack on HC emissions have been previously reported [17,18]. These have mostly been for steady, fully warm engine configurations but some have been concerned with cold starting [19]. The basis for all is the equation of state:

$$PV = nRT$$

where:

P pressure $\left[\dfrac{N}{m^2}\right]$

V volume $\left[m^3\right]$

n moles

R universal gas constant $\left[\dfrac{KJ}{KmolK}\right]$

T temperature $[K]$

By experimentally or mathematically defining the in-cylinder pressure, the number of moles of a substance resident in a fixed volume is governed by the local temperature. By rearranging the formula the incremental mass during a cycle can be computed.

The basic assumption can be improved on by more accurately describing the temperature of the crevice gas (usually assumed to be isothermal with the piston), by determining whether the in or out-fluxing material is fresh or burnt mixture and by accounting for the motion of the rings in their grooves.

The process just described covers simple flow through unrestricted orifices such as into the top land crevice. However, by using optical experimental equipment allowing optical analysis [18], jet like flows from the ring pack of a reciprocating engine were observed during blow down and exhaust strokes. Treating flow into the second land crevice volume as compressible flow through a restriction, it was found that these jet like flows could be predicted. This function of the ring gap was experimentally proved to be of consequence [20] in 1968. More recently the critical nature of second land flows has been thrown into doubt [21] as manufacturers tolerances have become increasingly tighter.

The flow into and out of the top ring groove is considered so close to the top land crevice, that in all computations the two crevices will be treated as one.

To predict cold start crevice HC contributions the transient piston dimensions must be known. In this work the dimensions of a piston at 200°C were measured and these were expressed as a percentage of the cold dimensions. The temperatures of the cylinder liners were then used to roughly estimate the piston temperatures and the crevice volumes then calculated.

The assumptions made when calculating the mass trapped in the crevices are that the crevice gas is isothermal with the liner, that the piston remains central in its bore, that the crevices only contain unburned mixture and that the rings always remain seated on the bottom of their respective grooves. Future improvements for the prediction of crevice volumes include an attempt to predict piston expansion under a given load using finite element analysis. Also a refinement of the assumption that crevice volumes only contain unburnt mixture will be undertaken.

One factor not accounted for in the model but which influences the two release mechanisms described above is that of poor fuel preparation due to low component temperatures. Previous research [22] has demonstrated there to be a discrepancy between the mass of fuel injected into the inlet manifold and the amount of fuel burnt in the cylinder during the initial stages of cold start. This discrepancy is explained by the build up of a film of liquid fuel on the cold surfaces of the inlet manifold which subsequently evaporates as the temperature rise.

5 Post flame oxidation

The final part of the in-cylinder HC model does not relate to a release mechanism as such but to the subsequent demise of HCs returning to the combustion chamber after flame extinguishment. Studies of post flame oxidation of fuel material [23,24] in a SI engine have highlighted its major significance in the final level of HCs in the exhaust gas.

Fuel mixture, trapped by the mechanisms described, returns to the combustion chamber during the blown down and exhaust strokes. As mixing occurs with the cooling burnt gas the fuel mixture is subjected to a range of elevated temperatures. Oxidation will take place if the temperature is still sufficiently high and there is an availability of free oxygen (O_2). The rate at which oxidation occurs is dependant on the temperature and concentration of free O_2. The temperature below which oxidation stops, or is sufficiently reduced, have been the subject of research [25]. Mechanisms, predicting the extent of oxidation from a series of reaction rate steps have been explored with varying degrees of experimental agreement. The more simplistic approach of a temperature below which oxidation effectively ceases was also investigated. The discrepancy between this approach and experimental data was found to be less than for the more complicated rate controlled approach.

The mechanism for post flame oxidation used in the model reported here is based upon the temperature of sudden freezing [25]. The burnt gas temperature at the combustion chamber extremities is assumed to be equal to the bulk gas temperature and the returning fuel mixture in the boundary layer is assumed to be at the same temperature as the cylinder liner. After limited mixing between the two occurs, the subsequent temperature is calculated as the mass weighted average of the contributing sources. Above this temperature, oxidation of HCs returning to the combustion chamber is assumed to be of a global nature with no spatial or temporal variations. Below this temperature, oxidation is assumed to have ceased.

6 Experimental

The test facility used for the experimental section of this work is centred around a four cylinder SI production engine. The engine is equipped with a management system for the control of fuel injection and ignition. The fuelling requirements of the engine are controlled to suit the needs of an underbody type TWC. The engine was run on reference fuel (CEC RF-08) for all emission tests. The engine is instrumented with various thermocouples, pressure transducers and flow measuring devices. A schematic of the engine test facility is shown in figure (2).

Stainless steel sample pipes were included in the exhaust system to allow extraction of exhaust samples for analysis. Samples were transported to the analysis equipment through a heated line maintained at 180°C. On-line measurements are taken with a standard PC controlled bench analyser. Measurements of THC (as C_3), CO, CO_2, O_2 and NO_x were logged at one second intervals during cold start tests. Instantaneous air-fuel ratio (AFR) was obtained using an exhaust gas calculation.

At fixed points during cold start tests samples of exhaust were extracted from bench analyser. These were transported in heated glass syringes to a gas chromatograph (GC) where the concentration of over 200 individual HC species were ascertained. A typical exhaust chromatogram is shown in figure (3). For the scope of this work 60 individual HC species have been identified by comparison with published retention indices [26,27,28]. The species are named in table (1). Using these individual concentrations, the reactivity of the exhaust was calculated. The calculations were conducted using the concentrations of all species in equivalent propane units which although allowing comparisons within this work, does not allow comparison with values calculated using the specific concentrations of each molecule.

HCs are detected using FID in both the bench analyser and the chromatograph. CO and CO_2 are detected using NDIR, O_2 is detected using a paramagnetic type detector and NO_x is detected by chemiluminescence. The detectors in both the bench analyser and the GC were routinely calibrated before each test. Further details concerning the test apparatus and operational methods used can be found in [1].

By its own nature, gas chromatography is an off line procedure. Typical analyses times during this work were in the order of 55 minutes. Data previously published [27] has shown certain species to be unstable when retained in sample containers, particularly members of the di-ene group. As these molecules are highly reactive and thus of interest in this work, it was decided not to store samples but to process them immediately after their extraction from the exhaust. This necessitated four cold starts to obtain one complete set of speciation data for the four points per cold start. The extraction times during a cold start test are labelled in figure (4). As no automated process was employed for the extraction of gas samples, a window of ±5 seconds was placed around the alignment of gas chromatography data to other logged data.

Testing was after a minimum soak period of 12 hours, during which the ambient temperature was kept to between 10 and 15°C. The throttle and dynamometer were set the previous day to achieve a 2000 rpm 2 bar BMEP steady condition. The engine was stopped and rotated while compressed air purged the exhaust, inlet manifold and cylinders of unwanted HCs. For the next test it was set to TDC firing on cylinder 4. Between tests the battery was fully recharged to ensure no prolonged cranking periods.

Immediately before each test, the detectors were calibrated and instrumentation activated and checked. Fresh fuel, which was kept in sealed drums, was placed in the fuel tank in an effort to reduce inaccurate results as a consequence of lighter fuel species having evaporated between tests. The data logging was initiated and when the gas sampled from the exhaust contained no HCs, the engine was started. A counter started at the time the key was turned was used to align gas extraction into syringes. As the samples were taken at the bench analyser, a characteristic transport time was also included.

7 Results and discussion

The characteristic THC emissions during a cold start are shown in figure (5). The data shown are an average of between 4 and 5 tests in each case. Included are the totals from chromatographic analyses. Although the individual species concentrations could be expressed in units specific to each molecule, to achieve some correlation between the two detectors they were left in units of C_3.

There is a large initial concentration at time t=0, both before and after the catalyst, which rapidly reduces until 35 seconds when the reduction becomes more gradual. This more gradual reduction continues until 200 seconds when the engine out concentration goes through a minimum at approximately 650 ppmC_3. During the period 35 to 200 seconds, the catalyst temperature increases to a significant temperature corresponding to the increasing difference between engine out and after catalyst emissions. The catalyst reaches full working efficiency between 230 and 250 seconds after engine firing at $\approx300°C$.

The initial 'spike' can be attributed to several factors. From observations of the cylinder pressure during cold starting, it was noticed that there were on average 1-2 unfired cycles before ignition. A proportion of fuel form these cycles passes straight to the exhaust system contributing significantly to the total. An analysis of the early emissions (figure (6)) reveals a predominance of the liquid fuel species such as toluene, xylene and 2,2,4-trimethylpentane giving plausibility to this explanation. The unburnt fuel from unfired cycles would be displaced from the exhaust system soon after firing by the large increases in volumetric flow and temperature. The duration and reduction in rate of diminishment of the 'spike' is indication that, to some degree, other factors are involved.

Liquid fuel films are established in the cold inlet ports during the first few seconds that subsequently evaporate over a short period. This causes the inlet mixture to be overly rich, increasing HC emissions. Combustion chamber crevice volumes are largest when cold and can contain a greater mass at a given pressure as their temperatures are lower during a cold start. Fuel is substantially more soluble in cold oil films in the combustion chamber which due to a higher viscosity, are also likely to be thicker. Clearly these factors will play some part in the elevated emissions.

Characteristics of each group represented in the exhaust were observed. Of the major groups observed three were aliphatic (alkanes, alkenes and alkynes) and one cyclic (aromatic). Before catalyst light off, HCs before and after the catalyst exhibit very similar characteristics with only slight discrepancies in absolute value. This is clearly seen in figure (7).

Initial emissions are dominated by alkanes and aromatic species which make up the majority of the fuel. By 30 seconds levels have dropped across the board, but the drop in alkanes is greater than for aromatics. Of the two major fuel groups, alkanes and aromatics, aromatics are the more soluble in the oil films. As more aromatics would dissolve into the oil films and reach the exhaust this observation would seem to provide proof that oil films have a part to play, if not in the size of the initial 'spike' then in its latter stages. A point to note is that the reduction in alkanes may also be brought about by the fact that a large portion of alkanes, such as 2-methylbutane, are more easily volatilised than the heavier aromatics and thus will burn more efficiently in the cold conditions.

As time progresses, from 30 seconds onwards the HC emissions adopt a more gradual reduction. This phase of the cold start would seem to be even more dependant on the oil film and crevice effect. In figure (8), we can see that this gradual reduction, corresponds to the increase in cylinder liner temperatures. The thermostat opens and the liners reach thermal equilibrium at ≈200-210 seconds. This point also sees a stabilisation in the THC levels. As postulated earlier, crevice and oil film effects are governed by liner temperature. It would appear then that a major part of the cold start HCs (25-200 seconds), are due to these effects.

Alkenes, such as ethene, propene and butene are virtually non-existent in the liquid fuel. They are found in the exhaust only as a product of oxidation of other species. From the increases in contribution of alkenes observed during warm-up seen in figure (7), the levels of post flame oxidation can be assumed to be increasing. As this contributes more reactive alkenes to the total, the reactivity of the exhaust increases. This can be attributed to the increasing temperature of the trapped mixture, which when returned to the combustion chamber diffuses more rapidly into the hotter burnt gas and retains a higher temperature thus promoting more complete oxidation.

The pressure, temperature and AFR data gathered during a series of tests was used as input for a numerical model. A comparison of the HC predictions and experimental measurements is shown in figure (9).

Predictions for the first few seconds are lower than observed. This is to be expected as fuel from unfired cycles is not taken into account. The values shown are after the effects of post flame oxidation. The correlation between prediction and experiment is satisfactorily close. It can be seen that the oil films account for slightly over half the total. This result does not completely agree with previously published data for steady state conditions [3,4] in which crevice volumes contribute the majority of HCs emitted. However, the testing conditions in this work are transient and thus different results would be expected.

By 25 secs the model over predicts the concentration of HCs. This continues as the engine condition nears its steady state. The contribution to the total from crevice volumes and oil layers reaches a 50% split by 200 seconds. Although the predicted emissions are too high, the model agrees closely with the experimental data of when the emissions level out. As the model strongly relies on cylinder liner temperature, this reaffirms the strong link between cylinder liner temperature and HC emissions.

8 Conclusions

Concentrations of time aligned individual HC species have been recorded during cold start tests. The data have been repeatable within ±14% (1 standard deviation) for the first sample event and ±8% (1 standard deviation) for the remaining events. These concentrations have been related to OFP and exhaust reactivity which show significant increases during cold start and warm-up. This highlights the need to optimise control strategies and component parameters.

The detailed development of a model to predict THC has been described. The model has built on the experience and work of previous researchers in the field of S.I. HC emissions.

The model uses experimental information, such as AFR and liner temperature, to predict the HC contributions from combustion chamber crevices, oil films and post flame oxidation. A comparison with experimental emissions shows reasonable agreement. It is envisaged that component designs over transient conditions will be explored with the model and designs for lower emissions be suggested.

9. References

1. Edwards, J. M. et al.: The concentration of and source of HC species in the exhaust of a SI engine during cold start and transient operation. EAEC 5th International Congress. Powertrain and the Environment. 1995.

2. Kaiser, E. W. et al.: Time-resolved measurement of speciated HC emissions during cold start of a spark ignited engine. SAE940963.

3. Wentworth, J. T. et al. The piston crevice volume effect on exhaust HC emissions. Comb Sci Tech 1971. Vol. 4, pp97-100.

4. Haskell, W. W. et al.: Exhaust HC emissions from gasoline engines- surface phenomena. SAE720255.

5. Carter, W. P. L.: Development of ozone reactivity scales for volatile organic compounds. J Air Waste Manage Assoc 1994, Vol. 44, pp881-899.

6. Schramm, J. et al.: A model for hydrocarbon emissions form SI engines. SAE902169.

7. Dent, J. C. et al.: A model for absorption and desorption of fuel vapour by cylinder lubricating oil film and its contribution to HC emissions. SAE830652.

8. Willcock, M.: The effect of piston design on HC emissions in a SI engine. PhD Thesis 1993.

9. Carrier, G. et al.: Cyclic absorption / desorption of a gas in a lquid wall film. Comb Sci Tech 1981. Vol. 25, pp9-19.

10. Korematsu, K. et al. Effects of fuel of fuel absorbed in oil films on HC emissions from SI engines. JSME 1990 Vol. 33, pp606-614.

11. Gatellier, B. et al.: HC emissions of SI engines influenced by fuel absorption-desorption in oil films. SAE920095.

12. Chappelow, C. C. et al.: Solubilities of gases in high-boiling HC solvents. AIChE Journal. 1974. Vol. 20, No. 6. pp1097-1104.

13. Wilke, R. et al.: Correlations of diffusion coefficients in dilute solutions. AIChE Journal. 1955. Vol. 1, No.5. pp264-270.

14. Radcliff, C.: Piston ring friction. The mission of tribology 1993.

15. Miyachika, M. et al.: A consideration of second land pressure and oil consumption of IC engines. SAE840099.

16. Willcock, M. et al.: An evaluation of cylinder liner surface finish and synthetic oil on exhaust HC species from a SI engine. 1993. CEC 4th symposium CEC/93/EF11

17. Kuo, T. W. et al.: Calculation of flow in the piston-cylinder-ring crevices of a homogenous charge engine and comparison with experiment. SAE890838.

18. Namazian, M. et al.: Flow in the piston-cylinder-ring crevices of a SI engine: Effect on HC enmissions, efficiency and power. SAE820088.

19. Kaplan, J. et al.: Modelling the SI engine warmup process to predict component temperatures and hydrocarbon emissions. SAE910392.

20. Wentworth, J.T. et al.: Piston and ring variables affect exhaust HC emissions. SAE680109.

21. Willcock, M. et al.: A comparison of HC emissions from different piston designs in a SI engine. SAE930714.

22. Sorrell, A. J. et al.: SI engine performance during warmup. SAE890567.

23. Caton,J. A. et al.: Models for heat transfer, mixing and hydrocarbon oxidation in a exhaust port of a SI engine. SAE800290.

24. Mendillo, J. V. et al.: HC oxidation in the por of a SI engine. SAE810019.

25. Wiess, P. et al.: Fast sampling valve measurements of HCs in the cylinder of a CFR engine. SAE810149.

26. Kent Hoekman, S.: Speciated measurements and calculated reactivities of vehicle exhaust emissions from conventional and reformulated gasolines. Environ Sci Technol 1992. Vol. 26, pp1206-1216.

27. Kent Hoekman, S.: Improved gas chromatography procedure for speciated hydrocarbon measurements of vehicle emissions. J Chromatog 1993. Vol. 639, pp239-253.

28. Siegle, W. O. et al.: Improved emissions speciation methodology for phase II of the Auto/Oil Air Quality Improvement Research Program- hydrocarbons and oxygenates. SAE930142.

Table (1): Hydrocarbons species

No.	Component Name	Relative Retention Time	Group
1	methane	0	alkane (1)
2	ethene	0.006	alkene (2)
3	ethane	0.01	1
4	propene	0.041	2
5	propane	0.045	1
6	propyne	0.068	alkyne (3)
7	2-methylpropane	0.106	1
8	1+2-butene	0.141	2
9	1,3-butadiene	0.148	2
10	n-butane	0.155	1
11	trans-2-butene	0.174	2
12	cis-2-butene	0.198	2
13	2-methylbutane	0.29	1
14	1-pentene	0.322	2
15	2-methyl-1-butene	0.336	2
16	n-pentane	0.345	1
17	2-methyl-1,3-butadiene	0.354	2
18	trans-2-pentene	0.363	2
19	cis-2-pentene	0.379	2
20	2-methyl-2-butene	0.389	2
21	trans-1,3-pentadiene	0.395	2
22	cyclopentadiene	0.414	2
23	2,2-dimethylbutane	0.42	1
24	cyclopentene	0.458	2
25	cyclopentane	0.482	1
26	2,3-dimethylbutane	0.487	1
27	2-methylpentane	0.498	1
28	3-methylpentane	0.531	1
29	n-hexane	0.572	1
30	3-methyl-cis-2-pentene	0.623	1
31	methylcyclopentane	0.638	1
32	2,4-dimethylpentane	0.648	1
33	benzene	0.701	aromatic (4)
34	cyclohexane	0.724	1
35	2-methylhexane	0.746	1
36	2,3-dimethylpentane	0.752	1
37	3-methylhexane	0.77	1
38	2,2,4-trimethylpentane	0.807	1
39	n-heptane	0.837	1
40	methylcyclohexane	0.9	1
41	2,5-dimethylhexane	0.928	1
42	2,4-dimethylhexane	0.935	1
43	2,3,4-trimethylpentane	0.984	1
44	toluene	1	4
45	2,3-dimethylhexane	1.013	1
46	2,2,5-trimethylhexane	1.085	1
47	n-octane	1.131	1
48	ethylbenzene	1.288	4
49	meta-xylene	1.314	4
50	para-xylene	1.318	4
51	styrene	1.37	4
52	ortho-xylene	1.386	4
53	n-nonane		1
54	n-propylbenzene	1.574	4
55	1-ethyl-3-methylbenzene	1.595	4
56	1-ethyl-4-methylbenzene	1.602	4
57	1,3,5-trimethylbenzene	1.618	4
58	1-ethyl-2-methylbenzene	1.65	4
59	1,2,4-trimethylbenzene	1.694	4
60	n-decane	1.732	1

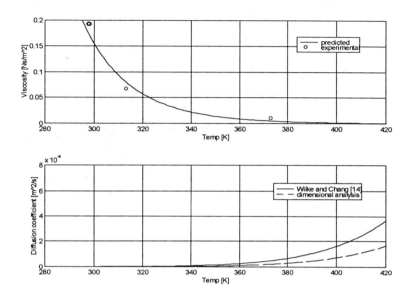

figure 1. Comparison of two methods for calculating the diffusion coefficient D.

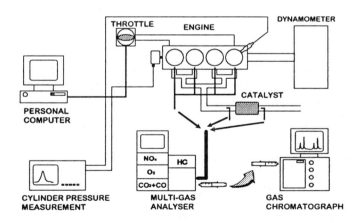

figure 2. Schematic of the engine test facility.

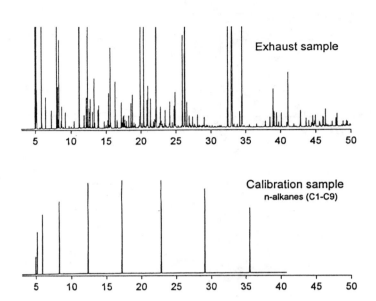

figure 3. Typical exhaust gas chromatograms.

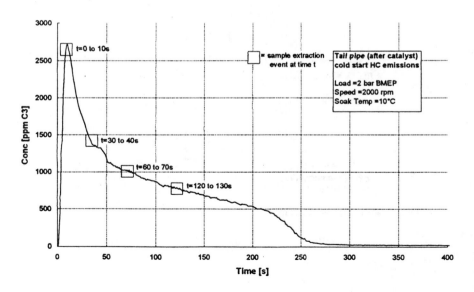

figure 4. Cold start tailpipe THC emissions with sample extraction events for speciation analysis.

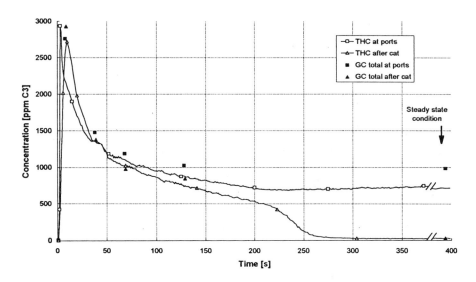

figure 5. Cold start THC emissions before and after the catalyst with GC totals in C_3 units.

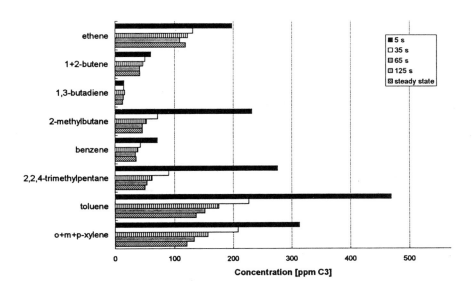

figure 6a. Cold start levels of selected species at the exhaust ports.

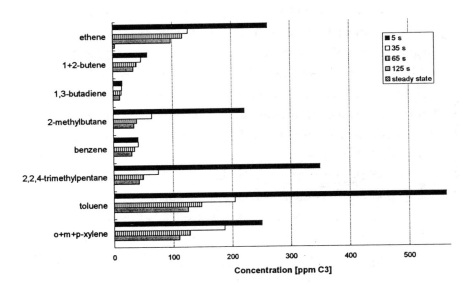

figure 6b. Cold start levels of selected species after the catalyst.

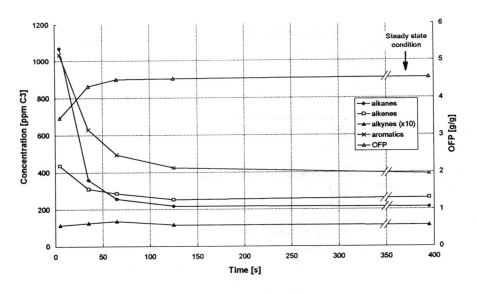

figure 7a. Variance of major HC groups at the exhaust ports during cold start.

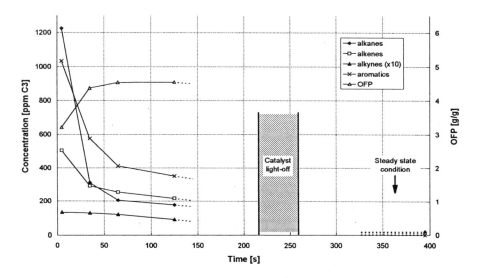

figure 7b. Variance of major HC groups after the catalyst during cold start.

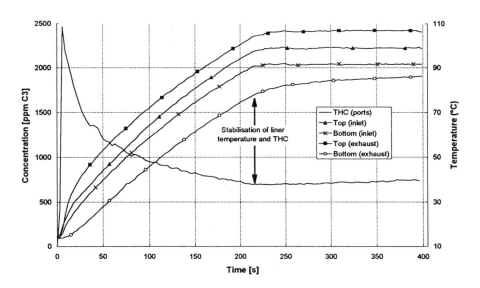

figure 8. Comparison of engine out HC emissions with cylinder liner temperatures during cold start.

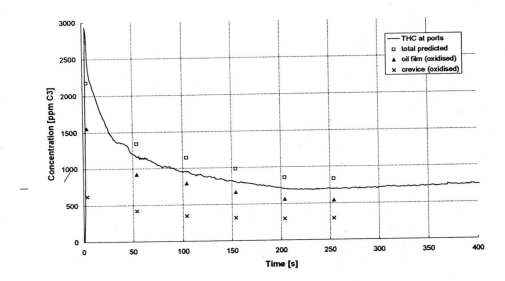

figure 9. Predicted and measured engine out HC emissions during cold start.

Emission of unregulated compounds from five different gasoline qualities

LAVESKOG MSAE
Motortestcenter, Swedish Motor Vehicle Inspection Company, Sweden

INTRODUCTION

Five different gasoline qualities have been tested. The project was part of a continuing search for less environmentally harmful Otto engine fuels in the range of what could still be called gasolines. SwEPA had three fuels tested and two oil companies one each. The gasolines ranged from pure alkylate to a gasoline typical of Sweden 1993-94. The tests were performed in a car with adaptive fuel metering system. The catalyst was removed in order to get sufficient quantities of unregulated pollutants especially for bioassays. The OICA type C ECE test cycle was used. Unregulated as well as regulated emissions were measured.

1 GASOLINE QUALITIES

The gasolines tested were all of qualities which are marketed in Sweden to different extents. The gasoline qualities were named A, T, 1, R, and 2.

Gasoline A was a pure alkylate chosen as a reference for the "cleanest" possible gasoline. This alkylate is produced by Neste in Finland, sold as "Chain saw gasoline" for mixing with 2 stroke oil and for use in 4-stroke lawn mowers and other utility 4-stroke engines.

Gasoline T, sold as "Aspen Taxi" is composed of 79% alkylate 16% ethanol and 5% xylenes. This gasoline is sold as a motor fuel for a fleet of taxi cars in Gothenburg and also as a fuel for outboard engines and snowmobiles (mostly competition).

Gasoline 1 is a stage two reformulated gasoline from Neste in Finland sold as a standard unleaded 95 RON gasoline having a ~ 90% share of the market in Finland ("city 2"). This gasoline has a reduced benzene concentration, reduced aromatic content and also a lowered final boiling point and sulphur content. It contains ~ 10% of MTBE.

Gasoline 2 is an early batch of "environment class 2"-gasoline ('OK 95T1) sold in Sweden since January 1994. It is manufactured by the OK Petroleum refinery in Gothenburg. When compared to "normal" gasoline of 1993-94 in Sweden, it has a slightly reduced benzene and aromatic content. The benzene concentration + (aromatics concentration /13) should be less than 5,5 % for 95 RON gasoline. The sulphur content is reduced, as well as the final boiling point and RVP. The composition limitations are regulated in Swedish tax laws and the entire composition is given in the proposed Swedish standard, SS 15 54 22.

Gasoline R is an unleaded 95 RON standard gasoline sold on the market in Sweden but with a more narrow specification, named "Blend 95", and used as a test gasoline for the Swedish automobile manufacturers.

2 TEST VEHICLE

The vehicle, lent to us by the emission department of Volvo in Gothenburg, was driven approx. 7000 km before the test and was prepared as follows. The engine oil filter was replaced by a new filter and, after oil drain, a fully synthetic OK 5W-40 oil was filled. The gasoline tank was filled with alkylate fuel which was additivated with an additive capable of cleaning the combustion chamber. The vehicle was driven on the road for about 1500 km with this special additive also added to standard 95 RON unleaded gasoline used later. Back at the laboratory the oil and oil filter were again exchanged and 5W-40 fully synthetic OK oil was filled. The entire exhaust system after the exhaust manifold was exchanged for a new exhaust system in stainless steel with only one muffler which was free of absorbent. The catalyst was replaced by a straight tube of stainless steel, the lambda sensor being retained. A flange for connection to the exhaust dilution system was welded to the tail pipe, thus avoiding materials other than stainless steel in the exhaust, dilution and sampling system. The fuel tank was thoroughly emptied and the line from the carbon canister vent line plugged as well as the line connection to the engine, thus avoiding interference from gasoline remaining in the fuel tank. The fuel lines were disconnected under the floor and plugged on the fuel tank side. The fuel pump was electrically disconnected and an external fuel system built up and connected to the engine side of the fuel lines under floor. The external fuel pump was electrically connected to the original pump relay in the vehicle. The fuel was taken from 20 l jerry cans and was thus easy to exchange.

2.1 Conditioning

Before the first test A-fuel was filled in the jerry can and the vehicle was conditioned for 10 minutes at 100 km/h, while connected to a direct outlet. After that, one OICA modified ECE test cycle was driven, with the exhaust pipe connected to the dilution tunnel. After soaking over night the first test on A-fuel was driven.

3 TESTING

The test cycle used was the OICA C modified ECE test cycle of 11 km. To make it possible to perform all the tests in the time available, the vehicle was forced soaked between the tests. The soaking was forced with 3 fans blowing into and under the engine compartment. During the first hour of soaking the temperature in the cell was lowered to +10 °C and then raised back to 20 °C. The engine oil and cylinder head temperatures were supervised and the test was started when both temperatures were within 1,5 °C of the cell temperature, which normally took around 4 hours from the end of the previous test.

3.1 Sampling

In order to minimise condensation in the sampling system, the tail pipe of the vehicle was connected to the point of dilution using only about 75 cm of piping, partly flexible. The diluted exhausts were transported in a 110 mm stainless steel pipe, of about 4 m in length, to the actual dilution tunnel, 250 mm in diameter and 3 m long. Sampling equipment for particulates as well as for aldehydes and light hydrocarbons was situated at the end of the tunnel. Regulated emissions were sampled and measured using a Horiba 9000 system.

Particulates were sampled on teflon coated glass fibre filters and semivolatiles in Porous Polyurethane Foam (PUF) which were in series after the filter. The normal flow through the filter and PUF was around 1 m^3/min out of the 6 m^3/min in the dilution tunnel. The sample flow was returned to the tunnel 0,75 m after the sampling point.

Aldehydes were sampled in 2,4 DNPH coated sampling tubes made by Micropore. Sampling was regulated to about 300 ml/min, in order not to overload the sampling tubes. Medium boiling point gasoline components were sampled in activated charcoal tubes, at the same rate of 300 ml/min. Gaseous compounds were sampled in polyester-aluminium foil laminated bags also at about 300 ml/min. The hydrocarbon sampling tubes as well as the gas balloons were kept at subzero temperatures and were sent for analysis every third day. Aldehyde sampling tubes were stored in closed bags made of laminated polyester and aluminium foil, in darkness in a freezer and were sent frozen for analysis at the end of the test period.

Regulated emissions and methane were sampled in polyester bags and analysed as bag values according to standard procedures. The gases filled in the bags were also analysed for some compounds with an FTIR instrument, however this instrument failed mid project.

4 EMISSSION RESULTS

The emissions are all measured and calculated as average values over the test cycle, EUDC included (a total of ~11 km).

Table 1 Fuel energy content, fuel consumption and regulated emissions.

Fuel	A	T	1	R	2
Qeff MJ/kg	44,1	40,4	41,5	41,9	42,3
Density kg/l	0,69	0,72	0,74	0,75	0,75
Fuel consumpt MJ/10km	30,0 ± 0,3	28,3 ± 0,3	28,9 ± 0,9	30,0 ± 0,3	29,8 ± 0,2
Fuel consumpt. l/10 km	0,99 ± 0,01	0,97 ± 0,01	0,94 ± 0,03	0,95 ± 0,01	0,94 ± 0,01
CO g/km	9,35 ± 0,48	7,69 ± 0,12	9,06 ± 0,37	9,17 ± 0,24	9,54 ± 0,24
HC g/km	1,70 ± 0,03	1,64 ± 0,01	1,64 ± 0,03	1,79 ± 0,04	1,74 ± 0,01
NOx g/km	1,83 ± 0,07	1,88 ± 0,04	2,19 ± 0,06	2,35 ± 0,04	2,32 ± 0,04
CH4 g/km	0,10 ± 0,01	0,08 ± 0,00	0,06 ± 0,00	0,05 ± 0,00	0,05 ± 0,00
CO2 g/km	193 ± 2	196 ± 2	200 ± 6	206 ± 2	204 ± 1
Total CO2 g/km	213 ± 2	213 ± 2	219 ± 7	226 ± 2	225 ± 1
Particulates mg/km	6,3 ± 0,3	7,6 ± 0,9	11,0 ± 0,4	12,2 ± 0,5	10,4 ± 0,9

4.1 Regulated emissions, Methane (CH_4), Fuel consumption (FC), CO_2 and particulates.

When driven on the five fuels, the emissions from the vehicle differ significantly in the case of CH_4 (two fuels), CO (one fuel) and of NO_X (three fuels), when comparing the test fuels to the reference R-fuel. The HC emissions are also somewhat lower for the two alkylate based fuels and the 1-fuel. When the fuel consumption was calculated according to the standard formula in the ECE regulations, where carbon fraction, density and energy content are taken into account, only the alkylate fuels differ significantly from the R-fuel.

The CO emissions are typical pre-catalyst emissions for a λ-sensor vehicle, around 9,3 g/km for A-, R- and 2-fuel respectively. Within the standard deviation, the 1-fuel has 9,0 g/km but the T-fuel has 7,7 g/km, this being far outside the standard deviation. A proposal from the automobile manufacturers is that the oxygen sensors are "fooled" when running on oxygen-containing fuels, and make the engine run somewhat lean. This is said, in turn, to give higher NO_X emissions. This is not true however, in this case, as the T-fuel gives around 20% lower NO_X emissions than the R-fuel, almost as low as the A-fuel. Besides this, the NO_X emissions, when driving on the 1-fuel, are more than 5% lower than the reference. The NO_X emissions when driving on the 2-fuel are somewhat lower than the reference but within the standard deviation. HC emissions do not differ more than 5%, with lower values for the T-fuel and the 1-fuel.

4.1.1 Fuel consumption

Concerning the fuel consumption (FC), it can be stated that the T-fuel gives around 3% larger calculated volumetric FC (compared to the R-fuel). The T-fuel has, according to the content of 15% ethanol, around 5% lower energy content than the A-fuel, but gives, in spite of this, lower (but not significantly so) volumetric fuel consumption than the A-fuel. However, as the T-fuel gives more than 15% reduction of CO emissions, which implies a 1% saving in fuel consumption. The low NO_X emission also indicates a possibly smoother burning, without for example trace knock, which could also save fuel to some extent. Seen on an energy basis, the T-fuel saves about 3% energy compared to the reference fuel, followed by the 1-fuel with a saving of around 2% of energy. Due to their high content of hydrogen, the A-fuel as well as the T-fuel emit significantly less CO_2, both of them about 5% less than the R-fuel.

4.1.2 Particulates

Particulate emissions are slightly higher for the R-fuel compared to the 1-fuel and the 2-fuel. This is expected, since the 1-fuel and the 2-fuel both have lower final boiling points and, according to the fuels analysis, also significantly lower PAH content. However the A-fuel and the T-fuel have particulate emissions almost 50% lower than the R-fuel. A larger difference had been expected for these fuels, based on earlier experiments with alkylate fuels. However the base level of particulates of this vehicle seems to be higher, probably due to the higher oil consumption of the engine in the present project.

4.2 Unregulated compounds

Unregulated pollutants are, in this context, chemical compounds in automotive exhausts which are not regulated in any regulations or laws. The vast majority of the compounds are part of the tens of thousand of chemical compounds together forming what is called HC (hydrocarbons) and the particulate fraction. When filtering automotive exhaust particulates, a lot of high molecular weight compounds pass through the filter since temperature, concentration and available time does not allow a complete condensation of particles. These compounds are named the "semivolatile phase" and are trapped on some type of absorbent. The boiling points of the trapped compounds later analysed are from about 200°C and upwards.

There are also some few compounds which are inorganic, such as ammonia and the individual nitrogen oxides including dinitrogenoxide, which are defined as unregulated compounds. The reason for measuring individual unregulated compounds is that there is always a suspicion that a fuel, fuel additive, engine construction or test condition will not only impair total HC or particulate emissions but also change the chemical composition, so as to make exhausts relatively more or less toxic.

The huge number of chemical compounds and their different chemical properties makes, especially for the particulate and semivolatile fraction, a complete chemical characterisation difficult or even impossible since some compounds, even in concentrations which are not measurable, exhibit a carcinogenic or mutagenic activity. Because of this, the biological test methods used on particulate and semivolatile extracts have, since the beginning of the 1980's, become more or less standard methods, especially for motor fuel testing. The methods now used are mutagenic testing according to Bruce Ames, with three strains of salmonella typhimurium, two of them both with and without metabolic activation. As a complement, during recent years a TCDD receptor binding assay, which also identifies extracts of probable mutagenic and cancerogenic properties, has also been used.

4.2.1 Gaseous compounds

The gasous compounds were analysed with GC as well as FTIR. The results of the GC analyses from the bags are however a factor 3 to 5 higher than expected (and possible) and are not reported here.

4.2.2 Aldehydes

Compared to the reference fuel all the other fuels have between 10 and 50% higher emissions of the aldehydes measured. Most increased are formaldehyde, acetaldehyde and the ketones, acetone and methylethyl ketone, the highest levels for the T- and A-fuels.

Table 2
Aldehyde and ketone emissions mg/km

Fuel	A		T		1		R		2	
Formaldehyde	63,9	± 5,7	64,4	± 3,2	51,8	± 2,8	40,5	± 1,9	44,6	± 1,0
Acetaldehyde	13,3	± 1,1	29,4	± 1,5	10,2	± 0,5	7,0	± 0,5	9,3	± 0,2
Acetone	10,8	± 0,9	9,38	± 0,84	6,57	± 0,52	3,08	± 0,31	4,02	± 0,07
Akrolein	5,52	± 0,51	5,37	± 0,28	5,33	± 0,42	4,58	± 0,20	5,35	± 0,24
Propylaldehyde	1,67	± 0,20	1,72	± 0,09	1,85	± 0,10	1,34	± 0,11	1,59	± 0,08
Krotonaldehyde	1,53	± 0,60	1,04	± 0,25	1,26	± 0,08	1,78	± 0,87	1,91	± 0,77
Metyletylketone	7,06	± 0,62	6,67	± 0,31	4,12	± 0,18	2,38	± 0,13	2,59	± 0,11
Butyraldehyde	1,80	± 0,23	1,62	± 0,17	1,37	± 0,15	0,96	± 0,03	1,23	± 0,29
Bensaldehyde	1,09	± 0,40	2,21	± 0,11	9,03	± 0,29	10,5	± 0,49	13,3	± 0,34
Metylbutyraldehyde	1,18	± 0,69	0,85	± 0,44	0,90	± 0,38	0,76	± 0,19	0,72	± 0,33
Valeraldehyde	0,02	± 0,02	0,02	± 0,02	0,22	± 0,10	0,08	± 0,09	0,19	± 0,12
2-metylbensaldehyde	0,08	± 0,05	0,97	± 0,11	2,27	± 0,25	2,82	± 0,05	3,37	± 0,10
3-metylbensaldehyde	0,14	± 0,16	3,37	± 1,13	4,08	± 1,08	7,50	± 1,57	7,59	± 4,49
Hexylaldehyde	0,13	± 0,09	0,17	± 0,06	0,11	± 0,02	0,14	± 0,06	0,20	± 0,07
Sum, aldehydes measured	108	± 10	127	± 8	99	± 6	83	± 6	96	± 7

The A- and T- fuel and to some extent 1-fuel had lower emissions of the aromatic aldehydes, bensaldehyde, 2 and 3-methyl bensaldehyde. The addition of ethanol and xylene to T-fuel gives a 20% higher total aldehyde emission compared to A-fuel, mostly depending on an almost 3-fold increase of acetaldehyde and an increase of aromatic aldehydes emissions, compared to the A-fuel. In this project phenols and cresols were not analysed. From earlier projects it is known that standard gasolines emit ~50 µg/km of phenols and cresols compared to 2 µg/km from pure alkylate fuel. (Emissions from a vehicle driven with 3 different fuels, to be published).

The increased formaldehyde formation from A-, T- and 1-fuel are probably due to cleavage of CH_3-groups from the isoparaffins, ethanol and MTBE being of high concentrations in these fuels. In the same manner acetaldehyde is produced from the ethanol in the T-fuel. The R- and 2-fuels contain more aromatics and thus the main precursors to formaldehyde are in lower concentrations.

Quite significant amounts (5-10%) of the increase in oxygenates emissions for A-, T- and 1-fuel are acetone and methyl-ethyl ketone, both of which are insignificant from the health effect point of view. The emission of 1 mg/km of bensaldehyde from the A-fuel is probably partly due to artefacts, since it seems unlikely that the engine can produce such quantities from isoparaffins.

4.2.3 Emissions of medium boiling point hydrocarbons, MTBE and ethanol

The emissions of medium boiling point hydrocarbons, of interest from a health point of view, and some other fuel components were sampled out of the diluted exhausts at the end of the dilution tunnel. The compounds were absorbed in glass ampoules containing activated charcoal at a rate of 300 ml/min.

Table 3 Emissions of medium boiling point hydrocarbons mg/km.

Fuel	A	T	1	R	2
Benzene	5 ± 1	11 ± 0,5	46 ± 2	95 ± 5	124 ± 9
Toluene	10 ± 5	17 ± 2	171 ± 7	238 ± 6	252 ± 13
Ethylbenzene	2 ± 1	18 ± 0,4	41 ± 2	51 ± 3	48 ± 2
Xylenes	12 ± 10	87 ± 2	153 ± 8	222 ± 8	227 ± 9
Styrene	<1	3 ± 0,1	8 ± 0,3	10 ± 1	8 ± 0,4
MTBE	13 ± 8	15 ± 8	145 ± 3	16 ± 6	7 ± 6
Ethanol	<1	60 ± 3	1 ± 1	<1	<1

The analyses were performed by GC analyses of extracts from the activated charcoal and thus interference is not excluded. MTBE emissions are of significance only in the case of the 1-fuel, which had 10% added. A comparison with benzene concentrations in the fuel and emissions reveals that benzene is about twice as stable as MTBE during the combustion process. Ethanol added at a concentration of 16 % to the T-fuel has an emission of 60 mg/km and is thus only a tenth as stable as benzene during the combustion process. The higher aromatic emissions of the T-fuel when compared to the A-fuel is a result of emissions of unreacted and partly reacted xylenes which had a concentration of ~6% in the fuel. The lower aromatic and benzene content in the 1-fuel gives lower emissions of aromatics than the 2- and R-fuels, which are quite equal.

4.2.4 Emissions of particulates and semivolatile compounds

Particulates and semivolatile compounds were sampled from diluted exhausts, particles on filters and the semivolatile phase in a porous polyurethane plug (PUF). Samples were stored in a freezer until extraction with acetone in Soxhlet extractors. After extraction, samples were divided in three parts and were used for analyses of PAH, mutagenic effect and TCDD receptor effect.

4.2.4.1 Emissions of PAH

PAH compounds are formed in the incomplete combustion of all organic compounds (except methanol). The emission of PAH, as well as soot, in the exhausts is enhanced by oxygen deficiency (fuel enrichment at cold start), aromatics in the fuel and also by PAH in the fuel. The fuels tested had a very large span of aromatic as well as PAH content. The concentration of PAH in the A-fuel as well as in the T-fuel was about a factor 100 lower than in the reference R-fuel. The aromatic content also varied from 0% in the A-fuel to 37,5% in the R-fuel. The Table below shows the total emissions of PAH in particulates and the semivolatile phase. For the convenience of the reader the PAH have been added together in three groups of compounds with different spread in mean molecular weight and boiling point.

Group 1. Napht.-Coro.
Naphthalene, Acenaphtylene, Acenaphtene, Fluorene, Phenantrene, Antracene, Fluorantene, Pyrene, Benso(a)fluorene, Benso(b)fluorene, Benso(a)antracene, Chrysene, Benso(b)fluorantene, Benso(k)fluorantene, Benso(e)pyrene, Benso(a)pyrene, Perylene, Indeno(1,2,3-cd)pyrene, Benso(g,h,i)perylene, Dibenso(a,h)antracene, Coronene

Group 2. Phen.-Coro.
Phenantrene, Antracene, Fluorantene, Pyrene, Benso(a)fluorene, Benso(b)fluorene, Benso(a)antracene, Chrysene, Benso(b)fluorantene, Benso(k)fluorantene, Benso(e)pyrene, Benso(a)pyrene, Perylene, Indeno(1,2-3-cd)pyrene, Benso(g,h,i)perylene, Dibenso(a,h)antracene, Coronene

Group 3. Benso(a)f.-Coro.
Benso(a)fluorene, Benso(b)fluorene, Benso(a)antracene, Chrysene, Benso(b)fluorantene, Benso(k)fluorantene, Benso(e)pyrene, Benso(a)pyrene, Perylene, Indeno(1,2,3-cd)pyrene, Benso(g,h,i)perylene, Dibenso(a,h)antracene, Coronene

Table 4
Polyaromatic hydrocarbon emissions, filter+ PUF, µg/km

Fuel	A	T	1	R	2
Sum Napht.-Coro., Filter+PUF	56 ± 22	43 ± 2	202 ± 29	441 ± 52	304 ± 64
Sum Phen.-Coro., Filter+PUF	19 ± 8	14 ± 2	68 ± 6	188 ± 19	152 ± 31
Sum Benso(a)f.-Coro., Fil.+PUF	3,8 ± 1,3	2,3 ± 0,5	12,0 ± 3,2	27,6 ± 3,9	23,1 ± 1,5

Polyaromatic hydrocarbon emissions, filter+ PUF, µg/km. Sample 12 omitted

Fuel	A	T	1	R	2
Sum Napht.-Coro., Filter+PUF	44 ± 12	43 ± 2	202 ± 29	441 ± 52	304 ± 64
Sum Phen.-Coro., Filter+PUF	14 ± 4	14 ± 2	68 ± 6	188 ± 19	152 ± 31
Sum Benso(a)f.-Coro., Fil.+PUF	3,1 ± 0,7	2,3 ± 0,5	12,0 ± 3,2	27,6 ± 3,9	23,1 ± 1,5

Table 5
Polyaromatics in fuel, µg/ml

Fuel	A	T	1	R	2
Sum Naptalene-Coronene	18	26	420	2630	1618
Sum Fenantrene-Coronene	0,47	0,76	26	85	78

As can be seen in Table 4 the emissions of PAH are much lower for the A- and T-fuels but also to a minor degree for the 1- and 2-fuels. When looking at the upper part of the table it can be seen that the A-fuel has an extremely large standard deviation and when looking into the primary values it can be seen that one value is far above the mean for the other three samples. It has been decided to exclude this value. It is probable that a carry over effect from the previous samples (of the 2-fuel) has taken place. This has been shown in earlier tests, despite conditioning drives in between the tests. In the lower part of Table 4 are shown the emissions of PAH in the three groups and in all the groups it can be seen that the emissions are around one tenth for the A-fuel and the T-fuel compared to the reference R-fuel. The 2-fuel also has lower emissions, with about one fourth for the group one compounds and lower, but not significantly so, for the other groups. The emissions from the 1-fuel has more than 50% lower concentrations of all groups, including the heavier compounds. Emissions however, are not directly proportional to the PAH content of the fuel. The 5% of xylenes in the T-fuel does not seem to have increased PAH emissions but the admixture of ethanol also could have played a role in this case.

4.2.4.2 Mutagenic effects

The mutagenic effects have been tested on extracts from particulates filtered on teflon coated filters and PolyUrethane Foam plugs (PUF:s) mounted after the filter. The compounds which are collected in the PUFs have, due to their low concentration, not condensed on particulates, or have evaporated, during sampling, from the particulates trapped on the filters.

Table 6
Mutagenetic effect, Rev/m, Filter.

Fuel	A	T	1	R	2
98-S9	0,8 ± 0,8	1,0 ± 0,2	1,7 ± 0,7	2,8 ± 1,2	1,6 ± 0,6
98+S9	0,7 ± 0,8	0,7 ± 0,5	1,9 ± 1,0	2,3 ± 0,9	2,7 ± 0,7
100-S9	1,4 ± 2,9	3,5 ± 2,8	3,5 ± 2,7	5,6 ± 2,6	2,5 ± 2,1
100+S9	1,4 ± 3,1	2,9 ± 2,9	6,0 ± 3,0	5,7 ± 2,3	2,9 ± 3,1
98 NR	0,4 ± 0,4	0,5 ± 0,5	0,9 ± 0,5	0,9 ± 0,5	1,0 ± 0,7
Sum	4,6 ± 0,0	8,6 ± 0,0	14,1 ± 0,0	17,3 ± 0,0	10,8 ± 0,0

Mutagenetic effect, Rev/m, PUF.

Fuel	A	T	1	R	2
98-S9	-0,0 ± 0,4	0,2 ± 0,3	0,7 ± 1,0	1,1 ± 1,1	0,7 ± 0,3
98+S9	0,3 ± 0,5	0,3 ± 0,3	0,2 ± 0,8	0,3 ± 1,4	1,3 ± 0,9
100-S9	2,2 ± 3,0	3,4 ± 2,9	2,2 ± 2,1	1,3 ± 3,5	1,1 ± 1,3
100+S9	1,7 ± 1,9	2,0 ± 2,6	2,8 ± 2,3	1,8 ± 3,7	2,4 ± 2,5
Sum	4,7 ± 5,1	6,5 ± 5,8	6,3 ± 6,0	5,0 ± 10,3	6,3 ± 5,0

In the above Table can be seen the sum of mutagenicity. This is by no means a scientifically proven measurement but is more a way of indicating the effects in a more readable form for the layman. The mutagenic effect is generally low and the results on separate strains show a low activity and thus many samples did not show a significant effect. For fuels showing no significant effect on all four tests the effect is given as "0". In other cases non significant samples are read as zero when calculating mean values. The spread is large and thus the results can only be used as relative ones, showing the order of how the fuels in the engine give rise to mutagenic compounds in the exhaust. The effect decreases in the order R, 2, 1, T and A.

4.2.4.3 TCDD receptor effect

The TCDD receptor effect has been tested on extracts from particulates collected on filters as well as from the semivolatile phase passing the filters and collected in PUF:s. A clear trend towards lower effect can be seen when going from the R-fuel to the 2-, 1-, T- and A-fuel, the A-fuel giving about 10 times less effect than the reference R-fuel. The 2-fuel gives about 25% less activity and the 1-fuel 50% less activity. The T-fuel, which in its base formulation is similar to the A-fuel, gives only 5 times lower activity than the R-fuel.

Table 7
TCDD receptor effect, l/km.

Fuel designation	A	T	1	R	2
Filter					
TCDD receptor l/km	37 ± 17	97 ± 129	118 ± 40	276 ± 71	205 ± 72
PUF					
TCDD receptor l/km	4 ± 6	5 ± 2	39 ± 8	129 ± 30	62 ± 12
Total					
TCDD receptor l/km	42 ± 22	102 ± 130	157 ± 45	405 ± 84	267 ± 62

However the standard deviation of the mean value of the filter activity for the T-fuel is extremely high and a check of the individual results shows one extreme outlier among the filter samples being 10 times higher than the rest of samples from the T-fuel. Concerning this sample (number 14), PAH analyses, mutagenic test and especially TCDD receptor analyses of the extract from the PUF sample show, however, no deviations on this sample compared to other samples from the T-fuel. Thus it is very probable that this result is incorrect and should be omitted. If this is done, the T-fuel and the A-fuel show the same low activity in the TCDD receptor test of particulates sampled out of the exhausts. When scrutinizing the results from the tests of the A-fuel, it is found that test number 12, which was performed after some tests on the 1-,2- and R-fuels, is three times higher than the mean of the other three samples, and is certainly far outside the standard deviation, (the same sample which was excluded from the PAH results). This is, as earlier mentioned, probably due to a carry-over effect from the previous tests, as has been mentioned previously. With the large differences in emissions between ordinary gasolines and alkylate fuel, the purging sequence of 10 minutes at 100 km/h plus one conditioning cycle is obviously not sufficient for conditioning engine, exhaust system and dilution tunnel. If this result, suspected of being too high, is also omitted the result will be as in the following Table.

Table 8
TCDD receptor effect, l/km.   Sample 12 and 14 omitted.

Fuel designation	A	T	1	R	2
Filter					
TCDD receptor l/km	28 ± 5	23 ± 4	118 ± 40	276 ± 71	205 ± 72
PUF					
TCDD receptor l/km	1 ± 1	4 ± 1	39 ± 8	129 ± 30	62 ± 12
Total					
TCDD receptor l/km	29 ± 4	27 ± 3	157 ± 45	405 ± 84	267 ± 62

The A- and T-fuels now show the same low activity, more than 10 times below the values from the R-fuel. What is also interesting to see is that the reformulated 2- and 1-fuels show an especially good reduction in the activity collected out of the semivolatile phase in the PUF:s.

5 CONCLUSIONS

The results from the tests show that there is a substantial possibility of reducing the effects on health from particulates and semivolatiles in the emissions from gasoline fuelled vehicles. It has also been found that small changes as the formulation of environmental class 2 gasoline will result in lower emissions and the reformulated quality (the 1-fuel) will definitely give much lower emissions. Low PAH concentration in the gasoline has, for this vehicle, been shown to be more important for low emissions of PAH than low aromatic content in the gasoline. However there is a potential for obtaining an even greater decrease in emissions by changing to fuels containing a larger quantitiy of pure isoparaffins.

6 ACKNOWLEDGEMENTS

Volvo is acknowledged for lending us the test vehicle, Saab Automobile for donating the reference fuel, Neste for donating the reformulated fuel 1, and the personal of MTC, Jan Hernebro, Gareth Tailor and Oskar Sander who devotedly performed their tasks. The project was financed by Swedish EPA, Aspen petroleum AB and OK petroleum AB.

Enhanced evaporative emissions requirements: a challenge to automotive industry, a chance for the environment

RAIMANN
Delphi Automotive Systems, Luxembourg

In the last decades the exhaust gas emission levels have gone down significantly due to more and more stringent regulatory requirements. In parallel, hydrocarbon emissions, that are emitted in the atmosphere as exhaust gases and fuel vapours, get more attention. The much more stringent US-96 evaporative legislation and the now proposed EC-2000 Evap requirements demonstrate the clear intention to reduce overall evaporative emissions as well to a considerable degree.

This publication deals with parameters lowering the evaporative emissions. After classifying the different sources the topic focusses on the tank system including carbon canisters, considering the following types of emissions: fuel vapour escaping from the vehicle during trips and soaking (US and Cal. Enhanced Evap-test, EC2000) and the hydrocarbons occurring when the cars are being refuelled. (Onboard Refueling Vapour Recovery, ORVR).

1   Introduction

Vehicle evaporative emissions may have different sources. The majority of hydrocarbon emissions is generated during vehicle driving or soaking (figure 1). Comparing the importance of the HC-sources, the hydrocarbons evaporating from solid materials like plastic have a significant influence when

---

[1] Based on a paper originally prepared by Dr. Olaf Weber

the vehicle is new. In addition all fuel ducting lines and hoses are to be designed to prevent HC-permeation. But the main target of investigations during vehicle development should be the tank vapours, the exhaust emissions during the so called "Running Loss" and the emissions when the car is being refuelled.

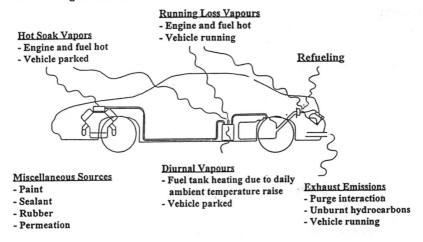

Figure 1: Classification of Evaporative Emissions

As the emission requirements on the exhaust side have been tightened to a large degree during the last decades and will continue to become more stringent in the future, the importance of reducing evaporative emission levels should be discussed. Figure 2, below, shows on the left the US-TLEV and LEV limits for exhaust HC emissions. The order of magnitude of the LEV-limit as well as the improvement from TLEV to LEV lies in the range of 0.05g/mi.

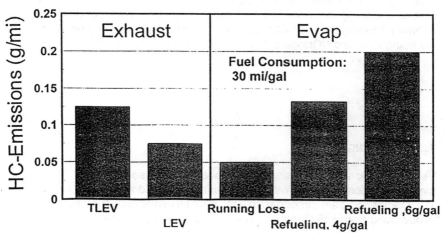

Figure 2: Comparison between Exhaust and Evaporative Emissions

ORVR-vapor generation rates are in the area of approx. 4 to 6 g of hydrocarbon vapour/gal of fuel dispensed (approx. 1-1.5 g/l). With the assumption that a vehicle's fuel consumption is app. 30 mi/gal, the refueling emissions will have values between 0.13 and 0.2 g/mile. If refueling emissions are eliminated, the benefit would be higher compared to the improvement on the exhaust side described above. The Running Loss limit is also in the same area of magnitude. As a conclusion, the evaporative legislation is useful with regard to exhaust emission levels, although different boundary conditions like test cycles, fuel composition and temperatures were not taken into account.

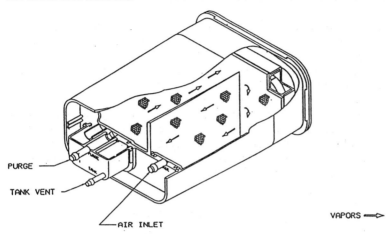

Figure 3: Example of a Carbon Canister

To prevent hydrocarbons from escaping a tank system, carbon canisters (example: figure 3) are used. A mixture of fuel vapour and air enters the canister and is adsorbed by the activated carbon. Pure air leaves the canister through the air tube. To purge the canister, air flows from the air tube to the purge tube. On their way through, the HC molecules can desorb and are transported to the internal combustion engine where they are burnt. Both the exotherme adsorption and the endotherme process depend on carbon quality, type of hydrocarbon, ambient pressure, humidity and temperature.

The determination of the canister size according to the legislation and customer test procedures is a real challenge, as a lot of boundary conditions are to be considered:

i.e.:
fuel composition[1]
tank volume
flow resistance of hoses
test cycle
geometry of tank

purge volume flow
geometry of carbon canister
carbon working capacity
temperatures

2. Enhanced Evaporative Emissions

The parameters influencing the evaporation during a diurnal have to be well understood to develop tank systems with low diurnal emissions. For certification testing there are different test procedures, for example the US and California Enhanced Evaporative Emission test and the proposed EC 2000 legislation. A maximum of 2 grams of hydrocarbon emissions for the US and Cal. long test as well as for the EC2000 are allowed during the highest diurnal sequence and the hot soak sequence together.

To simulate the behavior of the whole tank system including fuel and carbon canister a calculation program[2] (figure 4) has been created which allows accurate predictions to be made (figure 5). Evaporative (diurnal, running loss, hot soak) and refueling vapor generation are functions of fuel properties, fuel system parameters, and ambient conditions. Canister purging is a function of the amount of vapor in the canister, purge air volume, and temperature. Mathematical models have been developed for predicting the mass and composition of evaporative and refueling vapors as well as for the prediction of canister breakthrough emissions and canister purge. The calculation program merges the mathematical models working independently, so that all conceivable test sequences can be simulated.

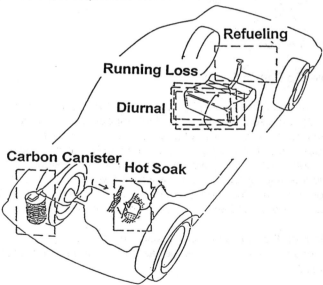

Figure 4: Various Mathematical Modules in Calculation Program "EVAP"[1]

To give an example, a comparison between the test results and the predicted results is presented in figure 5.

The figure shows the canister loading during the Enhanced Evap procedure. Before the test starts the canister is filled initially. Therefore the full loaded canister has the weight of 0g. During all stages of the Enhanced Evap there is only a small deviation between testing and calculation when the canister is loaded and secondly when purged. The so called back purge can also be observed when the tank system is cooled down and recondensation in addition to a volume increase in the tank sucks air through the canister.

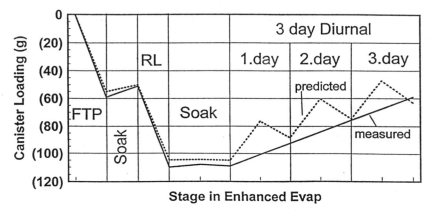

Figure 5: Enhanced Evap Results - Calculation and Test Results

The canister calculation module also gives the results of the breakthrough emissions but they can not be compared in this case because only the total vehicle evaporation emissions have been measured in test.

With this program it is possible to check out the tank system design as for example the necessary carbon canister capacity or the tank geometry. Influences on the vapor generation during the diurnal could be better understood and the development of a system is possible without expensive tests.

As a result, this kind of calculation program saves a lot of money and cost during the development process by predicting necessary purge rates, canister volumes, tank peak temperatures etc.

3   ORVR

The well known ORVR test procedure defines temperatures and fill rates:
filling rate: 4 - 10 gal/min   emission limit: 0.053 g/l of fuel disp.
tank temperature ca. 28°C   9 RVP fuel
dispensed fuel temp: ca. 19°C

To prevent the hydrocarbons from escaping the filler neck, a liquid seal is used due to its advantages concerning warranty and suitability to different fuel nozzle geometries.

The necessary ORVR tests are run in an ORVR shed, where the refueling emissions as well as the canister loading are determined. Similar to the Enhanced Evap procedure the canister size should be as small as possible. The vapor generated during refueling should be kept at a minimum. An improvement of the refueling system will lead to smaller vapour generation rates and therefore smaller and less expensive canisters.

Figure 7 shows the refueling results with a base system (left hand side) and the developed tank (right hand side). Canister size can be decreased by more than 20%, a much easier and flexible packaging is possible, and again, cost is reduced, while meeting the regulatory requirements of 0.2 g/gal (0.053 g/l).

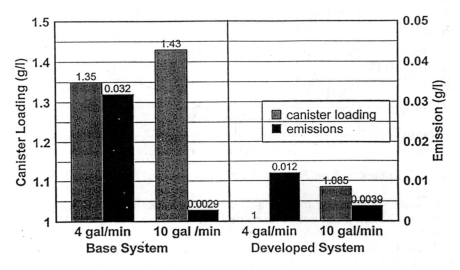

Figure 7: Non-Working and Working ORVR-System

4   Canister Design

The improvement on the tank system concerning vapour generation rates is only one possibility to make smaller canisters work. The canister design also has an impact on its performance (figure 8). Besides the well known influence of the L/D ratio of canisters, other design parameters are to be taken into account.

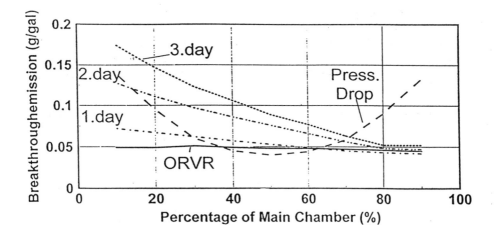

Figure 8: Influence of Canister Design on Breakthrough Emissions, Calculation results

The canister (see also figure 3) has a partitioned design: the main chamber to be entered by the vapour flow coming from the tank and a second chamber. On the x-axis (figure 8) the volume of the main chamber is enlarged by shifting the partition wall in the canister in the direction of the air outlet. The outer dimensions are held constant.

It can be observed that the breakthrough emissions of the canister in an Enhanced Evap test are decreased by this design change. The ORVR-performance is not affected. The relative pressure drop caused by the canister has the lowest value if both chambers have nearly the same size. This flow resistance will influence the fillability of the tank and has to be considered. For that reason all conceivable canister designs will be a compromise between L/D ratio, chamber sizes and carbon used.

5 Conclusion

After highlighting the importance of reducing evaporative emissions, the Enhanced Evap and ORVR test conditions and significant influencing parameters were discussed. The computer simulation model as an important tool in designing systems helps meeting the emissions standards and customer requirements as well as ensures a short and cost effective development process. In addition, the knowledge and test experience in ORVR-systems and canister design supports the future improvement of Evap Sub Systems.

6 References

1. Reddy, S.R. Evaporative Emissions from Gasoline and Alcohol-Containing Gasolines with Closely Matched Volatilities, SAE-Paper 861556, International Fuels and Lubricants Meeting and Exposition, Philadelphia, PA, October 6-9, 1986

2. Reddy, S.R. Evap-PC: A Computer Model for Predicting Vehicle Evaporative and Refueling Emissions, GM Research Report F7L-764, 1991

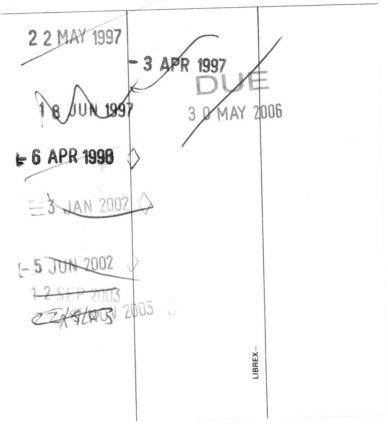